Bayesian Modeling in Bioinformatics

Chapman & Hall/CRC Biostatistics Series

Editor-in-Chief

Shein-Chung Chow, Ph.D.
Professor
Department of Biostatistics and Bioinformatics
Duke University School of Medicine
Durham, North Carolina, U.S.A.

Series Editors

Byron Jones
Senior Director
Statistical Research and Consulting Centre
(IPC 193)
Pfizer Global Research and Development
Sandwich, Kent, U. K.

Jen-pei Liu
Professor
Division of Biometry
Department of Agronomy
National Taiwan University
Taipei, Taiwan

Karl E. Peace
Georgia Cancer Coalition
Distinguished Cancer Scholar
Senior Research Scientist and
Professor of Biostatistics
Jiann-Ping Hsu College of Public Health
Georgia Southern University
Statesboro, Georgia

Bruce W. Turnbull
Professor
School of Operations Research
and Industrial Engineering
Cornell University
Ithaca, New York

Chapman & Hall/CRC Biostatistics Series

Published Titles

1. *Design and Analysis of Animal Studies in Pharmaceutical Development,*
   Shein-Chung Chow and Jen-pei Liu
2. *Basic Statistics and Pharmaceutical Statistical Applications,* James E. De Muth
3. *Design and Analysis of Bioavailability and Bioequivalence Studies,*
   Second Edition, Revised and Expanded, Shein-Chung Chow and Jen-pei Liu
4. *Meta-Analysis in Medicine and Health Policy,* Dalene K. Stangl and Donald A. Berry
5. *Generalized Linear Models: A Bayesian Perspective,* Dipak K. Dey,
   Sujit K. Ghosh, and Bani K. Mallick
6. *Difference Equations with Public Health Applications,* Lemuel A. Moyé
   and Asha Seth Kapadia
7. *Medical Biostatistics,* Abhaya Indrayan and Sanjeev B. Sarmukaddam
8. *Statistical Methods for Clinical Trials,* Mark X. Norleans
9. *Causal Analysis in Biomedicine and Epidemiology: Based on Minimal
   Sufficient Causation,* Mikel Aickin
10. *Statistics in Drug Research: Methodologies and Recent Developments,*
    Shein-Chung Chow and Jun Shao
11. *Sample Size Calculations in Clinical Research,* Shein-Chung Chow, Jun Shao, and Hansheng Wang
12. *Applied Statistical Design for the Researcher,* Daryl S. Paulson
13. *Advances in Clinical Trial Biostatistics,* Nancy L. Geller
14. *Statistics in the Pharmaceutical Industry, Third Edition,* Ralph Buncher
    and Jia-Yeong Tsay
15. *DNA Microarrays and Related Genomics Techniques: Design, Analysis, and Interpretation of Experiments,*
    David B. Allsion, Grier P. Page, T. Mark Beasley, and Jode W. Edwards
16. *Basic Statistics and Pharmaceutical Statistical Applications, Second Edition,*
    James E. De Muth
17. *Adaptive Design Methods in Clinical Trials,* Shein-Chung Chow and Mark Chang
18. *Handbook of Regression and Modeling: Applications for the Clinical and Pharmaceutical Industries,* Daryl
    S. Paulson
19. *Statistical Design and Analysis of Stability Studies,* Shein-Chung Chow
20. *Sample Size Calculations in Clinical Research, Second Edition,* Shein-Chung Chow,
    Jun Shao, and Hansheng Wang
21. *Elementary Bayesian Biostatistics,* Lemuel A. Moyé
22. *Adaptive Design Theory and Implementation Using SAS and R,* Mark Chang
23. *Computational Pharmacokinetics,* Anders Källén
24. *Computational Methods in Biomedical Research,* Ravindra Khattree and
    Dayanand N. Naik
25. *Medical Biostatistics, Second Edition,* A. Indrayan
26. *DNA Methylation Microarrays: Experimental Design and Statistical Analysis,*
    Sun-Chong Wang and Arturas Petronis
27. *Design and Analysis of Bioavailability and Bioequivalence Studies, Third Edition,*
    Shein Chung Chow and Jen-pei Liu
28. *Translational Medicine: Strategies and Statistical Methods,* Dennis Cosmatos and
    Shein-Chung Chow
29. *Bayesian Methods for Measures of Agreement,* Lyle D. Broemeling
30. *Data and Safety Monitoring Committees in Clinical Trials,* Jay Herson
31. *Design and Analysis of Clinical Trials with Time-to-Event Endpoints,* Karl E. Peace
32. *Bayesian Missing Data Problems: EM, Data Augmentation and Noniterative Computation,*
    Ming T. Tan, Guo-Liang Tian, and Kai Wang Ng
33. *Multiple Testing Problems in Pharmaceutical Statistics,* Alex Dmitrienko, Ajit C. Tamhane, and Frank Bretz
34. *Bayesian Modeling in Bioinformatics,* Dipak K. Dey, Samiran Ghosh, and Bani K. Mallick

Chapman & Hall/CRC Biostatistics Series

Bayesian Modeling in Bioinformatics

Edited by

Dipak K. Dey
University of Connecticut
Storrs, U.S.A.

Samiran Ghosh
Indiana University-Purdue University
Indianapolis, U.S.A.

Bani K. Mallick
Texas A&M University
College Station, U.S.A.

CRC Press
Taylor & Francis Group
Boca Raton London New York

CRC Press is an imprint of the
Taylor & Francis Group, an **informa** business

A CHAPMAN & HALL BOOK

CRC Press
Taylor & Francis Group
6000 Broken Sound Parkway NW, Suite 300
Boca Raton, FL 33487-2742

First issued in paperback 2019

© 2011 by Taylor & Francis Group, LLC
CRC Press is an imprint of Taylor & Francis Group, an Informa business

No claim to original U.S. Government works

ISBN-13: 978-1-4200-7017-0 (hbk)
ISBN-13: 978-0-367-38365-7 (pbk)

Library of Congress Cataloging-in-Publication Data

Dey, Dipak.
    Bayesian modeling in bioinformatics / Dipak K. Dey, Samiran Ghosh, Bani K. Mallick.
      p. cm. -- (Chapman & Hall/CRC biostatistics series ; 34)
    Includes bibliographical references and index.
    ISBN 978-1-4200-7017-0 (hardcover : alk. paper)
    1. Bioinformatics--Statistical methods. 2. Bayesian statistical decision theory. I. Ghosh, Samiran. II. Mallick, Bani K., 1965- III. Title. IV. Series.

QH324.2.D49 2010
570.285--dc22
                                         2009049470

Visit the Taylor & Francis Web site at
http://www.taylorandfrancis.com

and the CRC Press Web site at
http://www.crcpress.com

Contents

13 Proportional Hazards Regression Using Bayesian Kernel Machines   317

Arnab Maity and Bani K. Mallick

14 A Bayesian Mixture Model for Protein Biomarker Discovery   343

Peter Müller, Keith Baggerly, Kim Anh Do, and Raj Bandyopadhyay

List of Tables

List of Figures

Preface

Recent advances in genome sequencing (genomics) and protein identification (proteomics) have given rise to challenging research problems that require combined expertise from statistics, biology, computer science, and other fields. Similar challenges arise dealing with modeling genetic data in the form of sequences of various types, e.g., SNIP, RAPD, and microsatellite data. The interdisciplinary nature of bioinformatics presents many research challenges related to integrating concepts, methods, software, and multi-platform data. In addition to new tools for investigating biological systems via high-throughput genomic and proteomic measurements, statisticians face many novel methodological research questions generated by such data. The work in this volume is dedicated to the development and application of Bayesian statistical methods in the analysis of high-throughput bioinformatics data that arise from problems in medical research, in particular cancer and other disease related research, molecular and structural biology.

This volume does not aim to be comprehensive in all areas of bioinformatics. Rather, it presents a broad overview of statistical inference, clustering, and classification problems related to two main high-throughput platforms: microarray gene expression and phylogenic analysis. The volume's main focus is on the design, statistical inference, and data analysis, from a Bayesian perspective, of data sets arising from such high-throughput experiments. All the chapters included in the volume focus on Bayesian methodologies, parametric as well as nonparametric, but each is self contained and independent. There are topics closely related to each other, but we feel there is no need to maintain any kind of sequencing for reading the chapters. The chapters have been arranged purely in alphabetical order according to the last name of the first author and, in no way, should it be taken to mean any preferential order. Chapter 1 provides estimation and testing in time-course microarray experiments. This chapter discusses a variety of recently developed Bayesian techniques for detection of differentially expressed genes on the basis of time series of microarray data. Chapter 2 considers classification for differential gene expression using hierarchical Bayesian models. Chapter 3 deals with applications of mode-oriented stochastic search algorithms for discrete multi-way data for genome wide studies. Chapter 4 develops novel Bayesian nonparametric approaches to bioinformatics problems. Dirichlet process mixture models are introduced for multiple testing, high-dimensional regression, clustering, and functional data analysis. Chapter 5 develops measurement error and survival models for cDNA microarrays. Chapter 6 focuses on robust Bayesian inference for differential gene expression. Chapter 7 develops Bayesian hidden Markov modeling approach for CGH array data. Bayesian approaches to phylogenic analysis are developed in Chapter 8. Chapter 9 is concerned with gene selection for identification of biomarkers in high-throughput data. Chapter 10 develops sparsity priors for protein–protein interaction predictions. Chapter 11 describes Bayesian networks learning for gene expression data. Chapter 12 describes in-vitro to in-vivo factor profiling in genomic expressions. Proportional hazards regression using Bayesian kernel machines is considered in Chapter 13. Chapter 14 focuses on a Bayesian mixture model for protein biomarker discovery using matrix-assisted laser desorption and ionization mass spectrometry. Chapter 15 focuses on general Bayesian methodology for detecting differentially expressed genes. Bayes and empirical Bayes methods for spotted microarray data analysis are considered in Chapter 16. Finally Chapter 17 focuses on the Bayesian classification method for QTL mapping.

We thank all the collaborators for contributing their ideas and insights toward this volume in a timely manner. We fail in our duties if we do not express our sincere indebtedness to our referees, who were quite critical and unbiased in giving their opinions. We do realize that despite of their busy schedules the referees offered us every support in commenting on various chapters. Undoubtedly, it is the joint endeavor of the contributors and the referees that emerged in the form of such an important and significant volume for the Bayesian world.

We thankfully acknowledge David Grubbs from Chapman & Hall, CRC, for his continuous encouragement and support of this project. Our special thanks go to our spouses, Rita, Amrita, and Mou, for their continuous support and encouragement.

We are excited by the continuing opportunities for statistical challenges in the area of bioinformatics data analysis and modeling from a Bayesian point of view. We hope our readers will join us in being engaged with changing technologies and statistical development in this area of bioinformatics.

Dipak K. Dey
Samiran Ghosh
Bani Mallick

MATLAB® is a registered trademark of The Math Works, Inc. For product information, please contact:

The Math Works
3 Apple Hill Drive
Natick, MA
Tel: 508-647-7000
Fax: 508-647-7001
Email: info@mathworks.com
Web: http://www.mathworks.com

Symbol Description

σ^2 Variance of the instrumental error.

N Number of genes in a microarray.

M Total number of array in an experiment.

n Number of different time points in the temporal interval of the experiment.

K Maximum number of replicates in a time points of the experiment.

$\mathbf{z}_i$ The vector of measurements for the i-th gene.

$s_i(t)$ Underlying function of the i-th gene expression profile.

H Number of samples in the experiments.

Chapter 1

Estimation and Testing in Time-Course Microarray Experiments

C. Angelini, D. De Canditiis, and M. Pensky
(I.A.C.) CNR, Italy, and University of Central Florida, USA

1.1   Abstract

The paper discusses a variety of recently developed Bayesian techniques for detection of differentially expressed genes on the basis of time series microarray data. These techniques have different strengths and weaknesses and are constructed for different kinds of experimental designs. However, the Bayesian formulations, which the above described methodologies utilize, allow one to explicitly use prior information that biologists may provide. The methods successfully deal with various technical difficulties that arise in microarray time-course experiments such as a large number of genes, a small number of observations, non-uniform sampling intervals, missing or multiple data, and temporal dependence between observations for each gene.

Several software packages that implement methodologies discussed in the paper are described, and limited comparison between the techniques using a real data example is provided.

1.2   Introduction

Gene expression levels in a given cell can be influenced by different factors, namely pharmacological or medical treatments. The response to a given stimulus is usually different for different genes and may depend on time. One of the goals of modern molecular biology is the high-throughput identification of genes associated with a particular treatment or biological process of interest. The recently developed technology of microarrays allows one to simultaneously monitor the expression levels of thousands of genes. Although microarray experiments can be designed to study different factors of interest, in this paper, for simplicity, we consider experiments involving comparisons between two biological conditions (for example, "control" and "treatment") made over the course of time.

The data may consist of measurements of the expression levels of N genes in two distinct independently collected samples made over time [0,T] (the, so called, "two-sample problem"), or measurements of differences in the expression levels between the two samples (the "one-sample problem"). In both cases, the objective is first to identify the genes that are differentially expressed between the two biological samples and then to estimate the type of response. Subsequently, the curves may undergo some kind of clustering in order to group genes on the basis of their type of response to the treatment.

The problem represents a significant challenge since the number of genes N is very large, while the number of time points n where the measurements are made is rather small and, hence, no asymptotic method can be used. The data is contaminated by usually non-Gaussian heavy-tailed noise. In addition, non-uniform sampling intervals and missing or multiple data makes time series microarray experiments unsuitable to classical time-series and signal processing algorithms.

In the last decades the statistical literature has mostly addressed static microarray experiments, see Efron et al. (2001), Lonnstedt and Speed (2002), Dudoit et al. (2002), Kerr et al. (2000), Ishwaran and Rao (2003), among many others. The analysis of time-course microarray was mainly carried out by adapting some existing methodologies. For example, SAM version 3.0 software package—originally proposed by Tusher et al. (2001) and later described in Storey et al. (2003a, 2003b)—was adapted to handle time-course data by considering the time points as different groups. In a similar manner, the ANOVA approach by Kerr et al. (2000) and Wu et al. (2003) was applied to time-course experiments by treating the time variable as a particular experimental factor. Similar approaches have been considered by Di Camillo et al. (2005) and the Limma package by Smyth (2005), while the paper of Park et al. (2003) uses linear model regression combined with ANOVA analysis. All these methods can be useful when very short time-course experiments have to be analyzed (3-4 time points), however they have the shortcoming of applying statistical techniques designed for static data to time-course data, so that the results are invariant under permutation of the time points. The biological temporal structure of data is ignored.

Although microarray technology has been developed quite recently, nowadays, time-course experiments are becoming increasingly popular: a decreasing cost of manufacturing allows larger number of time points, moreover, it is well recognized that the time variable plays a crucial role in the development of a biological process. Indeed, Ernst and Bar-Joseph (2006) noted that currently time-course microarray experiments constitute about 30% of all microarray experiments (including both cell cycle and drug response studies). Nevertheless, there is still a shortage of statistical methods that are specifically designed for time-course microarray experiments.

This realization led to new developments in the area of analysis of time-course microarray data, see de Hoon et al. (2002), Bar-Joseph et al. (2003a, 2003b), and Bar-Joseph (2004). More recently, Storey et al. (2005), Conesa et al. (2006), Hong and Li (2006), and Angelini et al. (2007, 2009) proposed functional data approaches where gene expression profiles are expanded over a set of basis functions and are represented by the vectors of their coefficients thereafter. Tai and Speed (2006, 2007) considered a multivariate empirical Bayes approach, which applies statistical inference directly to the physical data itself. Another important class of methods which is based on the hidden Markov model approaches, was pioneered by Yuan and Kendziorski (2006).

The goal of the present chapter is to review the more recent techniques for analysis of the time-course microarray data, as well as the software designed for implementing them. The rest of the paper is organized as follows. Section 1.3 describes various possible designs of time-series microarray experiment with particular focus on the one-sample and the two-sample (or multi-sample) designs. Section 1.4 reviews the methods that are constructed specifically for the one-sample setup, while Section 1.5 is devoted to the methods for the two-sample and the multi-sample setups. Section 1.6 provides information on several software packages that implement methodologies described above. Section 1.7 reports results of the application of some recent techniques to a real data set. Finally, Section 1.8 concludes the chapter with a discussion.

1.3   Data Structure

The goal of the present paper is to review statistical methods for the automatic detection of differentially expressed genes in time series microarray experiments. However, the choice of the statistical methodology used for the analysis of a particular data set strongly depends on the way in which the data have been collected since different experimental designs account for different biological information and different assumptions.

First, one needs to distinguish longitudinal studies from independent and factorial ones. Pure longitudinal data is obtained when the same sample is monitored over a period of time, and different samples are separately and distinguishably recorded in the course of the experiment. Independent (or cross-sectional) experimental design takes place when at each time point independent samples are recorded, and replicates are indistinguishable from each other. Finally, factorial designs are those where two or more experimental factors are involved in the study. Also, one needs to distinguish between technical replicates and biological replicates, since the latter introduce additional variation in the data.

The choice of a statistical technique for data analysis is influenced by the number of time points at which observations are available (usually, 6 or more time points, or 3–5 time points for a "short" time series) and the grid design (equispaced or irregular). The presence of missing data also affects the choice of statistical methodology.

The objective of microarray experiments is to detect the differences in gene expression profiles in the biological samples taken under two or more experimental conditions. The differences between gene expression levels can be measured directly (usually, in logarithmic scale) leading to a "one-sample" problem where the differences between profiles are compared to an identical zero curve. If the measurements on each sample are taken independently, then one has a "two-sample" or a "multi-sample" problem where the goal is to compare the gene expression profiles for different samples. For example, if an experiment consists of a series of two-color microarrays, where the control (untreated) sample is directly hybridized with treated samples after various time intervals upon treatment, then one has the one-sample problem. On the other hand, if the two samples are hybridized on one-color arrays or independently hybridized with a reference on two color arrays, then it leads to the two-sample problem.

In what follows, we review statistical methods for the one-sample, two-sample, and multi-sample problems commenting on what types of experimental designs are accommodated by a particular statistical technique and how it deals with an issue of irregular design and missing data. However, regardless the type of experimental design, the microarray data are assumed to be already pre-processed to remove systematic sources of variation. For a detailed discussion of the normalization procedures for microarray data we refer the reader to, e.g., Yang et al. (2002), Cui et al. (2002), McLachlan et al. (2004), or Wit and McClure (2004).

1.3.1   Data Structure in the One-Sample Case

In general, the problem can be formulated as follows. Consider microarray data consisting of the records of N genes. The records are taken at n

different time points in $[0, T]$ where the sampling grid $t^{(1)}, t^{(2)}, \ldots, t^{(n)}$ is not necessarily uniformly spaced. For each array, the measurements consist of N normalized $\log_2$-ratios $z_i^{j,k}$, where $i = 1, \ldots, N$, is the gene number, index j corresponds to the time point $t^{(j)}$ and $k = 1, \ldots, k_i^{(j)}$, $k_i^{(j)} \geq 0$, accommodates for possible technical or biological replicates at time $t^{(j)}$. Note that usually, by the structure of the experimental design, $k_i^{(j)}$ are independent of i, i.e., $k_i^{(j)} \equiv k^{(j)}$ with $M = \sum_{j=1}^{n} k^{(j)}$ being the total number of records. However, since some observations may be missing due to technical errors in the experiment, we let $k_i^{(j)}$ to depend on i so that the total number of records for gene i is

$$M_i = \sum_{j=1}^{n} k_i^{(j)}. \tag{1.1}$$

The number of time points is relatively small ($n \approx 10$) and very few replications are available at each time point ($k_i^{(j)} = 0, 1, \ldots K$ where $K = 1, 2,$ or 3) while the number of genes is very large ($N \approx 10{,}000$).

1.3.2   Data Structure in the Multi-Sample Case

Consider data consisting of measurements of the expression levels of N genes in two or more distinct independently collected samples made over time $[0,T]$.

We note that, when the number of samples is two, the objective is first to identify the genes that are differentially expressed between the two samples and then to estimate the type of response. On the other hand, when more samples are considered one may be interested in detecting both genes that are differentially expressed under at least one biological condition or some specific contrasts, in a spirit that is similar to the multi-way ANOVA model and then to estimate the type of effect. In general, the problem can be formulated as follows. For sample $\aleph$, $\aleph = 1, \cdots, H$, data consists of the records $z_{\aleph i}^{j,k}$, $i = 1, \cdots, N$, on N genes, which are taken at time points $t_{\aleph}^{(j)} \in [0, T]$, $j = 1, .., n_{\aleph}$. For gene i at a time point $t_{\aleph}^{(j)}$, there are $k_{\aleph i}^{(j)}$ (with $k_{\aleph i}^{(j)} = 0, 1, \ldots, K$) records available, making the total number of records for gene i in sample $\aleph$ to be

$$M_{\aleph i} = \sum_{j=1}^{n} k_{\aleph i}^{(j)}. \tag{1.2}$$

Note that in this general setup it is required neither that the observations for the H samples are made at the same time points nor that the number of observations for different samples is the same. This model can be applied when observations on the H samples are made completely separately: the only requirement is that the samples are observed over the same period of time. The orders of magnitude of N, n, and K are the same as in the one-sample case.

1.4 Statistical Methods for the One-Sample Microarray Experiments

In this section we present the more recent and important methods for the one-sample microarray time-course experiments.

1.4.1 Multivariate Bayes Methodology

Tai and Speed (2006) suggest multivariate Bayes methodology for identification of differentially expressed genes. Their approach is specifically designed for longitudinal experiments where each replicate refers to one individual. The technique requires that for each gene the number of replicates $k_i^{(j)}$ is constant for all time points $k_i^{(j)} = k_i$, so that for each gene i there are k_i independent n-dimensional observations $\mathbf{z}_{i,k} = (z_i^{1,k}, \ldots, z_i^{n,k})^T \in R^n$, $k = 1, \ldots, k_i$. Each vector is modeled using multivariate normal distribution $\mathbf{z}_{i,k} \sim \mathcal{N}_n(I_i \boldsymbol{\mu}_i, \boldsymbol{\Sigma}_i)$ where I_i is the Bernoulli random variable, which is equal to 1 when gene i is differentially expressed and zero otherwise. For each gene i, the following hierarchical Bayesian model is built:

$$\boldsymbol{\mu}_i | \boldsymbol{\Sigma}_i, I_i = 1 \sim \mathcal{N}(\mathbf{0}, \eta_i^{-1} \boldsymbol{\Sigma}_i)$$
$$\boldsymbol{\mu}_i | \boldsymbol{\Sigma}_i, I_i = 0 = \mathbf{0}$$
$$\boldsymbol{\Sigma}_i \sim \text{InverseWishart}(\nu_i, \nu_i \boldsymbol{\Lambda}_i^{-1})$$
$$P(I_i = 0) = \pi_0.$$

Here ν_i, η_i and $\boldsymbol{\Lambda}_i$ are gene-specific parameters to be estimated from the data, while π_0 is a global user-defined parameter. To test whether a gene is differentially expressed, Tai and Speed (2006) evaluate posterior odds of $I_i = 1$

$$O_i = \frac{P(I_i = 1 | \mathbf{z}_{i,1}, \cdots, \mathbf{z}_{i,k_i})}{P(I_i = 0 | \mathbf{z}_{i,1}, \cdots, \mathbf{z}_{i,k_i})}. \tag{1.3}$$

Denote by $\bar{\mathbf{z}}_i$ and $\mathbf{S}_i$ the mean and the sample covariance matrix of the vectors $\mathbf{z}_{i,1}, \cdots, \mathbf{z}_{i,k_i}$, and let

$$\tilde{\mathbf{S}}_i = [(k_i - 1)\mathbf{S}_i + \nu_i \boldsymbol{\Lambda}_i]/(k_i - 1 + \nu_i) \quad \text{and} \quad \tilde{\mathbf{t}}_i = \sqrt{k_i} \tilde{\mathbf{S}}_i^{-1/2} \bar{\mathbf{z}}_i$$

be the moderated sample covariance matrix and multivariate t-statistic. Then, posterior odds (1.3) can be rewritten as

$$\frac{1 - \pi_0}{\pi_0} \left(\frac{\eta_i}{\eta_i + k_i} \right)^{n/2} \left(\frac{k_i - 1 + \nu_i + \tilde{\mathbf{t}}_i^T \tilde{\mathbf{t}}_i}{k_i - 1 + \nu_i + (k_i + \eta_i)^{-1} \eta_i \tilde{\mathbf{t}}_i^T \tilde{\mathbf{t}}_i} \right)^{(k_i + \nu_i)/2}. \tag{1.4}$$

Unknown gene specific parameters ν_i, η_i and $\boldsymbol{\Lambda}_i$ are estimated using a version of empirical Bayes approach described in Smyth (2004) and are plugged into expressions (1.4). After that, genes are ordered according to the values of posterior odds (1.4), and genes with the higher values of O_i are considered to be differentially expressed. Note that the user-defined parameter π_0 does not modify the order in the list of differentially expressed genes. The authors do not define the cut off point, so the number of differentially expressed genes has to be chosen by a biologist. Tai and Speed (2006) also notice that when all genes have the same number of replicates $k_i = k$, then the posterior odds is a monotonically increasing function of $\tilde{\mathbf{T}}_i^2 = \tilde{\mathbf{t}}_i^T \tilde{\mathbf{t}}_i$. Therefore, in this case they suggest to use the $\tilde{\mathbf{T}}_i^2$, $i = 1, \cdots, N$, statistics instead of posterior odds since they do not require estimation of η_i and lead to the same ranking as (1.4).

Since Tai and Speed (2006) apply Bayesian techniques directly to the vectors of observations, the method is not flexible. The analysis can be carried out only if the same number of replicates is present at every time point. If any observation is absent, missing data techniques have to be applied. Moreover, the results of analysis are completely independent of the time measurements, so that it is preferable to apply the technique only when the grid is equispaced. Finally, Tai and Speed (2006) do not provide estimators of the gene expression profiles.

1.4.2 Functional Data Approach

Although approach of Storey et al. (2005) is not Bayesian, we present it here since it is the first functional data approach to the time-series microarray data. Storey et al. (2005) treat each record $z_i^{j,k}$ as a noisy measurement of a function $s_i(t)$ at a time point $t^{(j)}$. The objective of the analysis is to identify and estimate the curves $s_i(t)$ that are different from the identical zero.

Since the response curve for each gene is relatively simple and only a few measurements for each gene are available, Storey et al. (2005) estimate each curve $s_i(t)$ by expanding it over a polynomial or a B-spline basis

$$s_i(t) = \sum_{l=0}^{L_i} c_i^{(l)} \phi_l(t). \tag{1.5}$$

The number of the basis functions is assumed to be the same for each gene: $L_i = L, i = 1, \cdots, N$. Therefore, each function is described by its $(L + 1)$ dimensional vector of coefficients $\mathbf{c}_i$. The hypotheses to be tested are therefore of the form $H_{0i} : \mathbf{c}_i \equiv \mathbf{0}$ versus $H_{1i} : \mathbf{c}_i \not\equiv \mathbf{0}$.

The vector of coefficients is estimated using the penalized least squares algorithm. In order to choose the dimension of the expansion (1.5), Storey et al. (2005) apply "eigen-gene" technique suggested by Alter et al. (2000). The basic idea of the approach is to take a singular value decomposition of the data and extract the top few "eigen-genes," which are vectors in the gene space. For those genes the number of basis functions is determined by generalized cross-validation. After that, this estimated value of L is used to fit the curves for all genes under investigation. The statistic for testing whether gene i is differentially expressed is analogous to the t or F statistics, which are commonly used in the static differential expression case:

$$F_i = (SS_i^0 - SS_i^1)/SS_i^1$$

where SS_i^0 is the sum of squares of the residuals obtained from the model fit under the null hypothesis, and SS_i^1 is the analogous quantity under the alternative hypothesis. The larger F_i is, the better the alternative model fit is over the null model fit. Therefore, it is reasonable to rank all the genes by the size of F_i, i.e., gene i is considered to be differentially expressed if $F_i \geq c$ for some cutoff point c. Since non-Gaussian heavy-tailed errors are very common for microarray data, the distribution of statistics F_i under the null hypothesis is assumed to be unknown. For this reason, the p-value for each gene is calculated using bootstrap resampling technique as the frequency by which the null statistics exceed the observed statistic.

However, although the p-value is a useful measure of significance for testing an individual hypothesis, it is difficult to interpret a p-value when thousands of hypotheses H_{0i}, $i = 1, \cdots, N$, are tested simultaneously. For this reason, the authors estimate q-values, which are the false discovery rate (FDR) related measure of significance. The q-values are evaluated on the basis of the p-values using an algorithm described in Storey and Tibshirani (2003a).

The approach of Storey et al. (2005) is very flexible: no parametric assumptions are made about distributions of the errors, moreover, methodology allows to take into account variations between individuals. However, this flexibility comes at a price: the technique requires resampling methods, which may be rather formidable to a practitioner.

1.4.3   Empirical Bayes Functional Data Approach

Angelini et al. (2007) propose an empirical Bayes functional data approach to identification of the differentially expressed genes, which is a good compromise between a rather inflexible method of Tai and Speed (2006) and a computationally expensive technique of Storey et al. (2005).

Similarly to Storey et al. (2005), in Angelini et al. (2007) each curve $s_i(t)$ is expanded over an orthonormal basis (Legendre polynomials or Fourier) as in (1.5) and is characterized by the vector of its coefficients $\mathbf{c}_i$, although the lengths $(L_i + 1)$ of the vectors may be different for different genes.

The genes are assumed to be conditionally independent, so that for each i, $i = 1, \ldots, N$ the data is modeled as

$$\mathbf{z}_i = \mathbf{D}_i \mathbf{c}_i + \boldsymbol{\zeta}_i \tag{1.6}$$

where $\mathbf{z}_i = (z_i^{1,1} \ldots z_i^{1,k_i^{(1)}}, \cdots, z_i^{n,1}, \ldots z_i^{n,k_i^{(n)}})^T \in R^{M_i}$ is the column vector of all measurements for gene i (see (1.1)), $\mathbf{c}_i = (c_i^0, \ldots, c_i^{L_i})^T \in R^{L_i+1}$ is the column vector of the coefficients of $s_i(t)$ in the chosen basis, $\boldsymbol{\zeta}_i = (\zeta_i^{1,1}, \ldots, \zeta_i^{1,k_i^{(1)}}, \cdots, \zeta_i^{n,1}, \ldots, \zeta_i^{n,k_i^{(n)}})^T \in R^{M_i}$ is the column vector of random errors and $\mathbf{D}_i$ is the $M_i \times (L_i + 1)$ block design matrix the j-row of which is the block vector $[\phi_0(t^{(j)}) \ \phi_1(t^{(j)}) \ \ldots \ \phi_{L_i}(t^{(j)})]$ replicated $k_i^{(j)}$ times.

The following Bayesian model is imposed on the data:

$\mathbf{z}_i \mid L_i, \mathbf{c}_i, \sigma^2 \sim \mathcal{N}(\mathbf{D}_i \mathbf{c}_i, \sigma^2 \mathbf{I}_{M_i});$

$L_i \sim Pois^*(\lambda, L_{\max}),$ Poisson with parameter λ truncated at $L_{\max}$;

$\mathbf{c}_i \mid L_i, \sigma^2 \sim \pi_0 \delta(0, \ldots, 0) + (1 - \pi_0) \mathcal{N}(0, \sigma^2 \tau_i^2 \mathbf{Q}_i^{-1}).$

Here, π_0 is the prior probability that gene i is not differentially expressed. Matrix $\mathbf{Q}_i$ is a diagonal matrix that accounts for the decay of the coefficients in the chosen basis and depends on i only through its dimension.

Parameter σ^2 is also assumed to be a random variable $\sigma^2 \sim \rho(\sigma^2)$, which allows to accommodate possibly non-Gaussian errors (quite common in microarray experiments), without sacrificing closed form expressions for estimators and test statistics. In particular, the following three types of priors $\rho(\cdot)$ are considered:

case 1: $\rho(\sigma^2) = \delta(\sigma^2 - \sigma_0^2)$, the point mass at σ_0^2. The marginal distribution of the error is normal.

case 2: $\rho(\sigma^2) = IG(\gamma, b)$, the Inverse Gamma distribution. The marginal distribution of the error is Student T.

case 3: $\rho(\sigma^2) = c_\mu \sigma^{(M-1)} e^{-\sigma^2 \mu/2}$, where M is the total number of arrays available in the experimental design setup. If the gene has no missing data ($M_i = M$), i.e. all replications at each time point are available, then the marginal distribution of the error is double exponential.

The procedure proposed in Angelini et al. (2007) is carried out as follows. The global parameters ν, λ and $L_{\max}$ are defined by a user. The global parameters π_0 and case-specific parameters, σ_0^2 for case 1, γ and b for case 2, and μ for case 3 are estimated from the data using records on all the genes that have no missing values. The gene-specific parameter τ_i^2 is estimated by maximizing the marginal likelihood $p(\mathbf{z}_i)$ while L_i is estimated using the mode or the mean of the posterior pdf $p(L_i|\mathbf{z}_i)$.

In order to test the hypotheses $H_{0i} : \mathbf{c}_i = 0$ versus $H_{1i} : \mathbf{c}_i \neq 0$ for $i = 1, \ldots, N$, and estimate the gene expression profiles, Angelini et al. (2007) introduce, respectively, the conditional and the averaged Bayes factors

$$BF_i(\hat{L}_i) = \frac{(1 - \pi_0)P(H_{0i}|\mathbf{z}_i, \hat{L}_i)}{\pi_0 P(H_{10i}|\mathbf{z}_i, \hat{L}_i)}, \quad BF_i = \frac{(1 - \pi_0)P(H_{0i}|\mathbf{z}_i)}{\pi_0 P(H_{10i}|\mathbf{z}_i)}.$$

Bayes factors can be evaluated explicitly using the following expressions

$$BF_i(\hat{L}_i) = A_{0i}(\hat{L}_i)/A_{1i}(\hat{L}_i), \quad BF_i = \left[\sum_{L_i=0}^{L_{\max}} g_\lambda(L_i)A_{0i}(L_i)\right] \Bigg/ \quad (1.7)$$

$$\times \left[\sum_{L_i=0}^{L_{\max}} g_\lambda(L_i)A_{1i}(L_i)\right] \quad (1.8)$$

where

$$A_{0i}(L_i) = F(M_i, \mathbf{z}_i^T \mathbf{z}_i), \quad A_{1i}(L_i)$$
$$= \tau_i^{-(L_i+1)}[(L_i+1)!]^\nu |\mathbf{V}_i|^{-1/2} F(M_i, H_i(\mathbf{z}_i)),$$

$$\mathbf{V}_i = \mathbf{D}_i^T \mathbf{D}_i + \tau_i^{-2}\mathbf{Q}_i, \quad H_i(\mathbf{z}_i) = \mathbf{z}_i^T \mathbf{z}_i - \mathbf{z}_i^T \mathbf{D}_i \mathbf{V}_i^{-1} \mathbf{D}_i^T \mathbf{z}_i,$$

and

$$F(A,B) = \begin{cases} \sigma_0^{-2A} e^{-B/2\sigma_0^2} & \text{in case 1,} \\ \frac{\Gamma(A+\gamma)}{\Gamma(\gamma)} b^{-A}(1+\frac{B}{2b})^{-(A+\gamma)} & \text{in case 2,} \\ \frac{B^{(M+1-2A)/4}\mu^{(M+1+2A)/4}}{2^{(M-1)/2}\Gamma((M+1)/2)} K_{((M+1-2A)/2)}(\sqrt{B\mu}) & \text{in case 3,} \end{cases}$$

Here $K_h(\cdot)$ is the Bessel function of degree h (see Gradshteyn and Ryzhik [1980] for the definition and Angelini et al. [2007]) for additional details).

The authors note that the two Bayes factors defined in (1.7) lead to an almost identical list of selected genes with the selection on the basis of BF_i being more theoretically sound but also slightly more computationally expensive.

Bayes factors BF_i, although useful for independent testing of the null hypotheses H_{0i}, $i = 1, \ldots, N$, do not account for the multiplicity when testing a large number of hypotheses simultaneously. In order to take into account multiplicity, the authors apply a Bayesian multiple testing procedure of Abramovich and Angelini (2006).

Finally, the coefficients $\mathbf{c}_i$ for the differentially expressed genes are estimated by the posterior means

$$\hat{\mathbf{c}}_i = \frac{(1-\pi_0)/\pi_0}{BF_i(\hat{L}_i)+(1-\pi_0)/\pi_0} \left(\mathbf{D}_i^T \mathbf{D}_i + \hat{\tau}_i^{-2}\mathbf{Q}_i\right)^{-1} \mathbf{D}_i^T \mathbf{z}_i.$$

Methodology of Angelini et al. (2007) is much more computationally efficient than that of Storey et al. (2005): all Bayesian calculations use explicit formulae and, hence, are easy to implement. In addition, by allowing a different number of basis functions for each curve, it avoids the need to pre-determine the group of the most significant genes in order to select the dimension of the fit. Also, by using hierarchical Bayes procedure for multiplicity control, Angelini et al. (2007) avoid a somewhat ad-hoc evaluation of the p-values.

On the other hand, the algorithm of Angelini et al. (2007) is much more flexible than that of Tai and Speed (2006): it takes into account the time measurements allowing not necessarily equispaced time point and different number of replicates per time point. It also avoids tiresome missing data techniques and both ranks the genes and provides a cutoff point to determine which genes are differentially expressed. However, we note that in Angelini et al. (2007) the replicates are considered to be statistically indistinguishable, so that the method cannot take advantage of a longitudinal experimental design.

1.5 Statistical Methods for the Multi-Sample Microarray Experiments

Many of the procedures suggested in literature are specifically constructed for the two-sample or multi-sample design, or can be easily adapted to it. For example, the functional data approach of Conesa et al. (2006) is directly applicable to the multi-sample design and Storey et al. (2005) accommodates the two-sample design, requiring some changes for the extension to the multi-sample design.

1.5.1 Multivariate Bayes Methodology in the Two-Sample Case

Tai and Speed (2006) reduce the two-sample case to the one-sample case by assuming that the number of time points in both samples is the same, and also that the number of replications for each time point is constant, i.e., $n_1 = n_2 = n$ and $k_{\aleph i}^{(j)} = k_{\aleph i}$, $\aleph = 1, 2$. It is also supposed implicitly that time measurements are the same on both samples, i.e., $t_1^{(j)} = t_2^{(j)}$, $j = 1, .., n$. Under these restrictions, Tai and Speed (2006) model the k-th time series for gene i in sample $\aleph$ as $\mathbf{z}_{i,k}^{\aleph} \sim \mathcal{N}(\boldsymbol{\mu}_{\aleph i}, \boldsymbol{\Sigma}_i)$, $\aleph = 1, 2$. The prior imposed on the vector $\boldsymbol{\mu}_i = \boldsymbol{\mu}_{2i} - \boldsymbol{\mu}_{1i}$ is identical to the prior in the Section 1.4.1, and the results of the analysis are very similar too. This model does not easily adapt to a general multi-sample setup.

1.5.2 Estimation and Testing in Empirical Bayes Functional Data Model

Angelini et al. (2009) propose an extension of the empirical Bayes functional data procedure of Section 1.4.3 to the case of the two-sample and multi-sample experimental design. For a gene i in the sample $\aleph$, it is assumed that evolution in time of its expression level is governed by a function $s_{\aleph i}(t)$ and each of the measurements $z_{\aleph i}^{j,k}$ involves some measurement error. For what concerns the two-sample problem, the quantity of interest is the difference between expression levels $s_i(t) = s_{1i}(t) - s_{2i}(t)$. In particular, $s_i(t) \equiv 0$ means that gene i is not differentially expressed between the two samples, while $s_i(t) \not\equiv 0$ indicates that it is. In the case of a multi-sample setup, a variety of statistical hypotheses can be tested. However, in the present paper we focus on the problem of identification of the genes that are differentially expressed under at least one biological condition.

Although mathematically the two-sample setup appears homogeneous, in reality it is not. In fact, it may involve comparison between the cells under different biological conditions (e.g., for the same species residing in different parts of the habitat) or estimating an effect of a treatment (e.g., effect of estrogen treatment on a breast cell). The difference between the two cases is that, in the first situation, the two curves $s_{1i}(t)$ and $s_{2i}(t)$ are truly interchangeable, while in the second case they are not: $s_{2i}(t) = s_{1i}(t) + s_i(t)$, where $s_i(t)$ is the effect of the treatment. In order to conveniently distinguish between those cases, Angelini et al. (2009) refer to them as "interchangeable" and "non-interchangeable" models. It is worth noting that which of the two models is more appropriate should be decided rather by a biologist than a statistician.

Similarly to the one-sample case, Angelini et al. (2009) expand each function $s_{\aleph i}(t)$ over a standard orthogonal basis on the interval $[0, T]$ (see eq. (1.5)), characterizing each $s_{\aleph i}(t)$ by the vector of its coefficients $\mathbf{c}_{\aleph i}$. The number of coefficients $L_i + 1$ can vary from gene to gene but it is required to be the same in all samples. Similarly to the one-sample case (see eq. (1.6)) one has

$$\mathbf{z}_{\aleph i} = \mathbf{D}_{\aleph i}\mathbf{c}_{\aleph i} + \boldsymbol{\zeta}_{\aleph i} \qquad (1.9)$$

where $\mathbf{z}_{\aleph i} = (z_{\aleph i}^{1,1} \ldots z_{\aleph i}^{1,k^{(1)}}, \cdots, z_{\aleph i}^{n,1}, \ldots z_{\aleph i}^{n,k^{(n)}})^T \in R^{M_{\aleph i}}$ is the column vector of all measurements for gene i in sample $\aleph$, $\mathbf{c}_{\aleph i}$ is the column vector of the coefficients of $s_{\aleph i}(t)$, $\boldsymbol{\zeta}_{\aleph i}$ is the column vector of random errors and $\mathbf{D}_{\aleph i}$ is the $M_{\aleph i} \times (L_i + 1)$ block design matrix the j-row of which is the block vector $[\phi_0(t_\aleph^{(j)}) \; \phi_1(t_\aleph^{(j)}) \; \cdots \; \phi_{L_i}(t_\aleph^{(j)})]$ replicated $k_{\aleph i}^{(j)}$ times. The following Bayesian model is imposed

$$\mathbf{z}_{\aleph i} \mid L_i, \mathbf{c}_{\aleph i}, \sigma^2 \;\sim\; \mathcal{N}(\mathbf{D}_{\aleph i}\mathbf{c}_{\aleph i}, \sigma^2 \mathbf{I}_{M_{\aleph i}})$$
$$L_i \;\sim\; Pois^*(\lambda, L_{\max}), \text{ Poisson with parameter } \lambda \text{ truncated at } L_{\max}.$$

The vectors of coefficients $\mathbf{c}_{\aleph i}$ are modeled differently in the interchangeable and noninterchangeable cases.

Interchangeable setup

Since the choice of which sample is the first, which is the second and so on, is completely arbitrary, it is reasonable to assume that a-priori vectors of the coefficients are either equal to each other or are independent and have identical distributions. The model naturally accommodates the H-sample design by eliciting the following delta-contaminated prior on the vectors $\mathbf{c}_{\aleph i}$, $\aleph = 1, \cdots, H$

$$
\begin{aligned}
\mathbf{c}_{1i}, \cdots, \mathbf{c}_{Hi} \mid L_i, \sigma^2 \sim{} & \pi_0 \mathcal{N}(\mathbf{c}_i \mid \mathbf{0}, \sigma^2 \tau_i^2 \mathbf{Q}_i^{-1}) \, \delta(\mathbf{c}_{1i} = \mathbf{c}_{2i} = \cdots \mathbf{c}_{Hi} = \mathbf{c}_i) \\
& + (1 - \pi_0) \textstyle\prod_{\aleph=1}^{H} \mathcal{N}(\mathbf{c}_{\aleph i} \mid \mathbf{0}, \sigma^2 \lambda_i^2 \mathbf{Q}_i^{-1}).
\end{aligned}
$$

where $\mathbf{0} = (0, \cdots, 0)^T$.

Gene i is considered to be differentially expressed if the null hypothesis $H_{0i} : \mathbf{c}_{1i} \equiv \mathbf{c}_{2i} \equiv \cdots \equiv \mathbf{c}_{Hi}$ is rejected.

Non-interchangeable setup

Consider the case when samples 1 and 2 are not interchangeable so that one cannot assume gene expression levels to be the same a-priori in both samples. It is intuitive to treat the expression level in sample 2 as the sum of the expression level in sample 1 and the effect of the treatment $\mathbf{d}_i$, i.e. $\mathbf{c}_{2i} = \mathbf{c}_{1i} + \mathbf{d}_i$ where $\mathbf{d}_i = \mathbf{c}_{2i} - \mathbf{c}_{1i}$. The prior imposed on the coefficients are a normal prior on $\mathbf{c}_{1i}$ and a delta contaminated prior on $\mathbf{d}_i$:

$$
\begin{aligned}
\mathbf{c}_{1i} \mid L_i, \sigma^2 &\sim \mathcal{N}(\mathbf{c}_{1i} \mid \mathbf{0}, \sigma^2 \tau_i^2 \mathbf{Q}_i^{-1}) \\
\mathbf{d}_i \mid L_i, \sigma^2 &\sim \pi_0 \delta(\mathbf{d}_i = \mathbf{0}) + (1 - \pi_0) \mathcal{N}(\mathbf{d}_i \mid \mathbf{0}, \sigma^2 \lambda_i^2 \mathbf{Q}_i^{-1}).
\end{aligned}
$$

Gene i is considered not to be differentially expressed if the null hypothesis $H_{0i} : \mathbf{d}_i \equiv 0$ is accepted. Of course, this setup does not allow a straightforward extension to the case of $H > 2$ samples.

In both setups, parameter σ^2 is treated as a random variable $\sigma^2 \sim \rho(\sigma^2)$ with the same three choices for $\rho(\sigma^2)$ that are considered in Section 1.4.3. Parameter π_0 is the prior probability that a gene is not affected by a particular biological condition in the interchangeable model, or by a particular treatment in the non-interchangeable model. Matrix $\mathbf{Q}_i$ accounts for the decay of the coefficients in the chosen basis.

In both setups, the method of Angelini et al. (2009) requires user-defined parameters ν, λ and $\mathrm{L_{max}}$ similarly to the one-sample situation. The authors recommend some empirical procedures for the evaluation of global parameter π_0, and case specific parameters σ_0^2 for case 1, γ and b for case 2 and μ for case 3. The gene-specific parameters τ_i^2 and λ_i^2 are estimated by maximizing $p(\mathbf{z}_{1i}, ..., \mathbf{z}_{Hi})$ ($H = 2$ in the non-interchangeable setup). The authors observe that in the interchangeable case this maximization step can be done as two independent one-dimensional optimization, while in the non-interchangeable setup this step requires a two-variable maximization procedure and therefore leads to a higher computational cost. Given the estimated values of τ_i^2 and λ_i^2, L_i is estimated by the mode or the mean of the posterior pdf $p(L_i \mid \mathbf{z}_{1i}, ..., \mathbf{z}_{Hi})$. Similarly to the one-sample setup, Angelini et al. (2009) propose to use the

Bayes factors BF_i for testing hypotheses H_{0i} and again the two kinds of Bayes factor are considered

$$BF_i(\hat{L}_i) = \frac{(1 - \pi_0)P(H_{0i}|\mathbf{z}_{1i}, ..., \mathbf{z}_{Hi}, \hat{L}_i)}{\pi_0 P(H_{10i}|\mathbf{z}_{1i}, ..., \mathbf{z}_{Hi}, \hat{L}_i)},$$

$$BF_i = \frac{(1 - \pi_0)P(H_{0i}|\mathbf{z}_{1i}, ..., \mathbf{z}_{Hi})}{\pi_0 P(H_{10i}|\mathbf{z}_{1i}, ..., \mathbf{z}_{Hi})}, \tag{1.10}$$

with $H = 2$ in the non-interchangeable setup. The multiplicity control is carried out similarly to the one-sample situation by using the Bayesian procedure of Abramovich and Angelini (2006).

For brevity, we do not report the formulae since they are quite complex mathematically and require some elaborate matrix manipulations, especially in the non-interchangeable setup (they can be found in Angelini et al. [2009]). Similarly to the one sample case, Angelini et al. (2009) also provide analytic expressions for the estimators of the gene expression profiles in both interchangeable and non-interchangeable setups.

Empirical Bayes functional data approach was also considered in the papers of Tai and Speed (2007) and Hong and Li (2006).

Tai and Speed (2008) extend their multivariate Bayes methodology to the multi-sample cross-sectional design where the biological conditions under investigation are interchangeable. The authors consider a Bayesian functional data approach. They expand the mean of the multivariate normal distribution, which models the data over a splines basis, and assume the number of basis functions to be a fixed constant chosen in advance. The model is similar to that of Angelini et al. (2009) with $\rho(\sigma^2)$ being Inverse Gamma (case 2 above). Also, the procedure only ranks the genes without providing any automatic cutoff point to select the differentially expressed genes.

Hong and Li (2006) propose a Bayesian functional data model for detecting differentially expressed genes in the case of cross-sectional design and non-interchangeable biological conditions. The model is somewhat similar to the non-interchangeable setup of Angelini et al. (2009) described above. Hong and Li (2006) expand the gene expression profiles over a polynomial or B-spline basis of fixed degree, which is globally estimated by cross validation. The authors impose a prior on $\mathbf{d}_i$ similar to (1.5.2) and treat $\mathbf{c}_{1i}$ as unknown nuisance parameters. They also elicit an Inverse Gamma prior on the error variance allowing experiment-specific, gene-specific, and time-specific errors $(\sigma^2)_{\aleph,i}^{(j)}$. The gene-specific parameters and the hyperparameters of the model are estimated using computationally expensive Monte Carlo EM algorithm.

In order to test hypotheses H_{0i}, Hong and Li (2006) evaluate posterior probabilities that $\mathbf{d}_i \neq \mathbf{0}$ and then rank the genes accordingly. Alternatively, they provide an automatic cutoff point by controlling the false discovery rate (FDR) with a procedure similar to Storey (2003c). Similarly to Angelini et al. (2009), Hong and Li (2006) method does not require the measurements for

the two samples to be taken at the same time points and provide estimators
for the gene expression profiles.

1.5.3 Bayesian Hierarchical Model with Subject-Specific Effect

Chi et al. (2007) develop a Bayesian hierarchical model for identification
of temporally differentially expressed genes under multiple biological conditions, which allows possible subjects-specific effects, i.e., the presence of biological replicates. In order to account for the subject-specific deviation from
the mean expression of the i-th gene in the $\aleph$-th sample of the individual s,
$s = 1, \ldots, S$, random effect $\gamma_{i\aleph s}$ is added to the right-hand side of the eq.
(1.9). The subject-specific effect is considered to be common across the genes
to avoid overparametrization, so that $\gamma_{i\aleph s} = \gamma_{\aleph s}$ for $i = 1, \cdots, N$.

Therefore, the data structure in matrix form is as follows:

$$\mathbf{z}_{\aleph i}^s = \mathbf{D}_{\aleph i}^s \mathbf{c}_{\aleph i} + \gamma_{\aleph s} \mathbf{1}_{M_{\aleph i}^s} + \boldsymbol{\zeta}_{i\aleph s} \tag{1.11}$$

where $\mathbf{1}_{M_{\aleph i}^s}$ is an $M_{\aleph i}^s \times 1$ vector of ones the length of which is determined
by the number of observations available and is allowed to differ across the
genes, the biological conditions, and the subjects to accommodate for possible
missing data. Note that if there is only one subject, i.e., $S = 1$, the structure
of the data is described by eq. (1.9).

Given the parameters, the data $\mathbf{z}_{\aleph i}^s$ is assumed to have a multivariate normal
distribution, i.e., $\boldsymbol{\zeta}_{i\aleph s} \sim \mathcal{N}(0, \sigma_i^2 I_{M_{\aleph i}^s})$ and $\gamma_{\aleph s} \sim \mathcal{N}(0, \tau^2)$ is independent from
$\boldsymbol{\zeta}_{i\aleph s}$. The variances of the measurement errors are allowed to differ across the
genes. Then, a double-layer hierarchical prior is elicited over the set of param-
eters $\boldsymbol{\theta} = \{\mathbf{c}_{\aleph i}, \aleph = 1, ..., H, i = 1, ...N; \sigma_i^2, i = 1, ...N; \tau^2\}$, and an empirical
Bayes approach is used to estimate the hyperparameters. The prior specifi-
cation leads to a conjugate posterior, which makes the computations more
efficient. In order to identify differentially expressed genes, first expectations
of the gene expression measurements $\mathbf{z}_{\aleph i}^{s\star} = E(\mathbf{z}_{\aleph i}^s | \boldsymbol{\theta})$ given the parameters
$\boldsymbol{\theta}$ are calculated. Then p-dimensional vectors $\boldsymbol{\psi}_{i,\aleph} = (\mathbf{D}_{\aleph i}^T \mathbf{D}_{\aleph i})^{-1} \mathbf{D}_{\aleph i}^T \mathbf{z}_{\aleph i}^{s\star}$ are
defined, where

$$\mathbf{D}_{\aleph i} = \left[\mathbf{D}_{\aleph i}^{1T}, \ldots, \mathbf{D}_{\aleph i}^{ST}\right]^T \quad \text{and} \quad \mathbf{z}_{\aleph i}^\star = \left[\mathbf{z}_{\aleph i}^{1\star}, \ldots, \mathbf{z}_{\aleph i}^{S\star}\right]^T .$$

and p is the number of covariates. For the i-th gene and the $\aleph$-th biological condition, each element of $\psi_{i,\aleph}$ represents the effect of the corresponding
covariate, and the ratio parameter based on these effects determines the maximum fold changes in the gene expression in response to each covariate. In
order to determine whether a gene is differentially expressed, a hypothesis
that a ratio parameter exceeds certain threshold (one, two, or three) is tested
using a version of the criterion proposed in Ibrahim et al. (2002).

1.5.4 Hidden Markov Models

Yuan and Kendziorski (2006) introduce hidden Markov chain based Bayesian models (HMM) for analysis of the time series microarray data. It is assumed that the vector of expression levels for gene i at time t under the biological conditions $\aleph = 1, \cdots, H$, is governed by an underlying state of a hidden Markov chain, i.e., $(\mathbf{z}_i^{(t)} | s_i^{(t)} = l) \sim f_{lt}(\mathbf{z}_i^{(t)})$ where l is the possible state of the HMM, and the length of the vector is equal to the number of measurements at time t under all H biological conditions. In the case of $H = 2$ biological conditions, the HMM has two states: genes are either equally expressed ($l = 1$) or differentially expressed ($l = 2$). In a general situation, the number of states of the HMM is equal to the number of nonempty subsets of the set with H elements.

The vectors $\mathbf{z}_i^{(t)}$ are assumed to have means $\boldsymbol{\mu}_i^{(t)}$, which are distributed according to some prior pdf. Unknown hyperparameters of the model are estimated by maximizing the marginal likelihood of the data on the basis of a version of EM algorithm. In the case of $H = 2$, the decision on whether gene i is differentially expressed is made by evaluating posterior odds of $P(s_i^{(t)} = 2)$; for general case, posterior probabilities can be evaluated for each configuration. In the particular version of the HMM model described in Yuan and Kendziorski (2006), the authors use Gamma distribution for both the data and the underlying means.

The paper provides a paradigm for modeling time series microarray data, which is very different from the methodologies described above. However, the weakness of the approach is that it tests whether a gene is differentially expressed at a particular time point not overall, thus, ignoring the temporal dependence between the expressions.

1.6 Software

Microarray data analysis is a very active area of research and several software packages have been recently released. Private companies who produce microarray equipment have their proprietary software, and several commercial software packages are now available. The use of the commercial software systems is however often limited by the cost of the license and the delay in incorporating innovative research techniques into the new releases. On the other hand, there are also a number of open source software packages for analysis of microarray data, the most widely used ones being distributed by *Bioconductor* (http://www.bioconductor.org/), an open source software project for the analysis and comprehension of genomic data (see Gentleman et al. [2004] and Gentleman et al. [2005]), which is based on R (http://www.r-project.org/).

Several packages include tools for analysis of time course microarray experiments; however, majority of them use methodology constructed for static microarray experiments and for a specific experimental design (see, for example, *SAM* version 3.0 software package originally proposed by Tusher et al. [2001] and later described in Storey et al. [2003a, 2003b], *MAANOVA* by Kerr et al. [2000] and Wu et al. [2003] and the *Limma* package by Smyth [2005]). On the other hand, recently other software packages have been developed that are specifically designed for analysis of time course microarray data. It should be noted that when the time series are very short ($n = 3$ or 4), the functional data analysis methodsm which are discussed in this review, are not recommended since simpler methods are more advantageous in this case, e.g., *MaSigPro*, the linear regression approach of Conesa et al. (2006), the *STEM* package by Ernst and Bar-Joseph (2006), or the *Timecourse* package of Tai and Speed (2006).

In what follows, we briefly present the software packages that implement some of the methodologies discussed in this review. We note that all these software packages require data to be already pre-processed to remove systematic sources of variation (see Gentleman et al. [2005] or Parmigiani et al. [2003] for discussion of the software).

EDGE

Software package *EDGE* (see Leek et al. [2006] for details) implements the functional data approach proposed in Storey et al. (2005). The software is distributed as a stand-alone package and can be freely downloaded from http://www.genomine.org/edge/ for academic non-commercial users. It relies on R language and *Bioconductor* in the background, however a user does not need to know R to use it but only to interact with a graphical user friendly interface.

Applications of *EDGE* are not limited to time-course microarray data: EDGE can also be employed for static experiments. Moreover, it also has some useful utilities that allow one to explore input data, such as drawing boxplots, checking for the presence of missing data, and imputing them with the "K nearest neighbor" (KNN) algorithm. In fact, *EDGE* does not allow missing data in the course of statistical analysis.

The software requires a user to define a number of permutations to be performed for evaluating the p-values. The shortcoming of this feature is that a relatively small number of permutations lead to a granularity problem (the p-values are quantized), while increasing number of permutations usually improves approximation of the p-values, but leads to a very high computational cost.

TIMECOURSE

The package *Timecourse* implements the multivariate Bayesian approach of Tai and Speed (2006) for both the one-sample and the two-sample designs. *Timecourse* is distributed as contributed package within Bioconductor (http://bioconductor.org/packages/2.1/bioc/html/timecourse.html), thus, its application requires the knowledge of the *R*-language. Although this can be advantageous to *R* users since *Timecourse* allows them to interact with other tools that are developed under Bioconductor project, this feature can be rather intimidating for a biologist or a person unfamiliar with *R* language. For example, the package neither allows missing data or unequal number of replicates, nor it provides any tools for handling these problems, although expert *R* users can import such tools from other *R* packages.

Since all calculations are based on explicit formulae, the computational cost of *Timecourse* is very low. However, the package only ranks the genes in the order of statistical significance without providing an automatic cutoff point to identify which genes are differentially expressed. Finally, the results of the analysis are completely independent of the actual time measurements, so that it is preferable to apply the software only when the grid is equispaced. On the other hand, the package can be successfully applied for short time-course experiments.

Current version of *Timecourse* does not include the functional approach described in Tai and Speed (2007), which we hope will be added in future releases.

BATS

The software package *BATS* (see Angelini et al. [2008]) implements the Bayesian functional approach proposed by Angelini et al. (2007). It is also distributed as a stand-alone software that can be freely downloaded from http://www.na.iac.cnr.it/bats/.

The software has a friendly graphical user interface. Although it is written in MATLAB®, it does not require a user to know MATLAB®.

Similarly to *EDGE*, *BATS* also has some helpful utilities, which allow to plot and filter the data, and compare the lists of differentially expressed genes obtained under different options. Due to flexibility of the methodology, the issue of missing data does not arise. Furthermore, by using the Bayesian methodology in combination with the functional data approach, *BATS* can successfully handle various other technical difficulties such as non-uniform sampling intervals, presence of multiple data as well as temporal dependence between observations for each gene, which are not fully addressed by *EDGE* and *Timecourse*. Since all calculations are based on explicit expressions, the computational cost of *BATS* is very low and is comparable to that of *Timecourse*.

The current version of BATS can be applied to a one-sample case only. The software allows different number of basis functions for each curve $s_i(t)$, thus improving the fit and avoiding the need to pre-determine the most significant genes to select the dimension of the approximation.

The two-sample and the multi-sample experimental designs will be accommodated in future releases by following methodology of Angelini et al. (2009) described above.

1.7   Comparisons between Techniques

Each of the methods presented above have been tested on simulated and real data (see Storey et al. [2005], Tai and Speed [2006, 2007], Hong and Li [2006] and Angelini et al. [2007, 2009]). However, each study considered only a single technique and did not have an objective of comparison between competitive approaches. To the best of the authors' knowledge, Angelini et al. (2007) and Mutarelli et al. (2008) are the only papers where a comparative study for the one-sample problem was attempted between the *EDGE* software, the R-package *Timecourse* and *BATS* software; in the following this comparative study is briefly presented (see Angelini et al. [2007] for details).

Since all three methods apply to different experimental designs, account for different biological information, and are valid under different assumptions, the methods were compared using a real data set that does not conform to specific assumptions. For comparisons, the cDNA microarray dataset of Cicatiello et al. (2004) was chosen (the data are available on the GEO repository - http://www.ncbi.nlm.nih.gov/geo/, accession number GSE1864) since it provides a "biology-guided" selection of significant genes that can be used as a "benchmark" in comparisons.

The objective of the experiment was to identify genes involved in the estrogen response in a human breast cancer cell line and to estimate the type of the response. Estrogen has a known role in promoting cell proliferation and thus in cancer development in hormone-responsive tissues such as breast and ovary. In the original experiment, ZR-75.1 cells were stimulated with a mitogenic dose of 17β-estradiol, after 5 days of starvation on an hormone-free medium, and samples were taken after $t = 1, 2, 4, 6, 8, 12, 16, 20, 24, 28, 32$ hours, with a total of 11 time points covering the completion of a full mitotic cycle in hormone-stimulated cells. For each time point at least two replicates were available (three replicates at $t = 2, 8, 16$).

The data was pre-processed and genes for which more than 20% of values were missing were removed from further investigation. As a result, the total number of analyzed genes was 8161 (among them about 350 contained at least one missing value). Since the *EDGE* software does not automatically account for missing values but only suggests a preliminary procedure (K-nearest-neighbors) for filling them in, the authors repeated the analysis both using this procedure and filtering out genes with missing values. *Timecourse* neither allows missing values nor suggests a specific procedure for treating them. Moreover, it requires that each time point has the same number of replicates (different number of replicates are allowed between different genes). In order to apply *Timecourse*, all the genes with missing observations were filtered out, and then the third observations, which was available at time points $t = 2, 8, 16$, were discarded.

TABLE 1.1:   Total Number of Genes Declared Affected by the Treatment and Overlap with Cicatiello et al. (2004)

Method	Selected genes	Overlap
All of the Angelini et al. 28 methods	574	270
At least one of the Angelini et al. 28 methods	958	309
Angelini et al., case 1, $\lambda = 9$ (default)	712	295
EDGE with default choices and q=0.05	767	186
EDGE with default choices and q=0.1	1178	219
Timecourse	500	174
Timecourse	1000	215

BATS was used with $L_{\max} = 6$, $\nu = 0$ and λ's ranging between 6 and 12 (7 values total) and three choices of $\rho(\sigma^2)$. In addition, in case 2, two possible techniques for estimation of hyperparameters of the inverse Gamma distribution were applied making the total of 28 combinations of techniques.

Table 1.1 displays the number of detected genes with different procedures and the overlap with the 344 genes detected as significant in Cicatiello et al. (2004).

It is evident that *BATS* has a noticeably wider overlap with the "biology guided" selection of differentially expressed genes of Cicatiello et al. (2004) than *EDGE* and *Timecourse*. In addition, *BATS* is the only software that allows an user to estimate the effect of the treatment as it is shown in Figure 1.1. Moreover, majority of genes selected by *EDGE*, R-package *Timecourse* and Cicatiello et al. (2004) were also selected by *BATS*.

1.8   Discussion

In this paper we discussed a variety of recently developed Bayesian techniques for detection of differentially expressed genes on the basis of time series microarray data. These techniques have different strengths and weaknesses and are constructed for different kinds of experimental designs. However, the Bayesian formulations that the above described methodologies utilize, allow one to explicitly use prior information that biologists may provide. The methods successfully deal with various technical difficulties that arise in microarray time-course experiments such as a small number of observations, non-uniform sampling intervals, missing or multiple data, and temporal dependence between observations for each gene.

We also describe several software packages that implement methodologies discussed in the paper, and provide a limited comparison between the techniques using a real data example. We hope that this paper provides a useful introduction to Bayesian methods for analysis of time-series microarray

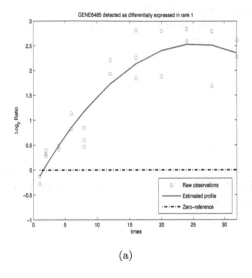

(a)

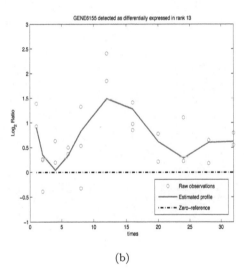

(b)

FIGURE 1.1: (a) Gene6485 (TFF1, a well-known target of the estrogen receptor) has been selected with rank 1 by BATS and included in the list of 574 genes selected by all the 28 combinations. This gene has been detected in Cicatiello et al. (2004) and by *Timecourse* and EDGE as well. (b) Gene6155 (MKI67, a gene involved in cell-cycle control but with a less clear association with estrogen action in literature) has been selected with rank 13 by BATS and included in the list of 574 genes selected by all the 28 combinations. This gene has not been detected by Cicatiello et al. (2004) or EDGE (with q-value=0.1), while it has been detected by *Timecourse* with rank 2. Profile showed here are obtained with BATS with choice case 1, $\lambda = 9$, $L_{max} = 6$ and $\nu = 0$.

experiments, though, try as we might, it cannot be made comprehensive since new statistical methodologies are coming up continuously as new microarray technologies are being developed.

Acknowledgments

This research was supported in part by the NSF grants DMS-0505133 and DMS-0652524, the CNR-Bioinformatics project, and the CNR DG.RSTL.004.002 project.

References

Abramovich, F. and Angelini, C. 2006. Bayesian maximum a posteriori multiple testing procedure. *Sankhya* 68: 436-460.

Alter, O., Brown, P.O., Botstein, D. 2000. Singular value decomposition for genome-wide expression data processing and modeling. *Proceedings of the National Academy of Sciences* 97: 10101-10106.

Angelini, C., De Canditiis, D., Mutarelli, M., and Pensky, M. 2007. A Bayesian Approach to Estimation and Testing in Time-course Microarray Experiments. *Statistical Applications in Genetics and Molecular Biology* 6, Iss.1 (Art.24). http://www.bepress.com/sagmb/vol6/iss1/art24

Angelini, C., Cutillo, L., De Canditiis, D., Mutarelli, M., Pensky, M. 2008. BATS: a Bayesian user-friendly software for Analyzing Time Series microarray experiments. *BMC: Bioinformatics*, 9, (Art. 415). http://www.biomedcentral.com/1471-2105/9/415

Angelini, C., De Canditiis, D., Pensky, M. 2009. Bayesian models for the two-sample time-course microarray experiments. *Computational Statistics & Data Analysis* 53: 1547–1565.

Bar-Joseph, Z. 2004. Analyzing time series gene expression data. *Bioinformatics* 20: 2493-2503.

Bar-Joseph, Z., Gerber, G., Jaakkila, T., Gifford, D., and Simon, I. 2003a. Comparing the continuous representation of time series expression profiles to identify differentially expressed genes. *Proc. Nat. Acad. Sci. USA* 100: 10146-10151.

Bar-Joseph, Z., Gerber, G., Jaakkila, T., Gifford, D., and Simon, I. 2003b. Continuous representation of time series gene expression data. *J. Comput. Biol.* 3-4: 341-356.

Chi, Y-Y, Ibrahim, J. G., Bissahoyo, A., and Threadgill, D. W. 2007. Bayesian Hierarchical Modeling for time course microarray experiments. *Biometrics* 63: 496-504.

Cicatiello, L., Scarfoglio, C., Altucci, L., et al. 2004. A genomic view of estrogen actions in human breast cancer cells by expression profiling of the hormone-responsive trascriptome. *Journal of Molecular Endocrinology* 32: 719-775.

Conesa, A., Nueda, M.J., Ferrer, A., and Talon, M. 2006. MaSigPro: a method to identify significantly differential expression profiles in time-course microarray-experiments.*Bioinformatics* 22: 1096-1102.

Cui, X., Kerr, M.K., and Churchill, G. A. 2003. Transformation for cDNA Microarray Data. *Statistical Applications in Genetics and Molecular Biology* 2, Iss.1 (Art.4). http://www.bepress.com/sagmb/vol2/iss1/art4

de Hoon, M.J.L., Imoto, S., and Miyano, S. 2002. Statistical analysis of a small set of time-ordered gene expression data using linear splines. *Bioinformatics* 18: 1477-1485.

Di Camillo, B., Sanchez-Cabo, F., Toffolo, G., Nair, S.K., Trajanosky, Z., and Cobelli, C. 2005. A quantization method based on threshold optimization for microarray short time series. *BMC Bioinformatics* 6.

Dudoit, S., Yang, Y.H., Callow, M. J., and Speed, T. P. 2002. Statistical methods for identifying differentially expressed genes in replicated cDNA microarray experiments. *Statistica Sinica* 12: 111-140.

Efron, B., Tibshirani, R., Storey, J.D., and Tusher, V. 2001. Empirical Bayes Analysis of a Microarray Experiment. *J. Amer. Statist. Assoc.* 96: 1151-1160.

Ernst, J. and Bar-Joseph, Z. 2006. STEM: a tool for the analysis of short time series gene expression data.*BMC Bioinformatics* 7: 191-192.

Gentleman, R. C., Carey, V. J., Bates, D. M. et al. 2004. *Bioconductor: open software development for computational biology and bioinformatics,Genome Biology.* http://bioconductor.org/packages/release/Software.html

Gentleman, R., Carey, V., Huber, W., Irizarry, R. and Dudoit, S. 2005. *Bioinformatics and Computational Biology Solutions Using R and Bioconductor.* Springer, series in Statistics for Biology and Health.

Hong, F. and Li, H. 2006. Functional hierarchical models for identifying genes with different time-course expression profiles. *Biometrics* 62: 534-544.

Kerr, M.K., Martin, M., and Churchill, G.A. 2000. Analysis of variance for gene expression microarray data. *Journal of Computational Biology* 7: 819-837.

Ibrahim, J.G., Chen, M.H., and Gray, R.J. 2002. Bayesian models for gene expression with DNA microarray data. *J. Amer. Statist. Assoc.* 97: 88-99.

Ishwaran, H., and Rao, J.S. 2003. Detecting differentially expressed genes in microarrays using Bayesian model selection. *J. Amer. Statist. Assoc.* 98: 438-455.

Leek, J.T., Monsen, E., Dabney, A.R., and Storey, J.D. 2006. EDGE: extraction and analysis of differential gene expression. *Bioinformatics* 22: 507-508.

Lonnstedt, I., and Speed, T. 2002. Replicated microarray data. *Statistica Sinica*, 12: 31-46.

McLachlan, G., Do, K.A., and Ambroise, C. 2004. *Analyzing microarray gene expression data.* Wiley series in Probability and Statistics.

Mutarelli, M., Cicatiello, L., Ferraro, L. et al. 2008. Time-course analysis of genome-wide gene expression data from hormone-responsive human breast cancer cells. *BMC Bioinformatics* 9: S12.

Park, T., Yi, S.G., Lee, S. et al. 2003. Statistical tests for identifying differentially expressed genes in time course microarray experiments. *Bioinformatics* 19: 694-703.

Parmigiani, G., Garrett, E.S., Irizarry, R., and Zeger, S.L. 2003. *The analysis of gene expression data: methods and softwarc.* Springer New York.

Smyth, G.K. 2004. Linear models and empirical Bayes methods for assessing differential expression in microarray experiments. *Statistical Applications in Genetics and Molecular Biology* 3, Iss.1 (Art.3). http://www.bepress.com/sagmb/vol3/iss1/art3

Smyth, G.K. 2005. Limma: linear models for microarray data. In *Bioinformatics and Computational Biology Solutions using R and Bioconductor*, eds. R. Gentleman, V. Carey, S. Dudoit, R. Irizarry, and W. Huber, 397-420. Springer New York.

Storey, J.D., and Tibshirani, R. 2003a. Statistical significance for genome wide studies. *Proc. Nat. Acad. Soc.* 100: 9440-9445.

Storey, J.D., and Tibshirani, R. 2003b. SAM thresholding and false discovery rates for detecting differential gene expression in DNA microarrays. In *The Analysis of Gene Expression Data: Methods and Software*, eds. G. Parmigiani, E.S. Garrett, R. A. Irizarry, and S.L. Zeger, 272-290. Statistics for Biology and Health, Springer.

Storey, J.D. 2003c. The positive false discovery rate: A Bayesian interpretation and the q-value. *Ann. Stat.* 31: 2013-2035.

Storey, J.D., Xiao, W., Leek, J.T., Tompkins, R.G., and Davis, R.W. 2005. Significance analysis of time course microarray experiments (with supplementary material). *Proc. Nat. Acad. Soc.* 102: 12837-12842.

Tai, Y. C. and Speed, T.P. 2006. A multivariate empirical Bayes statistic for replicated microarray time course data. *Ann. Stat.* 34: 2387-2412.

Tai, Y. C. and Speed, T.P. 2008. On gene ranking using replicated microarray time course data. *Biometrics*.

Tusher, V., Tibshirani, R., and Chu, C. 2001. Significance analysis of microarrays applied to the ionizing radiation response. *Proc. Nat. Acad. Soc.* 98: 5116-5121.

Yang, Y.H., Dudoit, S., Luu, P. et al. 2002. Normalization for cDNA microarray data: a robust composite method addressing single and multiple slide systematic variation. *Nucleic Acids Research* 30, no. 4 (February).

Yuan, M. and Kendziorski, C. 2006. Hidden Markov Models for microarray time course data in multiple biological conditions (with discussion). *J. Amer. Statist. Assoc.* 101: 1323-1340.

Wit, E., and McClure, J. 2004. *Statistics for Microarrays: Design, Analysis and Inference*. Wiley, Chichester, West Sussex, England.

Wu, H., Kerr, M. K., Cui, X., and Churchill, G.A. 2003. MAANOVA: A software package for analysis of spotted cDNA microarray experiments. In *The Analysis of Gene Expression Data: Methods and Software,* eds. G. Parmigiani, E.S. Garrett, R. A. Irizarry, and S.L. Zeger, 313-341. Statistics for Biology and Health. Springer.

Chapter 2

Classification for Differential Gene Expression Using Bayesian Hierarchical Models

Natalia Bochkina[1] and Alex Lewin[2]
[1]*School of Mathematics, University of Edinburgh, Mayfield Road, Edinburgh EH9 3JZ, UK;* [2]*Department of Epidemiology and Public Health, Imperial College, Norfolk Place, London W2 1PG, UK*

2.1 Introduction

High throughput DNA microarrays have become one of the main sources of information for functional genomics. Microarrays permit researchers to capture one of the fundamental processes in molecular biology, the transcription process from genes into mRNA (messenger RNA), that will be subsequently translated to form proteins. This process is called gene expression. By quantifying the amount of transcription, microarrays allow the identification of the genes that are expressed in different types of cells, different tissues and to understand the cellular processes in which they intervene.

Microarrays generally contain thousands of spots (or probes) at each of which a particular gene or sequence is represented. The data from a given array will share certain characteristics related to the manufacturing process of the particular array used and the extraction and handling of the biological sample hybridized to the array. The parallel structure of microarray experiments makes them particularly suited to analysis using Bayesian hierarchical models. In the hierarchical model framework unobserved quantities are organized into a small number of discrete levels with logically distinct and scientifically interpretable functions. At each level of the model information can be shared across parallel units. For example, gene expression experiments used by biologists to study fundamental processes of activation /suppression frequently involve genetically modified animals or specific cell lines, and such experiments are typically carried out only with a small number of biological samples. It is clear that this amount of replication makes standard estimates of gene variability unstable. By assuming exchangeability across the genes, inference is strengthened by borrowing information from comparable units.

Another strength of the Bayesian framework is the propagation of uncertainty through the model. Gene expression data is often processed through a series of steps, each time ignoring the uncertainty associated with the previous step. The end result of this process can be overconfident inference. In a Bayesian model it is straightforward to include each of these effects simultaneously, thus retaining the correct level of uncertainty on the final estimates. Further, when including in the model structured priors that are associated with classification, e.g. a mixture prior distribution, estimates of uncertainty of the classification can be obtained along with the fit of the model.

One of the most widely studied problems involving microarray data is the selection of groups of genes on the basis of their expression in mRNA samples derived under different experimental conditions. Many Bayesian models have been developed for differential expression between two conditions. We divide these models in two groups, with respect to the type of the prior distribution used for modeling differential expression: firstly a prior using separate distributions for data with and without differential expression, i.e. a *mixture prior*, and secondly a prior distribution without any structure which we refer to as an *unstructured prior* and which is also known as a noninformative or objective prior. Since it can be argued that any prior can be either noninformative or objective, we use the term unstructured instead. These are fit by Empirical Bayes and employ various strategies to estimate posterior probabilities of genes being differentially expressed or simply rank the genes in order of differential expression. Fully Bayesian models which can estimate the proportion of differentially expressed genes as part of the model include Broët et al. (2002), Reilly et al. (2003), Do et al. (2005), Lönnstedt & Britton (2005), Bhowmick et al. (2006) and Lewin et al. (2007). Baldi & Long (2001), Bhattacharjee et al. (2004) and Lewin et al. (2006) use unstructured priors on the differential expression parameters. A smaller number of models can be used to find genes which are differentially expressed amongst several experimental conditions (Parmigiani et al., 2002; Kendziorski et al., 2003; Broët et al., 2004; House et al., 2006; Gottardo et al., 2006; Bochkina & Richardson, 2007). These models have been reviewed in some detail in Lewin & Richardson (2007).

In this chapter we review our work using Bayesian hierarchical models to select groups of differentially expressed genes. We focus on different choices of structure used to model the differences in expression between conditions, and on several possible decision rules used to classify genes as differentially expressed or otherwise.

Section 2.2 introduces the modeling framework used in Lewin et al. (2006),

Bochkina & Richardson (2007) and Lewin et al. (2007), including some discussion of normalization between arrays and modeling gene variability. Section 2.3 discusses in detail several possible ways to infer differential expression. We consider both mixture and unstructured priors on the differential expression parameters. Bayesian inference in a mixture model is fairly straightforward, using posterior probabilities for each gene of allocation into the various mixture components. Different mixture components can be said to correspond to different hypotheses on the level of differential expression. The unstructured model does not provide a natural definition of the null hypothesis, therefore we must define it. We consider both point and interval null hypotheses, including the adaptive gene-dependent interval null proposed in Bochkina & Richardson (2007), as well as a fixed interval null. For classification in the mixture model and for the approaches using interval null hypotheses, we use a simple loss function defined in terms of misclassification between hypotheses. Consequently we can give Bayesian estimates of the false discovery rate (FDR). In other cases we use frequentist definitions of the FDR to define decision rules.

We use one particular data set to illustrate most of the methods in this chapter. This is a set of mRNA measurements for five wildtype mice and six mice with the gene IRS2 knocked out (Craig et al., 2007). The measurements were made using Affymetrix arrays and the data was preprocessed using the standard RMA algorithm (Irizarry et al., 2003). In Section 2.3.7 the advantages and disadvantages of the different classification approaches are discussed and illustrated on this data set.

Section 2.4 extends the problem to finding genes which are differentially expressed across several experimental conditions. In this case the choice of the null hypothesis is more complex, as it involves several comparisons of interest. In practice, if the null hypothesis is rejected, it is important to distinguish between difference patterns of differential expression, e.g. whether a gene is differentially expressed in all comparisons or in just one. We extend the decision-theoretic approach for testing a simple hypothesis to this more complex setup.

A major difference between the mixture and unstructured models priors is that using a parametric mixture prior we must perform model checks. This is often neglected in Bayesian analysis. Even when using the unstructured model for differential expression, the priors on the gene variances should still be checked. Section 2.5 describes predictive model checking for hierarchical models, demonstrating how this is done for both gene variances and differential expression parameters.

2.2 Bayesian Hierarchical Model
for Differential Expression

We start with a linear model, in which log gene expression y_{gsr} for gene g, experimental condition s and replicate r is modeled by a Normal distribution with additive effects for gene and array:

$$y_{gsr} \sim N(\mu_{gs} + \beta_{gsr}, \sigma_{gs}^2) \tag{2.1}$$

where μ_{gs} is the underlying gene expression level of gene g in condition s, β_{gsr} is the array effect, which depends on g through the overall expression level (see below), and σ_{gs}^2 is the gene-specific variance for condition s. We denote the number of genes by n and the number of replicates in condition s by R_s.

2.2.1 Normalization

Microarray data usually exhibit systematic differences between mRNA levels measured on different arrays. Very often the systematic effects for different genes are seen to depend on the overall level of expression of the genes. These effects can be accommodated in the linear model by the array effect parameters β_{gsr}. Ideally these should be estimated as part of the model, simultaneously with other model parameters. For example, Lewin et al. (2006) model the array effect as a smooth function of the expression level $\alpha_g \equiv \bar{\mu}_{g.}$, that is $\beta_{gsr} = f_{sr}(\alpha_g)$. For flexibility, f_{sr} is chosen to be a quadratic spline:

$$\beta_{gsr} = b_{sr0}^{(0)} + b_{sr0}^{(1)}(\alpha_g - a_0) + b_{sr0}^{(2)}(\alpha_g - a_0)^2$$

$$+ \sum_{k=1}^{K} b_{srk}^{(2)}(\alpha_g - a_{srk})^2 I[\alpha_g \geq a_{srk}] \tag{2.2}$$

where the polynomial coefficients $b_{srk}^{(p)}$ and knots a_{srk} are unknown parameters which are estimated as part of the model. The priors on the spline parameters are intended to be non-informative. The knots a_{srk} are Uniformly distributed on (a_0, a_{K+1}) where a_0 and a_{K+1} are fixed lower and upper limits (chosen to be wide enough not to affect the results). Polynomial coefficients $b_{srk}^{(p)}$ have independent $N(0, 10^2)$ priors.

An additive constraint is needed to identify the model given in equation (2.1), since constants can be transferred between the gene and array effects. Our choice is usually to normalize in a non-linear way within each condition by setting $\bar{\beta}_{gs.} = 0$ for all g, s, where the dot indicates that we are taking an average over the index r. This fully identifies the model. Normalizing using all genes within condition seems reasonable, as we do not expect systematic differences between genes on replicate arrays. It is sometimes necessary to further normalize between conditions by centering the differences between

μ_{gs} for different s. Alternatively, many people normalize across all arrays in one experiment, regardless of condition. This can be done in a similar manner by requiring $\bar{\beta}_{g..} = 0$ for all g. In this case the model is not fully identified, and the differences between μ_{gs} must be centered.

In practice the array effects often have low variability and it can be reasonable to treat them as fixed. This can be done using spline functions or locally weighted regression (loess), for example the loess estimates for the array effects (normalizing within each condition) are

$$\widehat{\beta}_{gsr} = \sum_{j \in \mathcal{N}(g)} w_{gj}(y_{jsr} - \bar{y}_{js.}) \tag{2.3}$$

where $\mathcal{N}(g)$ is a neighborhood of g with respect to expression level, and the points are weighted by distance: $w_{gj} = (1 - (d_{gj}/max_j(d_{gj}))^3)^3$ and $d_{gj} = |\bar{y}_{j..} - \bar{y}_{g..}|$. Model (2.1) can then be fit on data $y_{gsr} - \widehat{\beta}_{gsr}$, with no array effect terms in the model. Algorithmic methods are also available for normalizing arrays (Workman et al., 2002; Bolstad et al., 2003). These perform a transformation on the data without explicitly estimating array effects.

Since we used an algorithmic normalization for the main data set used in this chapter, we here illustrate array effects using a data set consisting of gene expression measures for three wildtype mice (Febbraio et al., 1999; Lewin et al., 2006). Figure 2.1 shows both posterior mean array effects and loess estimated array effects, as functions of expression level. The two estimating methods agree very well, except at very low expression levels, where there is very little data to estimate the curves.

2.2.2 Shrinkage of Gene Variances

Genes show wide variation in the amount of normal variability in expression, therefore we cannot assume equal variances for all genes. Often microarray experiments are performed with very small numbers of replicate arrays in each experimental condition. As a consequence, simple sample variance estimates can be very unstable. In a Bayesian hierarchical model, it is straightforward to stabilize the estimates of gene variances by assuming they are exchangeable within each condition.

Thus at the second level of our hierarchical model, the variances are assumed to come from a common distribution. This is essentially a random effects model for the variances, so we model variability around a "central" variance. We use either a log Normal

$$\sigma_{gs}^2 \sim \text{logNorm}(a_s, b_s) \tag{2.4}$$

or an Inverse Gamma distribution:

$$\sigma_{gs}^{-2} \sim \text{Gam}(a_s, b_s). \tag{2.5}$$

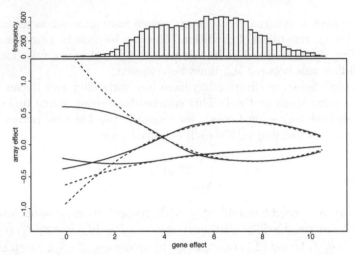

FIGURE 2.1:   Array effects as a function of expression level, for a wildtype mouse expression data set of 3 arrays, as presented in Lewin et al. (2006). Solid lines represent the spline array effects found as part of the Bayesian model, dashed lines represent the loess smoothed array effects. The histogram shows the distribution of the observed gene expression levels. Differences between the spline and loess curves are only apparent at low expression levels, where there is very little data.

When using the log Normal, the hyperparameters a_s, b_s have $N(0, 0.001)$ and $\text{Gam}(0.01, 0.01)$ priors respectively; for the Inverse Gamma both have $\text{Gam}(0.01, 0.01)$ priors. These distributions are proper approximations of improper Jeffreys priors, with density functions $f(x) = 1$ for $x \in (-\infty, +\infty)$ and $f(x) = 1/x$ for $x \in (0, +\infty)$ respectively.

Figure 2.2 compares the posterior mean gene variances found using the exchangeable model with the raw sample variances. The sample variances are each estimated using only 5 measurements (the number of wildtype mice) whereas the exchangeable variances use information from all genes (5 × 22690 measurements). In the exchangeable model the variances are shrunk towards the center of the variance distribution. The right hand plot of Figure 2.2 compares the posterior mean variances using the Inverse Gamma and log Normal priors. There is some difference in the degree of smoothing. Model checks can be used to assess the adequacy of the different choices of variance prior (see Section 2.5).

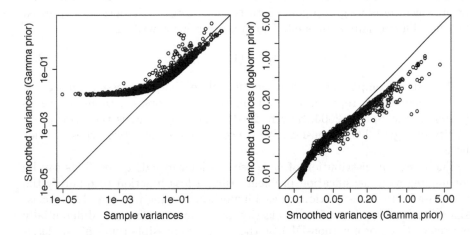

FIGURE 2.2: Left: sample variances and posterior mean variances (Inverse Gamma prior) for the IRS2 mouse data (wildtype only). Right: posterior mean variances from the Inverse Gamma and log Normal priors.

2.3 Differential Expression between Two Conditions

2.3.1 Classification Framework

It is useful to write the expression levels in two experimental conditions as

$$\mu_{g1} = \alpha_g - \delta_g/2,$$
$$\mu_{g2} = \alpha_g + \delta_g/2, \tag{2.6}$$

where α_g represents the overall expression level for gene g and δ_g represents the log differential expression. For two-color arrays the data can be given as log fold changes between the conditions (the data is paired) and in that case there is no α_g parameter. When the data is given separately for the two conditions, α_g must be modeled. It is usually treated as a fixed effect, so no information is shared between genes for this parameter.

In differential expression problems, the aim is to produce a list of genes which are expressed at different levels between different experimental conditions. Usually this corresponds to finding genes with sufficiently different means μ_{g1} and μ_{g2} of log expression, or, equivalently, with $|\delta_g|$ sufficiently large.

The fold change parameter δ_g can be given an unstructured prior $\delta_g \sim 1$ (e.g. Baldi & Long, 2001; Bhattacharjee et al., 2004; Lewin et al., 2006), which assumes no prior information on δ_g, or alternatively it can be given a mixture prior which aims to group genes into the categories differentially expressed and not differentially expressed. The mixture prior is written

$$\delta_g \sim w_0 f_0(\cdot|\phi_0) + (1 - w_0) f_1(\cdot|\phi_1) \tag{2.7}$$

with mixture component f_1 representing differentially expressed (DE) genes and component f_0 representing non-differentially expressed (non-DE) genes (Lönnstedt & Speed, 2003; Smyth, 2004). The weight w_0 is the prior probability of a gene being non-DE. We concentrate on these two types of prior here.

The posterior distribution of δ_g is used to identify DE genes. This can be formulated as a classification problem, using a loss function to quantify the penalties for misclassification, or as a hypothesis testing problem. In the classification framework we aim to classify genes into two groups, differentially expressed (DE) or not (non-DE), so there are two possible types of misclassification: genes that are truly not differentially expressed but are classified as DE (false positives), and genes that are truly DE but are classified as non-DE (false negatives). Thus there are two possible penalties for misclassification, one for false positives, one for false negatives. If the ratio of these two penalties is the same for all genes and is denoted by λ, the corresponding expected loss function given data is defined as

$$E(L|\mathbf{y}) \propto \sum_{g \in S_0} \mathbb{P}(\text{DE}^{(g)}|\mathbf{y}) + \lambda \sum_{g \in S_1} \mathbb{P}(\overline{\text{DE}}^{(g)}|\mathbf{y}), \tag{2.8}$$

where S_0 is the set of genes classified as non-differentially expressed, S_1 is the set of genes classified as differentially expressed. $\text{DE}^{(g)}$ means gene g is truly differentially expressed, $\overline{\text{DE}}^{(g)}$ means it is not. The corresponding optimal decision rule for classifying genes as differentially expressed (found by minimizing the expected loss) is

$$\mathbb{P}(\text{DE}^{(g)}|\mathbf{y}) > p_{cut}, \tag{2.9}$$

with $p_{cut} = \lambda/(1 + \lambda)$. This can be seen by considering that each gene contributes exactly one term to the expected loss (2.8), either $\mathbb{P}(\text{DE}^{(g)}|\mathbf{y})$ or $\lambda\mathbb{P}(\overline{\text{DE}}^{(g)}|\mathbf{y})$. In order to minimize the expected loss, classify gene g into S_1 if $\lambda\mathbb{P}(\overline{\text{DE}}^{(g)}|\mathbf{y}) < \mathbb{P}(\text{DE}^{(g)}|\mathbf{y})$, i.e. $\lambda(1 - \mathbb{P}(\text{DE}^{(g)}|\mathbf{y})) < \mathbb{P}(\text{DE}^{(g)}|\mathbf{y})$, which is equivalent to (2.9) with the specified p_{cut}. This classification approach works for the mixture model and in fact for any Bayesian model with non-zero prior mass of genes being in either of the DE or non-DE groups.

We refer to a data summary used for classification (in this case the posterior probability $p_g = \mathbb{P}(\text{DE}^{(g)}|\mathbf{y})$) as a decision function (as a function of data), and to an expression such as $p_g < p_{cut}$ or $p_g > p_{cut}$ as a decision rule. More generally, other types of data summaries p_g can be used as a decision function; we shall see an example of a decision function of a different type in Section 2.3.6.

Bayesian estimates of the false discovery (FDR) and false non-discovery (FNR) rates associated with the decision rule above are given (Newton et al., 2004) by

$$\widehat{FDR} = \frac{1}{|S_1|} \sum_{g \in S_1} \mathbb{P}(\overline{\mathrm{DE}}^{(g)} | \mathbf{y}),$$

$$\widehat{FNR} = \frac{1}{|S_0|} \sum_{g \in S_0} \mathbb{P}(\mathrm{DE}^{(g)} | \mathbf{y}), \tag{2.10}$$

where FDR is the expected ratio of the number of false positives to total number of declared positives, and FNR is the expected ratio of the number of false negatives to the total number of declared negatives (Benjamini & Hochberg, 1995; Storey, 2002). The false discovery rate is widely used in classical statistical analysis of gene expression data. It is useful to be able to give this estimate when comparing with different analysis methods and it has generally been found in simulation studies that (2.10) gives quite accurate estimates of the true FDR.

When a mixture prior is used for δ_g, the DE and non-DE classes naturally correspond to the different mixture components. With an unstructured prior, there is no such natural definition. In this situation it can be useful to formulate the problem in terms of hypothesis testing, where the null hypothesis corresponds to the non-DE genes (thus $H_0^{(g)}$ corresponds to $\overline{\mathrm{DE}}^{(g)}$ and $H_1^{(g)}$ to $\mathrm{DE}^{(g)}$). We consider both the usual point null hypothesis:

$$H_0^{(g)} : \delta_g = 0 \quad vs \quad H_1^{(g)} : \delta_g \neq 0, \tag{2.11}$$

and also an interval null hypothesis:

$$\tilde{H}_0^{(g)} : |\delta_g| \leqslant \theta_g \quad vs \quad \tilde{H}_1^{(g)} : |\delta_g| > \theta_g, \tag{2.12}$$

considered, for instance, in Lewin et al. (2006) and Bochkina & Richardson (2007). A decision function with fixed $\theta_g = \theta$ uses prior information about the threshold on δ_g which separates DE and non-DE genes (Lewin et al., 2006). If there is no a priori knowledge about the threshold, an adaptive variable-dependent threshold θ_g related to variability of gene g has been proposed (Bochkina & Richardson, 2007).

The loss function (2.8) and the estimates of FDR and FNR (2.10) are suitable for models with an unstructured prior on δ_g and an interval null hypothesis (with $\overline{\mathrm{DE}}^{(g)} \to H_0^{(g)}$ and $\mathrm{DE}^{(g)} \to H_1^{(g)}$). However they are not applicable to models where the prior (and thus the posterior) distribution for

δ_g has a zero mass of null hypothesis, e.g. the unstructured prior and point null hypothesis (2.11). In such cases, an appropriately chosen decision function p_g can be viewed as a test statistic from a frequentist point of view, and an estimate of FDR based on frequentist ideas can be used instead.

Following different choices of prior, null hypothesis and decision rule, we compare 5 setups:

1. Mixture prior, decision rule (2.9) (Lewin et al., 2007), i.e. with $p_g = \mathbb{P}(H_0^{(g)}|\mathbf{y})$;

2. Unstructured prior, interval null hypothesis (2.12) with a fixed interval $[-\theta, \theta]$ and decision rule (2.9) (Lewin et al., 2006), i.e. with $p_g = \mathbb{P}(|\delta_g| > \theta|\mathbf{y})$;

3. Unstructured prior, interval null hypothesis (2.12) with adaptive gene-dependent interval $[-\theta_g, \theta_g]$ and decision rule (2.9) (Bochkina & Richardson, 2007), i.e. with $p_g = \mathbb{P}(|\delta_g| > \theta_g|\mathbf{y})$;

4. Unstructured prior, point null hypothesis (2.11) and decision function $p_g = \mathbb{P}(|\delta_g| > \theta_g|\mathbf{y})$ (Bochkina & Richardson, 2007);

5. Unstructured prior, point null hypothesis (2.11) and decision function $p_g = \mathbb{P}(\delta_g > 0|\mathbf{y})$ (Bochkina & Richardson, 2007).

The choice of the null hypothesis determines the estimate of FDR (and FNR if possible) that we apply. For example, in the third and fourth setups we have the same prior and the same type of the decision rule but different null hypotheses and thus different estimates of FDR. The decision function in the last setup is equivalent to one-sided p-values (Bochkina & Richardson, 2007) which are considered in a particular (conjugate) case by Smyth (2004).

Below we describe these 5 setups in more detail. For illustration of the methods in this section we will use IRS2 microarray data referred to in the Introduction.

2.3.2   Mixture Model

A finite mixture distribution is a weighted sum of probability distributions,

$$\delta_g \sim \sum_{k=0}^{K-1} w_k f_k(\cdot|\phi_k), \tag{2.13}$$

where the weights sum to one ($\sum_{k=0}^{K-1} w_k = 1$). Each mixture component has a certain distribution f_k, with parameters ϕ_k. The weight w_k represents the probability of δ_g being assigned to mixture component k. In the context of differential expression, most mixture models used consist of two components ($K = 2$), one of which (f_0) can be thought of as representing the "null hypothesis" of there being no differential expression (Lönnstedt & Speed, 2003;

Parmigiani et al., 2002). The second component (f_1) corresponds to the alternative hypothesis that there is differential expression.

Two types of prior are usually considered for the non-DE genes. The first uses a probability mass at zero with density function $f_0(x) = \delta_0(x)$. Here $\delta_0(x)$ stands for Dirac function, $\delta_0(0) = \infty$ and $\delta_0(x) = 0$ for $x \neq 0$. This type of mixture is often known as a spike and slab model. The second choice is a "nugget" null, $f_0(x) = N(0, \varepsilon^2)$ with small values of $\varepsilon > 0$. The first type of mixture corresponds to the point null hypothesis (2.11) and the second to the null hypothesis which is an ε-"neighborhood" of the point null (for discussion, see e.g. Lönnstedt & Britton (2005)).

Usually, there is less information available to specify the density of the alternative hypothesis. For computational reasons, it can be convenient to choose a conjugate setup with equal variances in the two conditions (e.g. Lönnstedt & Speed, 2003):

$$f_1(x) = N(0, c\sigma_g^2), \tag{2.14}$$

where σ_g^2 is the variance in each of the two groups and $c > 0$. Alternatively, one could choose a uniform prior corresponding to the case of equal weight on all non-zero values of δ_g within chosen support:

$$f_1(x) = U[-\eta_-, \eta_+], \quad \eta_-, \eta_+ > 0. \tag{2.15}$$

Another possible approach to choose the alternative distribution is to allow for asymmetry in the prior distribution of the log fold change. For example, Lewin et al. (2007) proposed to use a mixture with the following three components: $\delta_g = 0$ (non-DE) with prior density f_0, $\delta_g > 0$ (DE, up regulated) with prior probability density function $f_+(x \mid \phi_+)$, and $\delta_g < 0$ (DE, down regulated) with prior probability density function $f_-(x \mid \phi_-)$. This mixture model can be written as:

$$\delta_g \sim w_0 f_0(x) + w_{+1} f_+(x \mid \phi_+) + w_{-1} f_-(x \mid \phi_-). \tag{2.16}$$

In addition to flexibility of modeling the prior distribution of δ_g, classification into these components is important for biological interpretation as the division of DE genes into up and down regulated which corresponds to opposite relationships to a studied biological process. The two components case, (2.13) with $K = 2$, can be viewed as a particular case of (2.16) with $w_{+1} f_+(x) = w_1 f_1(x \mid \phi_1) I(x \geqslant 0)$ and $w_{-1} f_-(x) = w_1 f_1(x \mid \phi_1) I(x \leqslant 0)$, with weights $w_{+}1$ and $w_{-}1$ such that f_+ and f_- are probability density functions sharing common parameters ϕ_1. Lewin et al. (2007) proposed to use gamma distributions for f_+ and f_-:

$$f_+(x|\phi_+) = \mathrm{Gam}(x|\lambda_+, \eta_+), \quad f_-(x|\phi_-) = \mathrm{Gam}(-x|\lambda_-, \eta_-), \tag{2.17}$$

where $\phi_+ = (\lambda_+, \eta_+)$ and $\phi_- = (\lambda_-, \eta_-)$. Since the choice of the parametric families f_k affects gene classification, mixed posterior predictive checks were used to choose the mixture model best fitting the data (see Section 2.5).

In the fully Bayesian mixture models described above, the decision whether a gene is differentially expressed is usually made using the posterior probability of a gene being allocated to the null mixture component. The mixture example given by (2.13) can also be written as

$$\delta_g | z_g \sim f_{z_g}(\phi_{z_g}), \quad \mathbb{P}(z_g = k) = w_k,$$

where z_g are allocation variables which label the mixture component to which gene g is assigned. Here the allocation variables z_g take values 0 and 1 for the two component mixture and values -1, 0 and $+1$ for the three component mixture. The posterior probability of gene g being in component k is $\mathbb{P}(z_g = k|\mathbf{y})$.

In the two-component mixture models, classification to the first component $z_g = 0$ corresponds to the event that hypothesis $H_0^{(g)}$ holds, and classification to the second component $z_g = 1$ corresponds to hypothesis $H_0^{(g)}$ rejected. Thus, the loss function for the two component mixture

$$E(L|\mathbf{y}) \propto \sum_{g \in S_0} \mathbb{P}(z_g \neq 0|\mathbf{y}) + \lambda \sum_{g \in S_1} \mathbb{P}(z_g = 0|\mathbf{y}) \tag{2.18}$$

is equivalent to the loss function (2.8) and thus is minimized by defining S_1 as the set of genes for which $\mathbb{P}(z_g \neq 0|\mathbf{y}) \geqslant \lambda/(1 + \lambda)$, i.e. genes are classified using a threshold on the posterior probabilities of classification in the mixture. This is not the only possible choice of loss function. Müller et al. (2007) discuss different loss functions and the decision rules they lead to.

Thus the Bayesian estimate of FDR (2.10) can be used for mixture models, based on the posterior probabilities of classification into components:

$$\widehat{FDR} = \frac{1}{|S_1|} \sum_{g \in S_1} \mathbb{P}(z_g = 0|\mathbf{y}), \tag{2.19}$$

(see Newton et al., 2004; Broët et al., 2004; Müller et al., 2007). An estimate of the false non-discovery rate can be defined similarly.

For mixtures of more than two components, one may consider different rules. For the 3 component mixture (2.16), in addition to two penalty terms in (2.18), an additional penalty term can be introduced which penalizes misclassification between non-null components f_+ and f_-:

$$E(L|\mathbf{y}) \propto \sum_{g \in S_0} \mathbb{P}(z_g \neq 0 \mid \mathbf{y}) + \lambda \sum_{g \notin S_0} \mathbb{P}(z_g = 0 \mid \mathbf{y}) \tag{2.20}$$

$$+ \tilde{\lambda} \left[\sum_{g \in S_+} \mathbb{P}(z_g = -1 \mid \mathbf{y}) + \sum_{g \in S_-} \mathbb{P}(z_g = +1 \mid \mathbf{y}) \right], \tag{2.21}$$

where S_-, S_0 and S_+ are defined by the decision rule: the groups of genes classified into the negative, zero and positive mixture components respectively,

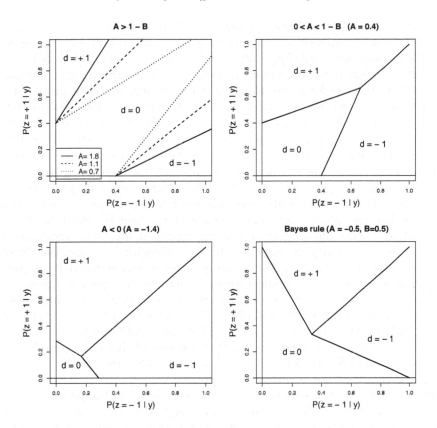

FIGURE 2.3: Decision areas for 3 component mixture in the $(\mathbb{P}(z_g = -1 \mid \mathbf{y}), \mathbb{P}(z_g = +1 \mid \mathbf{y}))$ space for different combination of penalty ratios. All plots (apart from Bayes rule) have $B = 0.4$.

labelled as $-1, 0, +1$ (Lewin et al., 2007). The parameter λ is, as before, the ratio of the penalties for false positives and false negatives (for classifying genes as null or non-null), and $\tilde{\lambda}$ is the ratio of the third penalty (for misclassifying between different non-null categories) to the penalty for false negatives.

The optimal decision rule can be written as:

1. $d_g = 0$ if $p_{g0} > 1 - B - (A + 1)\min(p_{g,+1}, p_{g,-1})$;

2. $d_y - +1$ if $p_{y,+1} > \max\{p_{y,-1}, Ap_{y,-1} + B\}$;

3. $d_g = -1$ if $p_{g,-1} > \max\{p_{g,+1}, Ap_{g,+1} + B\}$,

where $p_{gj} = P(z_g = j|x)$, $j \in \{-1, 0, +1\}$, $B = \lambda/(\lambda + 1)$, $A = \tilde{\lambda}/(\lambda + 1) - 1$. Different shapes of decision rules that arise from different values of A and B are shown in Figure 2.3.

When $\tilde{\lambda} = 0$ ($A = -1$) the loss function (2.20) reduces to the usual loss function for a 2-component mixture, leading to a simple rule classifying a gene as differentially expressed if $p_g = \mathbb{P}(z_g = 0 \mid \mathbf{y}) < 1/(\lambda + 1)$. This rule treats the left and right hand mixture components as one alternative distribution. This rule works well for genes which do not have high posterior probability of belonging to both left and right hand components. For genes with either $\mathbb{P}(z_g = -1 | \mathbf{y})$ or $\mathbb{P}(z_g = +1 | \mathbf{y})$ equal to zero, the 3-component rule also reduces to the 2-component rule. When $\lambda = \tilde{\lambda} = 1$ ($A = -0.5$, $B = 0.5$, bottom right plot in Figure 2.3), we obtain the Bayes rule, classifying each gene into the component it has maximum posterior probability of belonging to. When there are genes with non-negligible posterior probability of being in both left and right components, it is of interest to use the more general loss function. For example, consider a gene with posterior probabilities $1/3, 1/3 - \epsilon$ and $1/3 + \epsilon$ of being allocated into the under-expressed, null and overexpressed components respectively (for small $\epsilon > 0$). The rule with $\tilde{\lambda} = 0$ and the Bayes rule would declare this as differentially expressed, but it could be argued that this gene should be classified as non-differentially expressed, as it has essentially equal evidence for under and over-expression. The contribution to the loss function of this gene is $(2/3 + \epsilon)I[g \in S_0] + (\lambda(1/3 - \epsilon) + \tilde{\lambda}/3)I[g \in S_+] + (\lambda(1/3 - \epsilon) + \tilde{\lambda}(1/3 + \epsilon))I[g \in S_-]$. This gene will never be classified in S_-, as the third term is always bigger than the second. It will be classified in S_0 rather than in S_+ when $(1 - 3\epsilon)\lambda + \tilde{\lambda} > 2 + 3\epsilon$. Thus rules with $\lambda + \tilde{\lambda} > 2$ classify this gene as non-differentially expressed.

In Figure 2.3 we can see how the relationship between the penalty ratios 1, λ and $\tilde{\lambda}$ affects the classification areas. If $A > 1 - B$, i.e. if the penalty ratio $\tilde{\lambda}$ for the misclassification between components $+1$ and -1 is too large, namely greater than twice the sum of the other two penalty ratios, there is no possibility of jumping between the different alternative hypotheses (Figure 2.3, top left; the intersection of the classification lines with the axes is at point $B = 1/(\lambda + 1)$). If $A \in [0, 1 - B]$, i.e. $\tilde{\lambda}$ is large but not as large as before, then there is a possibility of jumping between the alternative components (Figure 2.3, top right; intersection with the axes is again at point B). If $A < 0$, i.e. if $\tilde{\lambda} < 1 + \lambda$, then more genes are classified into the alternative components (Figure 2.3, bottom left; intersection with the axes is at point $B/(1 - A)$). In all cases, the higher the value of λ, the more genes are classified into the null component (this can be seen for example in Figure 2.3, top left).

2.3.3   Unstructured Prior, Fixed Interval Null

In this and the following sections, we consider the unstructured (noninformative) prior for the log fold change parameter δ_g:

$$\delta_g \sim 1. \tag{2.22}$$

Function $f(x) = 1$, $x \in (-\infty, +\infty)$, is a density of the improper distribution which is invariant to shifts. However, it is easy to show that the posterior

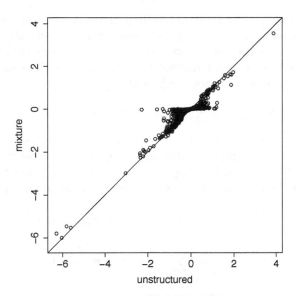

FIGURE 2.4: Posterior mean of δ_g under the mixture prior (y-axis) and under the unstructured prior (x-axis) for IRS2 data.

distribution of δ_g is proper under the likelihood (2.1) for a sufficient number of observations. The posterior mean estimates of δ_g under a mixture and the unstructured prior distributions are plotted in Figure 2.4. It can be shown that the unstructured mean estimate coincides with the empirical mean, and the mixture estimate shrinks small values towards zero. Suppose that the genes of interest are those whose fold change is above a certain (a priori fixed) value c, set according to a biologically interesting level of differential expression. Then differential expression is defined as δ_g being greater than $\log c$, corresponding to an interval null hypothesis with the interval fixed a priori (2.12) with $\theta = \log c$. Loss function (2.8) implies the optimal decision rule (2.9), namely $\mathbb{P}(|\delta_g| > \log c \mid \mathbf{y}) > p_{cut}$ in this case. The corresponding Bayesian estimates of FDR and FNR (2.10) in this case are given by

$$\widehat{FDR} = \frac{1}{|S_1|} \sum_{g \in S_1} \mathbb{P}(|\delta_g| \leqslant \log c | \mathbf{y}),$$

$$\widehat{FNR} = \frac{1}{|S_0|} \sum_{g \in S_0} \mathbb{P}(|\delta_g| > \log c | \mathbf{y}).$$

This approach was proposed by Lewin et al. (2006) and it combines statistical significance with biological significance, through the choice of c.

2.3.4   Unstructured Prior, Adaptive Interval Null

When the interval for an interval null hypothesis is required not to be fixed a priori, Bochkina & Richardson (2007) suggest to use an *adaptive interval null hypothesis* where the cutoff depends on gene-specific variability. This can be written as an interval null hypothesis (2.12) where the threshold θ depends on the variable-specific parameters σ_{g1}^2 and σ_{g2}^2:

$$\tilde{H}_0^{(g)} : |\delta_g| \leqslant \theta(\sigma_{gk}^2) \quad vs \quad \tilde{H}_1^{(g)} : |\delta_g| > \theta(\sigma_{gk}^2). \tag{2.23}$$

Following the strategy described in Section 2.3.1, for the loss function (2.8) corresponding to such a null hypothesis, the optimal decision rule for classifying genes as DE is given by $\mathbb{P}(|\delta_g| > \theta(\sigma_{gk}^2) \mid \mathbf{y}) > p_{cut}$ and the Bayesian estimates of the false discovery and non-discovery rates are given by

$$\widehat{FDR} = \frac{1}{|S_1|} \sum_{g \in S_1} \mathbb{P}(|\delta_g| \leqslant \theta(\sigma_{gk}^2) \mid \mathbf{y}),$$

$$\widehat{FNR} = \frac{1}{|S_0|} \sum_{g \in S_0} \mathbb{P}(|\delta_g| > \theta(\sigma_{gk}^2) \mid \mathbf{y}).$$

Threshold $\theta(\sigma_{gk}^2) = c_t \sqrt{\sigma_{g1}^2/R_1 + \sigma_{g2}^2/R_2}$ with some positive constant c_t suggested by Bochkina & Richardson (2007) is directly related to variability. It depends explicitly on the variance parameters of the model and is proportional to the conditional standard error of δ_g. The corresponding decision rule is based on

$$\mathbb{P}(|\delta_g|/v_g > c_t | \mathbf{y}), \tag{2.24}$$

where $v_g^2 = \sigma_{g1}^2/R_1 + \sigma_{g2}^2/R_2$. In the context of microarray data, this is the posterior probability of the between variability of $|\delta_g|$ exceeding the within variability v_g by c_t. It can also be interpreted as a fixed interval hypothesis for the standardized log fold change $t_g = \delta_g/v_g$ with a fixed threshold c_t. The authors suggest a default choice of $c_t = 2$, i.e. that the between variability $|\delta_g|$ exceeds its conditional standard error v_g by a factor of 2. Bochkina & Richardson (2007) suggested a parametrization of the threshold c_t by a parameter $\alpha \in (0, 1)$: $c_t = \Phi(1 - \alpha/2) := t^{(\alpha)}$ where α can be interpreted as the tail of the posterior distribution of δ_g if its observed value $\bar{y}_g$ were zero. Under the new parametrization, p_g is called a tail posterior probability. The authors also proposed an alternative type of threshold θ_g that depends on the variability of the data implicitly (not discussed here).

2.3.5   Unstructured Prior, Bayesian Decision Rule, Point Null

In this section we are interested in studying the frequentist performance of decision functions based on Bayesian posterior probabilities, particularly those based on interval null hypotheses. We concentrate on the decision function $\mathbb{P}(|\delta_g| > c_t v_g \mid \mathbf{y})$ based on the adaptive interval hypothesis defined in Section 2.3.4, which behaves as a frequentist test statistic in the sense that it has a gene-independent distribution under the point null hypothesis (Bochkina & Richardson, 2007).

We cannot estimate the FDR for the point null hypothesis using Bayesian methods since under the unstructured prior $\mathbb{P}(\delta_g = 0 \mid \mathbf{y}) = 0$ and thus the Bayesian estimate of FDR is 0. However, if we treat $p_g = \mathbb{P}(|\delta_g|/v_g > c_t|\mathbf{y})$ as a test statistic, it has a gene-independent distribution under the point null $H_0 : \delta_g = 0$, the frequentist FDR for the selection rule $p_g > p_{cut}$ is given by

$$\widehat{FDR}(p_{cut}) = \frac{w_0 \mathbb{P}(p_g > p_{cut}|H_0)}{\mathbb{P}(p_g > p_{cut})}, \qquad (2.25)$$

where w_0 is the probability that a gene is not differentially expressed (Storey, 2002). The probability $\mathbb{P}(p_g > p_{cut})$ in the denominator can be estimated as the proportion of genes whose posterior probability is above p_{cut}, and $\mathbb{P}(p_g > p_{cut}|H_0)$ is estimated by $1 - F_0(p_g)$ where $F_0(p)$ is the cumulative distribution function of p_g under the point null hypothesis. To estimate w_0, the approach of Storey (2002) can be adapted to obtain the following estimate:

$$\hat{w}_0(\lambda) = \frac{Card\{g : g \in \{1, \ldots, n\} \& p_g < \lambda\}}{n F_0(\lambda)}$$

with small values of $\lambda \in [0.1, 0.2]$, where $Card\{S\}$ is the number of elements in set S.

FDR estimates under the point null and under the adaptive interval hypotheses for IRS2 data are compared in Section 2.3.7.

2.3.6 Unstructured Prior, Point Null and Marginal P-Values

Bochkina & Richardson (2007) also consider a one-sided rule with threshold zero: $p(\delta_g, 0) = \mathbb{P}(\delta_g > 0 \mid \mathbf{y})$ under the unstructured prior for δ_g. It is shown to be equivalent to the corresponding one-sided p-value based on the moderated t-statistic of Smyth (2004) when the equal variance model is used.

More generally, $p(\delta_g, 0)$ is the complement to the right-sided p-value based on the marginal distribution of $\bar{y}_g = \bar{y}_{g2} - \bar{y}_{g1}$ where $\bar{y}_{gk} = \frac{1}{n_k} \sum_{r=1}^{n_k} y_{gkr}$, for any prior distribution for the variance parameters (Bochkina & Richardson, 2007). Similarly $1 - p(\delta_g, 0)$ and $2\max\{p(\delta_g, 0), 1 - p(\delta_g, 0)\} - 1$ are complements to the left- and two-sided p-value respectively. Thus, testing the null hypothesis $\delta_g = 0$ against alternatives $\delta_g > 0$, $\delta_g < 0$ or $\delta_g \neq 0$ using the posterior probabilities above is equivalent to testing the same hypotheses using corresponding p-values based on the marginal distribution of $\bar{y}_g$ under the null hypothesis.

This result is a consequence of a more general theorem given in Bochkina & Richardson (2007), that for any error distribution, if the location parameter is given the unstructured prior and the prior distributions of the remaining parameters and of the location parameter are independent, then we also have the correspondence between the posterior probability that the location parameter is greater than zero and the right-sided p-value under the hypothesis that the location parameter is zero.

This shows the equivalence between the frequentist testing using marginal p-values and the Bayesian approach where we compare the parameter of interest δ_g to its value under H_0.

2.3.7 Discussion

We compare the decision rules described in the sections above. We introduce notation to summarize all the posterior probabilities with unstructured distribution of δ_g: we denote $p(T_g, \theta) = \mathbb{P}(|T_g| > \theta \mid \mathbf{y})$, with the exclusion of the case with $\theta = 0$ and $T_g = \delta_g$ where $p(\delta_g, 0) = \mathbb{P}(\delta_g > 0 \mid \mathbf{y})$. Then, denoting $t_g = \delta_g / w_g$ the standardized difference parameter, we have that $p(\delta_g, c)$ is the posterior probability corresponding to the interval null hypothesis with fixed threshold c, $p(t_g, t^{(\alpha)})$ corresponds to the tail posterior probability. The mixture model was fit with the 3-component mixture prior

$$\delta_g \sim w_0 N(0, (0.07)^2) + w_{+1}\text{Gam}^{(+)}(5, \eta_+) + w_{-1}\text{Gam}^{(-)}(5, \eta_-), \quad (2.26)$$

which was found to be a good fit to the data (see Section 2.5.2).

The criterion with the fixed threshold divides genes in three biologically interpretable groups (see Figure 2.5b): genes whose "true" fold change is greater than the threshold $c = \log 2$ with posterior probability close to 1 (differentially expressed), a large group of genes with "true" fold change greater than $\log 2$ with posterior probability close to 0 (non-differentially expressed), and a small group of genes (compared to the other approaches), where there is substantial uncertainty whether their "true" fold change is above or below the threshold. Interestingly, the volcano plot for the posterior probability with the fixed interval threshold looks similar to the volcano plot for the posterior probability based on the mixture prior (Figure 2.5a).

In contrast, for the two-sided criterion $p(\delta_g, 0)$; (Figure 2.5d), the interval of observed fold changes where the posterior probability to be differentially expressed is close to zero, is very narrow, and genes with $\bar{y}_g$ around zero can have high posterior probability. Hence this posterior probability, as well as the corresponding p-value, is difficult to interpret on its own, and in practice a joint criterion with cutoffs both on the corresponding p-value and on the fold change is often applied.

On the other hand, the tail posterior probability $p(t_g, t^{(\alpha)})$, Figure 2.5c, is much less peaked around zero than $p(\delta_g, 0)$, taking small values for observed differences around zero and values close to one for large absolute differences, thus specifying groups of genes with low as well as high probability of differential expression, without relying on an arbitrary threshold. In addition, it

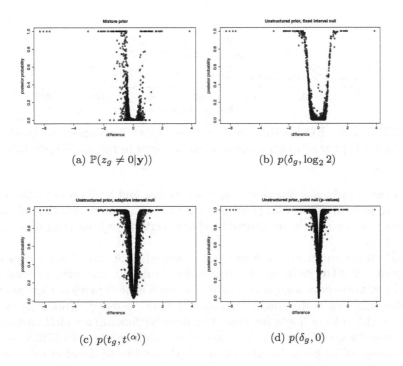

(a) $\mathbb{P}(z_g \neq 0|\mathbf{y}))$

(b) $p(\delta_g, \log_2 2)$

(c) $p(t_g, t^{(\alpha)})$

(d) $p(\delta_g, 0)$

FIGURE 2.5: Volcano plots $(\bar{y}_g, p_g)$ based on different posterior probabilities p_g for IRS2 data ($\alpha = 0.05$). In (d), $2 \max(p_g, 1 - p_g) - 1$ is plotted for the corresponding posterior probability.

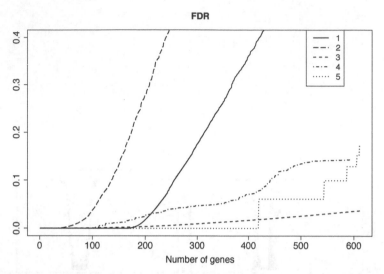

FIGURE 2.6: FDR estimates corresponding to 5 scenarios as labeled in Section 2.3.1, plotted against the number of genes in the list (IRS2 data).

varies less steeply as a function of $\bar{y}_g$ between low and high probabilities compared to $p(\delta_g, 0)$, with a large proportion of genes in between, thus allowing us to choose genes with the desired level of uncertainty about their differential expression.

FDR estimates under the 5 setups introduced in Section 2.3.1 are plotted in Figure 2.6. FDR estimates for the decision rules for the fixed interval null and the mixture have a good separation between gene lists with a low number of false positives and a higher number of false positives. This reflects the shape of the volcano plots for these two decision functions which have a flat part around zero difference values and then rapidly increase to high values of the corresponding posterior probabilities. All considered decision rules, apart from the fixed interval null, agree on the first approximately 100 genes to be differentially expressed with high probability. Comparing Bayesian (setup 3) and frequentist (setup 4) decision rules based on the posterior probability of the adaptive interval hypothesis, the Bayesian FDR estimate is lower than the corresponding frequentist FDR estimate under the point null hypothesis for gene lists larger than 100, and they coincide at being close to 0 for the gene lists of size less than 100. The number of differentially expressed genes at the 5 setups with a conservative choice of FDR = 1% are 191, 71, 306, 128 and 441 respectively.

Advantages and disadvantages of the mixture and unstructured priors are given schematically in Table 2.1. Note that given decision rules are determined by the chosen loss function (2.8), and a different choice of loss function would result in different decision rules. The mixture prior imbeds an a priori belief

TABLE 2.1:   Comparison of Mixture and Unstructured Model for δ_g

	Mixture prior	Unstructured prior
Model specification	need to specify priors f_1 and f_0	simple: $\delta_g \sim 1$
Null hypothesis	corresponds to null component	need to specify
Decision rule	$\mathbb{P}(z_g \neq 0 \mid \mathbf{y}) > p_{cut}$, interpretation of $p_{cut}/(1 - p_{cut})$ as posterior odds	need to specify, Bayesian or frequentist
FDR (FNR)	Bayesian estimate	Bayesian estimate if interval null; frequentist estimate of FDR under certain conditions if point null
Model checks for prior of δ_g	need to perform to choose appropriate f_0 and f_1	no need
Applicability	any cost of FP and FN can be accommodated	usually high cost of FP, low cost of FN

that there are (at least) two groups of genes: DE and non-DE (DE genes can be divided into further groups), which becomes a part of the model. The part of the information needed to fully specify the mixture prior and which is not usually available is the distribution of δ_g for DE genes (and even the distribution of non-DE genes, as we shall see later). Thus, mixture models require model checks to choose the prior distributions best explaining given data. If we take the unstructured prior approach, we do not need to perform the model checks for the distribution of δ_g as in the mixture model approach. However, in this case we need to specify a decision rule to classify genes, which in the cases considered here corresponds to specifying the length of the interval null hypothesis or appropriate choice of test statistic.

2.4   Testing Hypotheses for Multiclass Data

In this section we consider an approach suitable to discovering genes of interest for data sets with more complex design than considered in the previous section, for example to compare the actions of several drugs and a control sample simultaneously, or to compare different tumor samples. As with the mixture models described previously, it can be useful to describe the classification of genes in terms of null and alternative hypotheses. There are a number of different choices of alternative hypothesis for multi-class data.

Consider a data set with M conditions (classes):

$$y_{gsr} \sim N(\mu_{gs}, \sigma_{gs}^2), \quad s = 1, \ldots, M,$$

and the contrasts $\delta_g = (\delta_{g1}, \ldots, \delta_{gm})$ which we want to compare to zero are given by the matrix C:

$$\delta_g = C\mu_g,$$

where $\mu_g = (\mu_{g1}, \ldots, \mu_{gm})$ is a vector of means for gene g.

Suppose that we are interested in testing a compound hypothesis about two parameters δ_{g1} and δ_{g2} (or, more generally, about a subset of parameters δ_{gi}, $i \in S \subset \{1, \ldots, m\}$) for each gene g:

$$H_0^{(g,c)} : \quad \delta_{g1} = \delta_{g2} = 0. \tag{2.27}$$

A common approach is to consider an ANOVA-type alternative, i.e. a negation of $H_0^{(g,c)}$. In this case, F-statistic can be applied (Smyth, 2004; Broët et al., 2004). If the null hypothesis is rejected, one is left with a bouquet of different types of rejections, i.e. rejected for different combinations of conditions where a gene is differentially expressed. In practice, it is however of interest to discover a pattern of differential expression for each gene, e.g. in which conditions it is differentially expressed and in which conditions it is not. Separating these different types by taking the intersection of the single-hypothesis decision rules, e.g. $d_{g1} = 1$ and $d_{g2} = 1$ for genes differentially expressed in both comparisons, does not take into account possible dependence between parameters which is commonly present in Bayesian hierarchical models.

For the compound null hypothesis (2.27), there are 3 available patterns for rejecting the null hypothesis: a gene is differentially expressed only in comparison 1 $\{\delta_{g1} \neq 0, \delta_{g2} = 0\}$, only in comparison 2 $\{\delta_{g1} = 0, \delta_{g2} \neq 0\}$, or that a gene is differentially expressed in both comparisons $\{\delta_{g1} \neq 0, \delta_{g2} \neq 0\}$. Thus, it would be useful to be able to make a decision to which pattern each gene belongs to.

A simple way to take into account dependence is to consider a joint loss function for several parameters of interest for each gene:

$$L(\delta, P_d) = \sum_{g \in P_d} \sum_{P_t} \lambda_{P_d}(P_t) I\{\delta_{\mathbf{g}} \sim P_t\},$$

where P_d stands for the decided (estimated) pattern of differential expression (e.g. "DE, DE"), P_t for the true pattern of the parameters, and $I\{\delta_{\mathbf{g}} \sim P_t\}$ is the indicator which equals 1 if δ_g follows pattern P_t, and 0 otherwise. $\lambda_{P_d}(P_t)$ is the penalty for discovering pattern P_d instead of true pattern P_t; if the patterns P_d and P_t match the penalty is zero.

Bochkina & Richardson (2007) performed structured testing of the interval version of hypothesis (2.27) in a sequential way. First, for each comparison of interest they tested the pairwise hypothesis (2.12) and thus identified comparisons with differentially expressed genes, two in their data set. The next

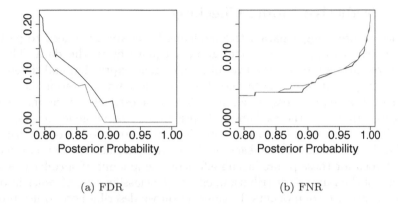

(a) FDR (b) FNR

FIGURE 2.7: False discovery (FDR) and non-discovery (FNR) rates for different thresholds, simulated data, as presented in Bochkina & Richardson (2007). Joint probability: light, minimum of pairwise probabilities: dark.

biological question, to find genes differentially in both comparisons, was answered using the following loss function:

$$L(\delta, d^c, \theta) = \sum_g (1 - d_g^c) I\{|\delta_{g1}| > \theta_g \ \& \ |\delta_{g2}| > \theta_g\}$$
$$+ \lambda_c \sum_g d_g^c I\{|\delta_{g1}| \leqslant \theta_g \ or \ |\delta_{g2}| \leqslant \theta_g\},$$

where $d_g^c = 1$ if a gene is classified as differentially expressed in both comparisons, and $d_g^c = 0$ otherwise. λ_c is the relative penalty for false positives in the compound hypothesis testing with respect to false negatives. This loss function corresponds to testing structured alternative hypothesis:

$$H_0^{(g,s)} : \delta_{g1} = 0 \ or \ \delta_{g2} = 0, \ vs \ H_1^{(g,s)} : \delta_{g1} \neq 0 \ and \ \delta_{g2} \neq 0,$$

with the optimal decision rule

$$p_g^J = P\{|\delta_{g1}| > \theta_g \ \& \ |\delta_{g2}| > \theta_g \mid \mathbf{y}\} > 1/(1 + \lambda_c). \tag{2.28}$$

The authors showed on simulated data that using the joint decision function results in a lower FDR and a similar FNR compared to combining individual gene lists if the parameters δ_{gk}/θ_g are correlated (Figure 2.7). This may occur, for instance, if an adaptive interval hypothesis is used with θ_g depending on the variance parameter which is shared between the conditions compared.

2.5 Predictive Model Checking

Since we are using a parametric modeling framework, we need to perform some checks to ensure the model is an adequate fit to the data. The unstructured priors used on the overall level of gene expression and differential expression parameters impose very little on the data: posterior mean estimates of these parameters are close to the simple estimates using means over replicates. However, the exchangeable prior used on the gene variances and the mixture prior for differential expression parameters do impose a structure on the data, leading to shrinkage of the model parameters. The parametric form chosen for these priors has an effect on the amount of shrinkage, and in the case of the mixture distribution, effects the classification of genes into the different mixture components. In this section we describe how to use predictive model checks to assess these two parts of the model (the priors for the gene variances and fold change parameters).

Our approach to model checking is to make predictions about new data points using the fitted model. In order to do this, there are two issues to consider. One is how to make the comparison between the predictions and the observed data. The other is how to make predictions from the model in the first place.

Checking Function

We want a measure of fit for each gene, rather than just one summarizing the fit of the whole model. We can then look at the distribution of these measures, which makes the checks much more sensitive to the shape of the prior distribution compared to a single p-value per model. It also allows us to consider some genes as outliers, rather than rejecting a model based on the measures of fit for a handful of genes.

There are a number of ways to obtain a measure of fit for each gene (Gelman et al., 1996). The most straightforward way is by using a Bayesian p-value. This has the advantage that the distribution should be near Uniform if the model fits the data well. We use the p-value as a convenient measure of difference between model and data. Departures from uniformity can give us useful information about the way in which the model can be improved (as will be seen in Section 2.5.2).

Prediction Scheme

Ideally we would like to perform cross-validation, in which part of the data is left out whilst fitting the model and making predictions. The predictions would then be compared with the data which was left out. However, with thousands of genes this would not be practical, as the model would have to be estimated thousands of times.

Instead we use a posterior predictive distribution (Gelman et al., 1996), for which data points are not removed from the data set. This means that the model can be run just once, and predictions made for each gene simultaneously. However, care must be taken with this approach, as posterior predictive p-values can be very conservative, as the predictions from the model have been influenced by the data to be compared with. If the influence on the model of one particular data point is large, a very small p-value can never be obtained for that data point, so the distribution of p-values is not Uniform, even when fitting the correct model (Bayarri & Berger, 2000).

The key here is the amount of influence the data for one gene has on the predictive distribution. We use the example of modeling variances exchangeably to illustrate this point. Figure 2.8 shows a directed acyclic graph (DAG) for the part of the model involving gene variances (here S_{gs} is the sample variance $\frac{1}{(R_s-1)} \sum_r (y_{gsr} - \bar{y}_{gs.})^2$ and R_s is the number of replicate arrays for condition s). The naive way to define a posterior predictive distribution in our hierarchical model would be to predict a new sample variance $S_{gs}^{(pred)}$ for each gene conditional on the model variance σ_{gs}^2 for that gene. However the predicted $S_{gs}^{(pred)}$ will be strongly dependent on the observed sample variance $S_{gs}^{(obs)}$, as can be seen from the conditional independence structure of the DAG.

A better way is to use the so-called *mixed predictive* distribution (Gelman et al., 1996; Marshall & Spiegelhalter, 2003, 2006). This is found by predicting new first level parameters $\sigma_s^{2\,(mix)}$ conditional on the second level parameters a_s, b_s, and predicting new data points $S_s^{(mix)}$ conditional on the predicted first level parameters, hence the name "mixed" predictive. The $S_s^{(mix)}$ depend on the observed sample variances only through the global parameters a_s, b_s, and thus the influence of each gene on the predicted sample variances is vastly reduced. This is emphasized by the fact that the mixed predictive sample variances do not have a g index. The mixed predictive distribution is in fact a posterior predictive distribution, conditioned on different model parameters from the naive posterior predictive distribution.

2.5.1 Checking Gene Variances

For the purposes of model checking (and fitting) it is convenient to write the first level of the model in terms of the observed sample means and variances:

$$\bar{y}_{gs.}|\mu_{gs}, \sigma_{gs}^2 \sim N(\mu_{gs}, \sigma_{gs}^2/R_s)$$

$$S_{gs}|\sigma_{gs}^2 \sim Gam(\frac{1}{2}(R_s-1), \frac{1}{2}(R_s-1)\sigma_{gs}^{-2}) \tag{2.29}$$

As seen in the previous section, S_{gs} is dependent on just one set of gene-specific parameters, σ_{gs}^2, and these are dependent on the hyperparameters a_s, b_s. Therefore the mixed predictive density of $S_s^{(mix)}$ is

$$f(S_s^{(mix)}|\mathbf{y}^{obs}) = \int f(S_s^{mix}|a_s, b_s)\pi(a_s, b_s|\mathbf{y}^{obs})da_s db_s, \tag{2.30}$$

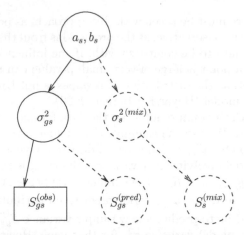

FIGURE 2.8: Directed acyclic graph for the part of the model involved in predicting sum of squares data. Solid lines show the actual model being fitted, while dotted lines show the predicted quantities.

where $\pi(a_s, b_s | \mathbf{y}^{obs})$ is the posterior density of a_s, b_s and $f(S_s^{(mix)} | a_s, b_s)$ can be found by predicting new parameters $\sigma_s^{2\,(mix)}$:

$$\sigma_s^{2\,(mix)} \sim \pi(\sigma_s^{2\,(mix)} | a_s, b_s),$$

$$S_s^{(mix)} | \sigma_s^{2\,(mix)} \sim Gam\left(\frac{1}{2}(R_s - 1), \frac{1}{2}(R_s - 1)\sigma_s^{-2\,(mix)}\right) \qquad (2.31)$$

where $\pi(.|a_s, b_s)$ is the prior density on σ_{gs}^2.

The $S_s^{(mix)}$ can be simulated from the same Monte Carlo Markov Chain (MCMC) run as the model parameters. This is done by predicting a new $\sigma_s^{2\,(mix)}$ at each MCMC iteration conditional on the current values of a_s, b_s, and predicting $S_s^{(mix)}$ conditional on $\sigma_s^{2\,(mix)}$, as shown in (2.31). If the MCMC chain has converged, the simulated values of $S_s^{(mix)}$ will be a sample from the mixed predictive distribution. The Bayesian p-values are then defined as

$$\mathbb{P}(S_s^{(mix)} > S_{gs}^{(obs)} | \mathbf{y}^{obs}). \qquad (2.32)$$

These can be calculated by averaging an indicator function over MCMC iterations.

Figure 2.9 (left hand plot) shows the mixed predictive p-values for sample variances for the model using the Inverse Gamma prior for exchangeable gene variances given by equation 2.5 (with unstructured prior on the δ_g). This model gives a reasonable fit for the sample variances. As a comparison, we also show an example which does not fit well. The right hand plot shows the predictive p-values for the same model except that all genes have the same variances: $\sigma_{gs}^2 = \sigma_s^2$ with $\sigma_s^{-2} \sim Gam(0.01, 0.01)$.

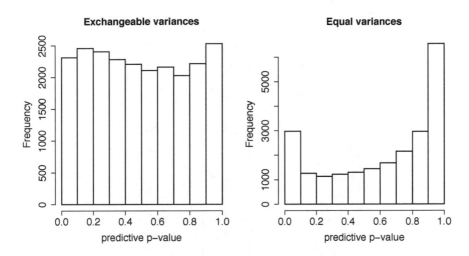

FIGURE 2.9: Bayesian predictive p-values for checking the gene variance prior, for two models applied to the IRS2 data (wildtype data only). Left: exchangeable variances, right: equal variances.

In this non-hierarchical model the predictive distribution is not called a mixed predictive, but it is in fact a global distribution for the sample variances, and so will suffer very little conservatism due to using the data to fit the model and compare to predictions. When the σ_{gs}^2 are integrated out of the hierarchical model the global distribution for the sample variances $(S_{gs}|a_s, b_s)$ is much wider than the global Inverse Gamma $(S_{gs}|\sigma_s^2)$ used in the non-hierarchical model. This is the reason for the better fit of the model.

2.5.2 Checking Mixture Prior

In this section we will use the notation $\tilde{y}_g$ to denote the sample log fold change $\bar{y}_{g2\cdot} - \bar{y}_{g1\cdot}$. These are the quantities which will be predicted from the model in order to check the mixture prior. As before, we want to predict new $\tilde{y}_g$ from the global prior without using gene-specific parameters. However, the situation is more complicated than for the sample variances as there are several sets of gene-specific parameters which may affect the $\tilde{y}_g$. Figure 2.10 shows the DAG for part of the model involving $\tilde{y}_g$. The parameter α_g is not shown as it is a fixed effect and so re-predicting this parameter would have negligible influence on $\tilde{y}_g$. Our investigations have shown that re-predicting the gene variances σ_{gs}^2 also has little effect on the Bayesian p-values for $\tilde{y}_g$,

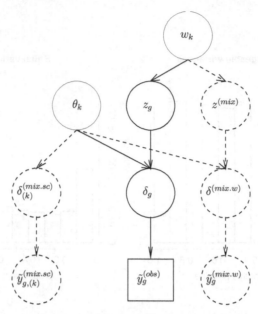

FIGURE 2.10:   Directed acyclic graph for predicting the data means. Solid lines show the actual model being fitted, while dotted lines show the predicted quantities.

so we condition on these in all results shown here. Therefore we have left the σ^2_{gs} out of the DAG as well.

We are left with the gene-specific parameters δ_g and z_g. In the same way as we had to predict new $\sigma_s^{2\,(mix)}$ in order to reduce the influence of the observed sample variances on $S_s^{(mix)}$, here we will have to predict new δ in order to reduce the influence of the observed fold changes on the predicted $\tilde{y}_g$.

Looking at the DAG, it would seem that we would want to re-predict the allocation parameters z as well, since this would take us back to the global mixture distribution (with parameters w_k and ϕ_k). In this case we would obtain the predicted quantities $\tilde{y}_g^{(mix.w)}$ shown in the DAG. The predictive distribution for $\tilde{y}_g^{(mix.w)}$ can be written

$$f(\tilde{y}_g^{(mix.w)}|\mathbf{y}^{obs}) = \int f(\tilde{y}_g^{(mix.w)}|\phi,\mathbf{w},\mathbf{y}^{obs})\pi(\phi,\mathbf{w}|\mathbf{y}^{obs})d\phi d\mathbf{w},$$

where $\pi(\phi,\mathbf{w}|\mathbf{y}^{obs})$ is the posterior density of $\phi,\mathbf{w}$ and $f(\tilde{y}_g^{(mix.w)}|\phi,\mathbf{w},\mathbf{y}^{obs})$

can be found by predicting new parameters:

$$\delta^{(mix.w)} \sim \sum_k w_k f_k(\cdot|\phi_k)$$

$$\bar{y}_{gs.}^{(mix.w)} \sim N(\alpha_g \mp \delta^{(mix.w)}/2, \sigma_{gs}^2/R_s)$$

$$\tilde{y}_g^{(mix.w)} = \bar{y}_{g2.}^{(mix.w)} - \bar{y}_{g1.}^{(mix.w)} \qquad (2.33)$$

This prediction scheme enables us to assess whether the overall shape of the mixture distribution is a good fit for the data. Note that $\tilde{y}_g^{(mix.w)}$ has a g index only because of the dependence on α_g and σ_{gs}^2. There is no g index in the mixture contribution to the log fold changes, thus individual genes will have little influence on the predictions.

However, if we are using the mixture to classify genes into separate components, it would be good to be able to predict from each component separately. We can do this by choosing to go back only as far as the three separate distributions. This means using the predicted $\tilde{y}_{g,(k)}^{(mix.sc)}$ separately for each k, where the index $k = -1, 0, 1$ labels the mixture components. The predictive distribution for $\tilde{y}_{g,(k)}^{(mix.sc)}$ is

$$f(\tilde{y}_{g,(k)}^{(mix.sc)}|\mathbf{y}^{obs}) = \int f(\tilde{y}_{g,(k)}^{(mix.sc)}|\phi_k, \mathbf{y}^{obs})\pi(\phi_k|\mathbf{y}^{obs})d\phi_k,$$

where $\pi(\phi_k|\mathbf{y}^{obs})$ is the posterior density of ϕ_k and $f(\tilde{y}_{g,(k)}^{(mix.sc)}|\phi_k, \mathbf{y}^{obs})$ can be found by predicting new parameters:

$$\delta_k^{(mix.sc)} \sim f_k(\cdot|\phi_k)$$

$$\bar{y}_{gs.(k)}^{(mix.sc)} \sim N(\alpha_g \mp \delta_k^{(mix.sc)}/2, \sigma_{gs}^2/R_s)$$

$$\tilde{y}_{g,(k)}^{(mix.sc)} = \bar{y}_{g2.(k)}^{(mix.sc)} - \bar{y}_{g1.(k)}^{(mix.sc)} \qquad (2.34)$$

Again, the g index in $\tilde{y}_{g,(k)}^{(mix.sc)}$ appears only because of the dependence on α_g and σ_{gs}^2, thus individual genes will have little influence on the predictions for each mixture component.

There is an additional issue when comparing the predictive distributions for the separate mixture components with the observed data. At any given MCMC iteration, the information for ϕ_k comes only from genes with $z_g = k$ at that iteration. Therefore the comparison for each component should only be made for genes which are classified with reasonably high probability to that component. We define conditional Bayesian p-values:

$$p_{gk} \equiv \mathbb{P}(\tilde{y}_{g,(k)}^{(mix.sc)} > \tilde{y}_g^{(obs)}|\mathbf{y}^{obs}, z_g = k) \qquad (2.35)$$

and look at the distributions of p_{gk} (separately for each k) only for genes allocated with $\mathbb{P}(z_g = k|\mathbf{y}^{obs})$ above some threshold. This reduces the contribution of the mis-classified genes.

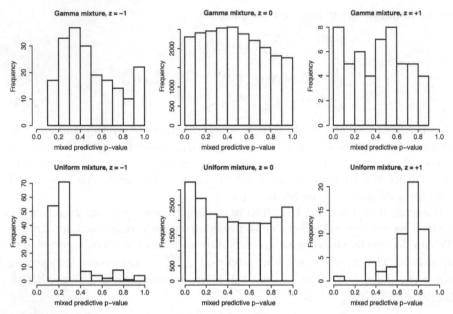

FIGURE 2.11: Bayesian mixed predictive p-values for checking separate mixture components, for two models applied to the IRS2 data. The three columns show histograms of the p-values for components $z = -1, 0, 1$ separately. Each histogram uses only genes allocated to that component with posterior probability greater than 0.5. Top row shows results for the Gamma model in Equation 2.36, bottom row shows results for the Uniform model in Equation 2.37.

As an example, Figure 2.11 shows the mixed predictive p-values checking separate mixture components for two different mixture models fitting the IRS2 data described in the introduction. The first model is the one used to produce the results in Section 2.3.7, that uses the mixture prior

$$\delta_g \sim w_0 N(0, (0.07)^2) + w_{+1} \text{Gam}^{(+)}(5, \eta_+) + w_{-1} \text{Gam}^{(-)}(5, \eta_-). \quad (2.36)$$

The variances have an exchangeable Inverse Gamma prior. This is a model we consider to fit the data reasonably well. To compare, we also show an example of a model which does not fit well. This is the same as the previous model, but using the following mixture prior:

$$\delta_g \sim w_0 \delta_0 + w_{+1} \text{Unif}(0, 3) + w_{-1} \text{Unif}(-3, 0). \quad (2.37)$$

The histograms of p-values are plotted separately for each mixture component. Each histogram uses only genes allocated to that component with posterior probability greater than 0.5.

In both cases the outside mixture components have a deficit of extreme p-values (small values for the left-hand component, large values for the right).

This is inevitable, due to the overlap between the mixture components. Genes which would have extreme p-values for the outside components are those which are more likely to be allocated to the central component, and hence do not contribute to the histograms for the outside component p-values.

However, the shape of the p-value distributions can give us information on how the model can be improved, since the shape indicates the difference between the shapes of the $\tilde{y}_{g,(k)}^{(mix.sc)}$ and $\tilde{y}_g^{(obs)}$ distributions. For the Uniform model the histograms for the outside components are skewed, suggesting that any Uniform distribution would be the wrong shape for this data (indeed we find this to be true for models with different parameters in the Uniform distributions).

In addition, the Uniform model appears to have an excess of extreme p-values for the central distribution. The Gamma model has flatter distributions of p-values for all three components, suggesting this model is an adequate fit to the data.

Acknowledgments

The authors would like to thank Sylvia Richardson, who created the Bayesian Gene Expression project supported by BBSRC "Exploiting Genomics" grant 28EGM16093 (www.bgx.org.uk) through which the work in this chapter was carried out. She contributed to all the work reviewed in this chapter. We are very grateful for all her help and guidance in our work.

NB would also like to acknowledge the financial support of a Wellcome Trust Cardio-Vascular grant 066780/Z/01/Z. The authors gratefully acknowledge the Wellcome Trust Functional Genomics Development Initiative (FGDI) thematic award "Biological Atlas of Insulin Resistance (BAIR)", PC2910 DHCT, which has supported the generation of the data used in this chapter.

References

Baldi, P. & Long, A. D. (2001). A Bayesian framework for the analysis of microarray data: regularized t-test and statistical inferences of gene changes. *Bioinformatics* 17 509–519.

Bayarri, M. J. & Berger, J. (2000). P-values for composite null models. *Journal of the American Statistical Association* 95 1127–1142.

Benjamini, Y. & Hochberg, Y. (1995). Controlling the false discovery rate: a practical and powerful approach to multiple testing. *Journal of the Royal Statistical Society, Series B* 57 289–300.

Bhattacharjee, M., Pritchard, C. C., Nelson, P. S. & Arjas, E. (2004). Bayesian integrated functional analysis of microarray data. *Bioinformatics* 20 2943–2953.

Bhowmick, D., Davison, A. C., Goldstein, D. R. & Ruffieux, Y. (2006). A Laplace mixture model for identification of differential expression in microarray experiments. *Biostatistics* 7 630–641.

Bochkina, N. & Richardson, S. (2007). Tail posterior probability for inference in pairwise and multiclass gene expression data. *Biometrics* 63 1117–1125. Advanced access: doi:10.1111/j.1541-0420.2006.00807.x.

Bolstad, B., Irizarry, R., Astrand, M. & Speed, T. (2003). A comparison of normalization methods for high density oligonucleotide array data based on variance and bias. *Bioinformatics* 19 185–193.

Broët, P., Lewin, A., Richardson, S., Dalmasso, C. & Magdelenat, H. (2004). A mixture model based strategy for selecting sets of genes in multiclass response microarray experiments. *Bioinformatics* 20(16) 2562–2571.

Broët, P., Richardson, S. & Radvanyi, F. (2002). Bayesian hierarchical model for identifying changes in gene expression from microarray experiments. *Journal of Computational Biology* 9 671–683.

Craig, A., Mandek, K., Charalambous, C., Bochkina, N., Cloarec, O., Dumas, M., Mangion, J., Simmgen, M., MacIver, F., Richardson, S., Tatoud, R., White, M., Scott, J., Aitman, T., Nicholson, J. & Withers, D. (2007). Integrated analysis of gene expression and metabolism in type 2 diabetes. Submitted to PLOS Biology.

Do, K.-A., Mueller, P. & Tang, F. (2005). A Bayesian mixture model for differential gene expression. *Applied Statistics* 54 627–644.

Febbraio, M., Abumrad, N. A., Hajjar, D. P., Sharma, K., Cheng, W., Pearce, S. F. A. & Silverstein, R. L. (1999). A null mutation in murine cd36 reveals an important role in fatty acid and lipoprotein metabolism. *Journal of Biological Chemistry* 274 19055–19062.

Gelman, A., Meng, X.-L. & Stern, H. (1996). Posterior predictive assessment of model fitness via realized discrepancies. *Statistica Sinica* 6 733–807.

Gottardo, R., Raftery, A. E., Yeung, K. Y. & Bumgarner, R. E. (2006). Bayesian robust inference for differential gene expression in microarrays with multiple samples. *Biometrics* 62 10–18.

House, L., Clyde, M. & Huang, T., Y.-C. (2006). Bayesian identification of differential gene expression induced by metals in human bronchial epithelial cells. *Bayesian Analysis* 1 105–120.

Irizarry, R. A., Hobbs, B., Collin, F., Beazer-Barclay, Y. D., Antonellis, K. J., Scherf, U. & Speed, T. P. (2003). Exploration, normalization, and summaries of high density oligonucleotide array probe level data. *Biostatistics* 4 249–264.

Kendziorski, C., Newton, M., Lan, H. & Gould, M. N. (2003). On parametric empirical Bayes methods for comparing multiple groups using replicated gene expression profiles. *Statistics in Medicine* 22 3899–3914.

Lewin, A., Bochkina, N. & Richardson, S. (2007). Fully Bayesian mixture model for differential gene expression: simulations and model checks. *Statistical Applications in Genetics and Molecular Biology* 6 Article 36.

Lewin, A. & Richardson, S. (2007). Bayesian methods for microarray data. In D. Balding, B. Martin & C. Cannings, eds., *Handbook of Statistical Genetics, 3rd edition*. Chichester: Wiley.

Lewin, A., Richardson, S., Marshall, C., Glazier, A. & Aitman, T. (2006). Bayesian modelling of differential gene expression. *Biometrics* 62 1–9.

Lönnstedt, I. & Britton, T. (2005). Hierarchical Bayes models for cDNA microarray gene expression. *Biostatistics* 6 279–291.

Lönnstedt, I. & Speed, T. (2003). Replicated microarray data. *Statistica Sinica* 12 31–46.

Marshall, E. C. & Spiegelhalter, D. J. (2003). Approximate cross-validatory predictive checks in disease mapping models. *Statistics in Medicine* 22 1649–1660.

Marshall, E. C. & Spiegelhalter, D. J. (2006). Identifying outliers in Bayesian hierarchical models: a simulation approach. Technical Report.

Müller, P., Parmigiani, G. & Rice, K. (2007). FDR and Bayesian multiple comparison rules. *Bayesian Statistics* 8 349–370.

Newton, M., Noueiry, A., Sarkar, D. & Ahlquist, P. (2004). Detecting differential gene expression with a semiparametric hierarchical mixture model. *Biostatistics* 5 155–176.

Parmigiani, G., Garrett, E. S., Anbazhagan, R. & Gabrielson, E. (2002). A statistical framework for expression-based molecular classification in cancer. *Journal of the Royal Statistical Society B* 64 1–20.

Reilly, C., Wang, C. & Rutherford, M. (2003). A method for normalizing microarrays using the genes that are not differentially expressed. *Journal of the American Statistical Association* 98 868–878.

Smyth, G. K. (2004). Linear models and empirical Bayes methods for assessing differential expression in microarray experiments. *Statistical Applications in Genetics and Molecular Biology* 3 3.

Storey, J. D. (2002). A direct approach to false discovery rates. *Journal of the Royal Statistical Society, Series B* 64 479–498.

Workman, C., Jensen, L., Jarmer, H., Berka, R., Gautier, L., Nielsen, H., Saxild, H., Nielsen, C., Brunak, S. & Knudsen, S. (2002). A new non-linear normalization method for reducing variability in DNA microarray experiments. *Genome Biology* 3(9) 0048.1–0048.16.

Chapter 3

Applications of the Mode Oriented Stochastic Search (MOSS) Algorithm for Discrete Multi-Way Data to Genomewide Studies

Adrian Dobra[1], Laurent Briollais[2], Hamdi Jarjanazi[2], Hilmi Ozcelik[2], and Hélène Massam[3]

[1]*Department of Statistics, University of Washington, Seattle.* [2] *Samuel Lunenfeld Research Institute, Mount Sinai Hospital, Toronto.* [3]*Department of Mathematics and Statistics, York University, Toronto*

Abstract   We present a Bayesian variable selection procedure that is applicable to genomewide studies involving a combination of clinical, gene expression and genotype information. We use the Mode Oriented Stochastic Search (MOSS) algorithm of Dobra and Massam (2010) to explore regions of high posterior probability for regression models involving discrete covariates and to perform hierarchical log-linear model search to identify the most relevant associations among the resulting subsets of regressors. We illustrate our methodology with simulated data, expression data and SNP data.

Key words: Bayesian analysis; contingency tables; expression data; log-linear models; model selecton; SNP data; stochastic search; variable selection.

3.1 Introduction

High-throughput sequencing studies together with clinical and physiological data produce large amounts of biological information that is used for molecular phenotyping of many diseases. Due to the extremely small sample size relative to the total number of possible covariates, it is imperative that the key feature in the development of predictive models based on a combination of gene expression, genotype and phenotype data should be the selection of a small number of predictors. Many variable selection approaches proposed in the literature involve using univariate rankings that individually measure the dependency between each candidate predictor and the response – see, for example, Golub et al. (1999); Nguyen and Rocke (2002); Dudoit et al. (2002); Tusher et al. (2001). Besides the need to address complex issues related to assessing the statistical significance of a huge number of null hypothesis tests associated with each individual predictor (Benjamini and Hochberg (1995); Efron and Tibshirani (2002); Storey and Tibshirani (2003)), there is no theoretical justification of the implicit claim that the resulting set of predictors can be converted into a good classification model for two reasons: (i) the variable selection criteria are not necessarily related to the actual classification method; (ii) there is no clear way to include other candidate variables if the covariates already selected do not lead to the desired classification peformance.

The alternative is to take into consideration *combinations* of predictors, that leads to an exponential increase of the number of candidate models. Exhaustively enumerating all these models is computationally infeasible for genomewide studies, thus a lot of efforts have been invested in finding stochastic search algorithms that are capable of quickly discovering regions of interest in the model space. In the context of linear regression, the stepwise methods of Furnival and Wilson (1974) can only be used for very small datasets due to their inability to escape local modes created by complex patterns of collinear predictors. A significant step forward were Markov chain Monte Carlo (MCMC) algorithms that explore the models space by sampling from the joint posterior distribution of the candidate models and regression parameters – see, for example, George and McCulloch (1993, 1997); Green (1995); Raftery et al. (1997); Nott and Green (2004). Excellent review papers about Bayesian variable selection for Gaussian linear regression models are Carlin and Chib (1995), Chipman et al. (2001) and Clyde and George (2004). Lee et al. (2003) make use of MCMC techniques in the context of probit regression to develop cancer classifiers based on expression data. Theoretical considerations related to the choice of priors for regression parameters are discussed in Fernández et al. (2003) and Liang et al. (2008).

Of particular interest for us is the analysis of genomewide single-nucleodite polymorphisms (SNPs) data – see, for example, Christensen and Murray (2007). While some relative consensus seems to emerge on the design of such studies, emphasizing the advantage of multistage designs in terms of cost-efficiency, their analysis raises many methodological questions, yet not answered — see Thomas et al. (2005). Hoggart et al. (2008) address the problem of multiple testing in SNP data arising from multiple studies and genotyping platforms. Although Schaid (2004) points out that single-marker analysis exploits a fraction of the information that exists in multilocus genotype data, only a few papers proposed analytic approaches that go beyond this exhaustive single-marker association testing. Marchini et al. (2005); Zhang and Liu (2007) showed the feasibility of genome-scale testing of the joint effect of two markers, but their extension to higher dimensionality problems still needs to be demonstrated. Wang and Abbott (2008) describe a multiple regression approach in which the predictors are the principal components of the sample covariance matrix of the SNP genotype scores. Clark et al. (2007) identify combinations of SNPs and environmental factors associated with disease status by extending the logic trees of Ruczinski et al. (2003) to a Bayesian framework. Swartz et al. (2007) discuss the use of model selection and Bayesian methods in the context of genomewide association studies (GWAS, henceforth) with the aim of selecting the best subsets of markers associated with the outcome of interest. Cordell and Clayton (2002) give a stepwise logistic regression procedure applicable for small, tightly linked subsets of the genome, while Lunn et al. (2006) extend this approach to a Bayesian framework that accounts for uncertainty in variable selection and missing data. Verzilli et al. (2006) proposed Bayesian graphical models for GWAS discrete data. Model selection and fitting is based on the MCMC approach of Madigan and York (1995). This methodology focused only on decomposable graphical models restricted to SNPs physically close to each other, thereby potentially ignoring the complex nature of association patterns in GWAS. Therefore there is a compelling need for developing appropriate methodologies for GWAS, allowing an efficient evaluation of a large number of multi-marker models, where the markers can be linked or unlinked.

MCMC methods can have a slow convergence rate due to the high model uncertainty resulting from the small number of available samples. This uncertainty is readily apparent when there are many models each with a very small posterior probability and the total posterior probability of these models approches one. Any MCMC approach might fail to discover better models in a reasonable number of iterations because it spends most of its time moving around models that are ultimately not relevant. MCMC algorithms are required only if the model parameters cannot be integrated out. If the marginal likelihood of each model can be evaluated exactly or at least approximated, more efficient stochastic search methods that do not attempt to sample from the posterior distribution over all candidate models can be developed. A very good example is Yeung et al. (2005) who develop a multi-class classification method by introducing a stochastic search algorithm called *iterative Bayesian model averaging*. While this method performs very well in the context of gene selection in microarray studies, it is still based on an univariate ordering of the candidate predictors. Hans et al. (2007) make another step forward and propose the *shotgun stochastic search* (SSS) algorithm that is capable of quickly moving towards high-probable models while evaluating and recording complete neighborhoods around the current most promising models.

The aim of this chapter is to describe the use of the Mode Oriented Stochastic Search (MOSS, henceforth) algorithm of Dobra and Massam (2010) to perform variable selection in regression models involving discrete covariates. We further employ MOSS to determine the most probable hierarchical log-linear models associated with the small subsets of regressors identified at the variable selection step. As such, our proposed methodology represents a unified framework of the variable selection MCMC methods of Lunn et al. (2006) and of the graphical models selection approach of Verzilli et al. (2006). However, our model determination step is not restricted to a subset of decomposable log-linear models as proposed in Verzilli et al. (2006). The use of MOSS makes our methods scale to genomewide data due to its abililTy to rapidly reach regions of high posterior probability in the target models space.

Following the ideas in Pittman et al. (2004) and Hu et al. (2009), we transform the observed data in multi-way contingency tables. While there is an inherent loss of information associated with our proposed discretization, we are able to treat covariates coming from various sources (e.g., expression levels, SNPs or clinical data) in a coherent manner. In addition, we reduce the influence of the particular choice of the normalization method employed for the initial pre-processing of the available information.

The structure of this chapter is as follows. In Section 3.2 we recall MOSS introduced by Dobra and Massam (2010) in the context of hierarchical log-linear models. We show the connections between MOSS and SSS, and describe how to use MOSS to perform small subsets regression selection. In Section 3.3 we briefly discuss the conjugate priors for log-linear parameters from Massam et al. (2009) and give the formula for the marginal likelihood of a regression involving the response variable and a reduced number of predictors. In Section 3.4 we show how to transform the observed data into a multi-way contingency table. In Section 3.5 we describe our Bayesian model averaging algorithm for performing variable selection and for developing efficient multi-class classifiers. Sections 3.6, 3.7 and 3.8 illustrate the use of our proposed methods for simulated datasets, expression data and SNP data, respectively. In Section 3.9 we make some concluding remarks.

3.2 MOSS

We denote by (n) the available data. The Bayesian paradigm to model determination involves choosing models m with high posterior probability

$\Pr(m|(n))$ selected from a set $\mathcal{M}$ of competing models. We associate with each candidate model $m \in \mathcal{M}$ a neighborhood $\text{nbd}(m) \subset \mathcal{M}$. Any two models $m, m' \in \mathcal{M}$ are connected through at least a path $m = m_1, m_2, \ldots, m_l = m'$ such that $m_j \in \text{nbd}(m_{j-1})$ for $j = 2, \ldots, l$. The neighborhoods are defined with respect to the class of models considered. There are two types of models of interest for the development of our framework:

(i) Regressions. As described in Hans et al. (2007), the neighborhood of a regression m containing k regressors is obtained by: (i) addition moves: individually including in m any variable that is not in m. The resulting neighbors contain $k + 1$ regressors; (ii) deletion moves: individually deleting from m any variable that belongs to m. The resulting neighbors contain $k - 1$ regressors; and (iii) replacement moves: replacing any one variable in m with any one variable that is not in m. The resulting neighbors contain k regressors. Replacement moves are extremely important especially if we restrict the search to models having at most k' regressors, where k' is small (say, at most 5) with respect to the total number of available predictors. Since a variable cannot be added to a regression with k' predictors, a stochastic search algorithm would have to move to a smaller dimensional model before being able to consider a new covariate for inclusion in the current model. Here we assumed that no interactions among two or more predictors are considered for inclusion in a regression model. Since the search involves only main effects, each term can be included or removed independently. The removal of a main effect involved in an interaction requires the deletion of the corresponding interaction term, hence other types of moves would have to be developed – see Hans et al. (2007).

(ii) Hierarchical log-linear models. The neighborhood of a hierarchical model m consists of those hierarchical models obtained from m by adding one of its dual generators (i.e., minimal interaction terms not present in the model) or deleting one of its generators (i.e., maximal interaction terms present in the model). For details see Edwards and Havranek (1985) and Dellaportas and Forster (1999).

The MC^3 algorithm proposed by Madigan and York (1995) constructs an irreducible Markov chain m_t, $t = 1, 2, \ldots$ with state space $\mathcal{M}$ and equilibrium distribution $\{\Pr(m|(n)) : m \in \mathcal{M}\}$. If the chain is in state m_t at time t, a candidate model m' is drawn from a uniform distribution on $\text{nbd}(m_t)$. The chain moves in state m' at time $t + 1$, i.e. $m_{t+1} = m'$ with probability

$$\min \left\{ 1, \frac{\Pr(m_{t+1}|(n))/\#\text{nbd}(m_{t+1})}{\Pr(m_t|(n))/\#\text{nbd}(m_t)} \right\}, \qquad (3.1)$$

where $\#\text{nbd}(m)$ denotes the number of neighbors of m. Otherwise the chain does not move, i.e. we set $m_{t+1} = m_t$.

Jones et al. (2005) and Hans et al. (2007) build on the MC^3 approach to construct the shotgun stochastic search (SSS) algorithm specifically designed to explore regions of high posterior probability in the candidate model space. SSS originates from earlier ideas proposed in Dobra et al. (2004). It visits the space by evaluating the posterior probability $\Pr(m'|(n))$ of each model m' in the neighborhood of the current model m and by choosing a new current model by sampling from these candidates with probabilities proportional with $\Pr(m'|(n))$. The efficiency of SSS comes from the fact that it focuses on rapidly moving towards models that maximize $\Pr(m|(n))$, $m \in \mathcal{M}$. As opposed to SSS, MC^3 might spend most of its time exploring models with low posterior probability if there are many such models and their total posterior probability dominates the space. This is precisely what happens when the sample size remains fixed while the number of candidate models increases rapidly which leads to a high degree of model uncertainty.

Dobra and Massam (2010) define the mode oriented stochastic search (MOSS, henceforth) algorithm that identifies models in

$$\mathcal{M}(c) = \left\{ m \in \mathcal{M} : \Pr(m|(n)) \geq c \cdot \max_{m' \in \mathcal{M}} \Pr(m'|(n)) \right\}, \qquad (3.2)$$

where $c \in (0, 1)$. As proposed in Madigan and Raftery (1994), models with a low posterior probability compared to the highest posterior probability model are discarded. This choice drastically reduces the size of the target space from $\mathcal{M}$ to $\mathcal{M}(c)$. For suitable choices of c, $\mathcal{M}(c)$ can be exhaustively enumerated.

MOSS makes use of a current list $\mathcal{S}$ of models that is updated during the search. Define the subset $\mathcal{S}(c)$ of $\mathcal{S}$ in the same way we defined $\mathcal{M}(c)$ based on $\mathcal{M}$. Define $\mathcal{S}(c')$ with $0 < c' < c$ so that $\mathcal{S}(c) \subset \mathcal{S}(c')$. Let q be the probability of pruning the models in $\mathcal{S} \setminus \mathcal{S}(c)$. A model m is called *explored* if all its neighbors $m' \in \text{nbd}(m)$ have been visited. A model in $\mathcal{S}$ can be explored or unexplored. MOSS proceeds as follows:

PROCEDURE MOSS(c,c',q)

    (a) Initialize the starting list of models $\mathcal{S}$. For each model $m \in \mathcal{S}$, calculate and record its posterior probability $\Pr(m|(n))$. Mark m as unexplored.

    (b) Let $\mathcal{L}$ be the set of unexplored models in $\mathcal{S}$. Sample a model $m \in \mathcal{L}$ according to probabilities proportional with $\Pr(m|(n))$ normalized within $\mathcal{L}$. Mark m as explored.

    (c) For each $m' \in \text{nbd}(m)$, check if m' is currently in $\mathcal{S}$. If it is not, evaluate and record its posterior probability $\Pr(m'|(n))$. If $m' \in \mathcal{S}(c')$, include m' in $\mathcal{S}$ and mark m' as unexplored. If m' is the model with the highest posterior probability in $\mathcal{S}$, eliminate from $\mathcal{S}$ the models in $\mathcal{S} \setminus \mathcal{S}(c')$.

    (d) With probability q, eliminate from $\mathcal{S}$ the models in $\mathcal{S} \setminus \mathcal{S}(c)$.

(e) If all the models in $\mathcal{S}$ are explored, eliminate from $\mathcal{S}$ the models in $\mathcal{S} \setminus \mathcal{S}(c)$ and STOP. Otherwise go back to step (b).

END.

At step (c) the models $m' \in \mathrm{nbd}(m)$ can be considered in any possible order. As opposed to MC^3 or SSS, MOSS edobra623652nds by itself without having to specify a maximum number of iterations to run. In a MOSS search, the model explored at each iteration is selected from the most promising models identified so far. In MC^3 or SSS, this model is selected from the neighbors of the model evaluated at the previous iteration. This feature allows MOSS to move faster towards regions of high posterior probability in the models space.

Choosing c in the intervals $(0, 0.01]$, $(0.01, 0.1]$, $(0.1, 1/3.2]$, $(1/3.2, 1]$ means eliminating models having decisive, strong, substantial or "not worth more than a bare mention" evidence against them with respect to the highest posterior probability model – see Kass and Raftery (1995). We recommend using a value for c' as close to zero as possible. The role of the parameter c' is to limit the number of models that are included in $\mathcal{S}$ to a manageable number. We also suggest running the algorithm with several choices of c, c' and q to determine the sensitivity of the set of models selected.

3.3 Conjugate Priors for Hierarchical Log-Linear Models

Massam et al. (2009) developed and studied the conjugate prior as defined by Diaconis and Ylvisaker (1979) (henceforth abbreviated the DY conjugate prior) for the log-linear parameters for the general class of discrete hierarchical log-linear models. We outline the notation and most relevant results from Massam et al. (2009). We also give a formula for the marginal likelihood of a regression model induced by these conjugate priors.

3.3.1 Model Parameterization

Let V be the set of criteria defining the contingency table. Denote the power set of V by $\mathcal{E}$ and take $\mathcal{E}_\ominus = \mathcal{E} \setminus \{\emptyset\}$. Let $X = (X_\gamma, \gamma \in V)$ such that X_γ takes its values (or levels) in the finite set I_γ of dimension $|I_\gamma|$. When a fixed number of individuals are classified according to the $|V|$ criteria, the data is collected in a contingency table with cells indexed by combination of levels for the $|V|$ variables. We adopt the notation of Lauritzen (1996) and denote a cell by $i = (i_\gamma, \gamma \in V) \in \mathcal{I} = \times_{\gamma \in V} I_\gamma$. The count in cell i is denoted $n(i)$ and the probability of an individual falling in cell i is denoted $p(i)$.

For $E \subset V$, cells in the E-marginal table are denoted $i_E \in \mathcal{I}_E = \times_{\gamma \in E} I_\gamma$.

The marginal counts are denoted $n(i_E)$. For $N = \sum_{i \in \mathcal{I}} n(i)$, $(n) = (n(i), i \in \mathcal{I})$ follows a multinomial $\mathcal{M}(N, p(i),\ i \in \mathcal{I})$ distribution with probability density function

$$P\big((n)\big) = \binom{N}{(n)} \prod_{i \in \mathcal{I}} p(i)^{n(i)} . \tag{3.3}$$

Let i^* be a fixed but arbitrary cell that we take to be the cell indexed by the "lowest levels" of each factor. We denote these lowest levels by 0. Therefore i^* can be thought to be the cell $i^* = (0, 0, \ldots, 0)$. We define the log-linear parameters to be

$$\theta(i_E) = \sum_{F \subseteq E} (-1)^{|E \setminus F|} \log p(i_F, i_{F^c}^*) \tag{3.4}$$

which, by the Moebius inversion, is equivalent to

$$p(i_E, i_{E^c}^*) = \exp \sum_{F \subseteq E} \theta(i_F) . \tag{3.5}$$

Remark that $\theta_{\emptyset}(i) = \log p(i^*)$, $i \in \mathcal{I}$. We denote $\theta_{\emptyset}(i^*) = \theta_{\emptyset}$ and $p(i^*) = p_{\emptyset} = \exp \theta_{\emptyset}$. It is easy to see that the following lemma holds.

LEMMA 3.1
If for $\gamma \in E, E \subseteq V$ we have $i_{\gamma} = i_{\gamma}^* = 0$, then $\theta(i_E) = 0$.

3.3.2 The Multinomial for Hierarchical Log-Linear Models

Consider the hierarchical log-linear model m generated by the class $\mathcal{A} = \{A_1, \ldots, A_k\}$ of subsets of V, which, without loss of generality, can be assumed to be maximal with respect to inclusion. We write $\mathcal{D} = \{E \subseteq_{\ominus} A_i, i = 1, \ldots, k\}$ for the indexing set of all possible interactions in the model, including the main effects. It follows from the theory of log-linear models (for example, see Darroch and Speed (1983)) and from Lemma 3.1 that the following constraints hold:

$$\theta(i_E) = 0, \quad E \notin \mathcal{D} \tag{3.6}$$

Therefore, in this case, for $i_E \in \mathcal{I}_E^*$, (3.5) becomes

$$\log p(i_E, i_{E^c}^*) = \log p(i(E)) = \theta_{\emptyset} + \sum_{F \subseteq E, F \in \mathcal{D}, i_F \in \mathcal{I}_F^*} \theta(i_F) . \tag{3.7}$$

and after the change of variable $(n) = (n(i), i \in \mathcal{I}^*) \mapsto (n(i_E), E \in \mathcal{E}_{\ominus})$, the multinomial distribution for the hierarchical log-linear model becomes the distribution for the random variable $Y = (n(i_D), D \in \mathcal{D}, i_D \in \mathcal{I}_D^*)$ with density

$$
f_{\mathcal{D}}(y; \theta) \propto \exp \left\{ \sum_{D \in \mathcal{D}} \sum_{i_D \in \mathcal{I}_D^*} \theta(i_D) n(i_D) \right.
$$

$$
\left. - N \log \left(1 + \sum_{E \in \mathcal{E}_\Theta, i_E \in \mathcal{I}_E^*} \exp \sum_{F \subseteq_D E} \theta(i_F) \right) \right\} \tag{3.8}
$$

with respect to a measure $\mu_\infty(y)$. It is important to note here that

$$
\theta_{\mathcal{D}} = (\theta(i_D), D \in \mathcal{D}, i_D \in \mathcal{I}_D^*) \tag{3.9}
$$

is the canonical parameter and

$$
p_{\mathcal{D}} = (p(i(D)), D \in \mathcal{D}, i_D \in \mathcal{I}_D^*), \tag{3.10}
$$

is the cell probability parameter of the multinomial distribution of m. The other cell probabilities $p(i(E)), E \notin \mathcal{D}$ are not free and are a function of $p_{\mathcal{D}}$.

3.3.3   The Diaconis–Ylvisaker Conjugate Prior

The natural exponential family form of the distribution of the marginal counts $Y = (n(i_E), E \in \mathcal{D}, i_E \in \mathcal{I}_E^*)$ of a contingency table with cell counts $n(i), i \in \mathcal{I}$ is given in Equation (3.8). The conjugate prior for θ as introduced by Diaconis and Ylvisaker (1979) is

$$
\pi_{\mathcal{D}}(\theta_{\mathcal{D}} | s, \alpha) = I_{\mathcal{D}}(s, \alpha)^{-1} h(\theta_{\mathcal{D}}), \tag{3.11}
$$

where $I_{\mathcal{D}}(s, \alpha) = \int_{\mathbb{R}^{d_{\mathcal{D}}}} h(\theta_{\mathcal{D}}) d\theta_{\mathcal{D}}$ is the normalizing constant of $\pi_{\mathcal{D}}(\theta_{\mathcal{D}} | s, \alpha)$, the dimension of the parameter space $d_{\mathcal{D}}$ is $\sum_{D \in \mathcal{D}} \prod_{\gamma \in D} (|I_\gamma| - 1)$ and

$$
h(\theta_{\mathcal{D}}) = \exp \left\{ \sum_{D \in \mathcal{D}} \sum_{i_D \in \mathcal{I}_D^*} \theta(i_D) s(i_D) \right.
$$

$$
\left. - \alpha \log \left(1 + \sum_{E \in \mathcal{E}_\Theta} \sum_{i_E \in \mathcal{I}_E^*} \exp \sum_{F \subseteq_D E} \theta(i_F) \right) \right\}. \tag{3.13}
$$

The corresponding hyper-parameters are:

$$
(s, \alpha) = (s(i_D), D \in \mathcal{D}, i_D \in \mathcal{I}_D^*, \alpha), \quad s \in \mathbb{R}^{d_{\mathcal{D}}}, \quad \alpha \in \mathbb{R}. \tag{3.14}
$$

Massam et al. (2009) give a necessary and sufficient condition for the distribution in Equation (3.11) to be proper as well as two methods to choose hyper-parameters (s, α) such that $I_{\mathcal{D}}(s, \alpha) < +\infty$.

From the similarity between the form (3.8) of the distribution of the marginal cell counts Y and the form (3.11) of the prior on θ, one can think of the hyper-parameters s as the marginal cell entries of a fictive contingency table whose cells contain positive real numbers. Consequently, α can be taken to be the grand total of this fictive table. The lack of prior information can be expressed through a non-informative prior specified by taking all the fictive cell entries to be equal to $\frac{\alpha}{|\mathcal{I}|}$ so that

$$s(i_D) = \sum_{j \in \mathcal{I}, j_D = i_D} \frac{\alpha}{|\mathcal{I}|}. \qquad (3.15)$$

In the case of decomposable log-linear models, this approach to constructing a conjugate prior is equivalent to eliciting hyper-Dirichlet priors – see, for example, Dawid and Lauritzen (1993) and Madigan and York (1997). While the hyper-Dirichlet priors are restricted to decomposable log-linear models, the properties of the Diaconis–Ylvisaker conjugate priors extend naturally to graphical and hierarchical log-linear models.

Given the prior $\pi_D(\theta_D | s, \alpha)$, from the form (3.8) of the distribution of the marginal cell counts Y, the posterior distribution of θ_D given $Y = y = (n(i_D), D \in \mathcal{D}, i_D \in \mathcal{I}_D^*)$ is

$$\pi_D(\theta_D | y, s, \alpha) = \frac{1}{I_D(s + y, \alpha + N)} \exp\left\{ \sum_{D \in \mathcal{D}} \sum_{i_D \in \mathcal{I}_{D^*}} \theta(i_D)(s(i_D) + n(i_D)) \right.$$

$$\left. -(\alpha + N) \log\left(1 + \sum_{E \in \mathcal{E}_\Theta, i_E \in \mathcal{I}_E^*} \exp \sum_{F \subseteq_D E} \theta(i_F)\right) \right\}. \qquad (3.16)$$

If we look at the posterior $\pi_D(\theta_D | y, s, \alpha)$ as a function of (s, α), $s(i_D) + n(i_D)$ represent the (i_D)-counts in the D-marginal entries of a table having the same structure as (n) and whose entries are obtained by augmenting the actual cell counts in (n) with the fictive cell counts from the prior. The grand total of this augmented table is $\alpha + N$.

3.3.4 Computing the Marginal Likelihood of a Regression

Let $Y = X_r$, $r \in V$ be a response variable and X_A, $A \subset V \setminus \{r\}$ are a set of explanatory variables. Denote by $(n)_{A \cup \{r\}}$ and $(n)_A$ the marginals of the full contingency table (n). Here (n), $(n)_{A \cup \{r\}}$ and $(n)_A$ are cross-classifications involving X_V, $X_{A \cup \{r\}}$ and X_A, respectively. Dobra and Massam (2010) explored the connection between log-linear models and the regressions derived from them. They expressed the regression parameters of the conditional $[Y|X_A]$ as a function of a log-linear model for Y and X_A. To assure the consistency of the distributions for regression parameters associated with various subsets of explanatory variables, we assume a saturated log-linear model for the full table (n). After collapsing across variables $X_{V \setminus (A \cup \{r\})}$, we obtain a saturated log-linear model for $(n)_{A\cup\{r\}}$. This means that the DY conjugate prior for the saturated log-linear model for (n) reduces to the DY conjugate prior for the saturated log-linear model for $(n)_{A\cup\{r\}}$ by setting some of the θ parameters to zero. A similar statement can be made about the posterior distribution in Equation (3.16).

The properties shared by the DY priors and posteriors imply that the marginal likelihood of the regression $[Y|X_A]$ is the ratio between the marginal likelihoods of the saturated models for $(n)_{A\cup\{r\}}$ and $(n)_A$ – see Dawid and Lauritzen (1993); Geiger and Heckerman (2002); Dobra and Massam (2010):

$$\Pr(r|A) = \frac{I_{\mathcal{D}^{A\cup\{r\}}}((s+y)^{A\cup\{r\}}, \alpha + N)}{I_{\mathcal{D}^A}((s+y)^A, \alpha + N)} \cdot \frac{I_{\mathcal{D}^A}(s^A, \alpha)}{I_{\mathcal{D}^{A\cup\{r\}}}(s^{A\cup\{r\}}, \alpha)}. \quad (3.17)$$

Dobra and Massam (2010) show that, for any set $B \subset V$, we have

$$I_{\mathcal{D}^B}(s^B, \alpha) = \Gamma(\alpha_\emptyset^B) \prod_{D\in\mathcal{D}^B} \prod_{i_D \in \mathcal{I}_B^*} \Gamma(\alpha^B(i_D, i_{D^c}^*)),$$

where

$$\alpha^B(i_D, i_B^*) = \sum_{\substack{B \supseteq F \supseteq D}} \sum_{\substack{j_F \in \mathcal{I}_F^* \\ (j_F)_D = i_D}} (-1)^{|F\setminus D|} s(j_F),$$

$$\alpha_\emptyset^B = \alpha + \sum_{D\subseteq B} (-1)^{|D|} \sum_{i\in\mathcal{I}_D^*} s(i_D).$$

3.4 Discretizing Ordered Variables

The methods developed in this chapter perform well for datasets involving covariates of almost any type. The only constraints we impose on the nature of the observed covariates is that they have to be either categorical or their values can be ordered. Variables of the latter type are transformed in categorical variables as follows. Let $\{x_1, x_2, \ldots, x_n\}$ be the possible values of x. Denote by $\tau_1 < \tau_2 < \ldots < \tau_{k-1} < \tau_k = \infty$ a set of levels of x called *splits*. Here τ_1 is the minimum value of x and $k \geq 3$. We replace each value x_i, $1 \leq i \leq n$, by $j \in \{1, \ldots, k-1\}$ if $\tau_j \leq x_i < \tau_{j+1}$. This implies that x is transformed in a categorical covariate with $(k-1)$ levels. A sequential discretization of all the observed variables leads to a multi-way contingency table (n) with the same dimension as the original data. The table (n) is invariant to monotonic transformations of covariates – see Hu et al. (2009).

The splits used for discretization can be chosen either based on background information that might be available or can be defined to maximize the individual predictive ability of each covariate with respect to a categorical response variable y. Let x_τ be the categorical version of a predictor x, where τ are a set of splits for x. Following Pittman et al. (2004) and Hu et al. (2009), we use percentiles as a possible choice of splits for continuous covariates such as gene expression levels. A segregating SNP site has three possible genotypes: 0/0, 0/1 and 1/1, where 0 is the wild type and 1 is the mutant allele. Diallelic SNPs can be represented as three-category discrete variables, or can be dichotomized as presence of 0 vs. absence of 0, or as presence of 1 vs. absence of 1.

We choose the splits τ having the largest marginal likelihood in Equation (3.17) of the regression of y on x_τ. Ideally we would like to consider combinations of choices of splits for all available predictors and choose those splits that lead to large values of the marginal likelihood of regressions of y on one, two or more categorical predictors. Each predictor will have several categorical versions associated with various splits sets. Unfortunately the datasets resulting from genomewide studies precludes us to perform such large scale computations, and we need to choose splits sequentially for each covariate.

The conjugate priors of Massam et al. (2009) are appropriate for the analysis of general multi-way tables in which each variable is allowed to have any number of categories. In order to increase the mean number of observations per cell we make the constraint that the data needs to be transformed in a dichotomous rather than a polychotomous contingency table. The response variable y is either dichotomous by design or has been dichotomized based on background information. Each predictor is replaced by a dichotomous version constructed with respect to $20, 21, \ldots, 80$-th percentiles. The smallest and largest percentile to be used as a split candidate is chosen such that at least one sample falls in each of the two categories. However, we emphasize that our theoretical framework can be employed to develop multi-class predictive models – see Yeung et al. (2005) for a related approach.

3.5   Bayesian Model Averaging

Let (n) be the multi-way contingency table obtained by discretizing the observed covariates as described in Section 3.4. As before, the variables cross-classified in (n) are X_γ, $\gamma \in V$. Denote by $Y = X_r$, $r \in V$, a response variable of interest. We use MOSS to identify regressions $[Y|X_A]$, $A \subset V \setminus \{r\}$ that maximize the marginal likelihood $\Pr(r|A)$ in Equation (3.17). We assume that all the models are apriori equally likely, so that the posterior probability of each regression is proportional with its marginal likelihood. However, we are only interested in regressions involving a relatively small number of predictors (less than five) because the available number of samples is usually small. One should evaluate the mean number of observations per cell in a marginal $(n)_{A \cup \{r\}}$, where A has the maximum number of predictors allowed. If the resulting mean is too low, the maximum number of predictors should be decreased.

MOSS identifies a set of regressions $\mathcal{S}$ for some $c \in (0, 1)$. The cutoff c is chosen so that $\mathcal{S} = \mathcal{S}(c)$ can be exhaustively enumerated. A regression $m_j \in \mathcal{S}(c)$ is $[Y | X_{A_j}]$ with $A_j \subset V \setminus \{r\}$ and $j \in B$. Here B is a set of indices. The regression of Y on the remaining variables is a weighted average of the regressions in $\mathcal{S}$, where the weights represent the posterior probability of each regression (see, for example, Yeung et al. (2005)):

$$\Pr(Y = y | (n)) = \sum_{j \in B} \Pr(Y = y | (n)_{A_j}) \cdot \Pr(m_j | (n)). \qquad (3.18)$$

Since we assumed that all models are apriori equally likely, the posterior probability of each regression is equal to its marginal likelihood normalized over all the models in $\mathcal{S}$:

$$\Pr(m_j | (n)) = \frac{\Pr(r | A_j)}{\sum\limits_{l \in B} \Pr(r | A_l)}.$$

Madigan and Raftery (1994) show that the weighted average of regressions in (3.18) has a better predictive performance than any individual model in $\mathcal{S}$. The revelance of each predictor X_j can be quantified by its posterior inclusion probability defined as the sum of the posterior probabilities of all the models that include X_j.

There is an additional step we perform to assure the most parsimonious versions of regressions $[Y | A_j]$ are employed in Equation (3.18). Dobra and Massam (2010) developed a version of MOSS for hierarchical log-linear models with the DY conjugate priors. We use their stochastic search procedure to determine the hierarchical log-linear model with the highest posterior probability for each marginal $(n)_{A_j \cup \{r\}}$, where A_j identifies a regression in $\mathcal{S}$. Dobra and Massam (2010) also determine the regression induced by a hierarchical log-linear model for a given response variable Y. It is possible that some variables X_γ, $\gamma \in A_j$ interact with Y indirectly through other variables in A_j and in this case $[Y | X_{A_j}]$ reduces to $[Y | X_{E_j}]$ for $E_j \subset A_j$. Each regression $[Y | X_{A_j}]$ in Equation (3.18) is the regression induced by the highest posterior probability log-linear model for the marginal $(n)_{A_j \cup \{r\}}$. Since the same choice of conjugate priors is employed in the stochastic search for regressions and in the subsequent stochastic search for log-linear models, the overall consistency of our variable selection method coupled with log-linear model determination is guaranteed.

Dobra and Massam (2010) describe an algorithm called the Bayesian iterate proportional fitting (Bayesian IPF, henceforth) for sampling from the joint posterior in Equation (3.16) of parameters θ corresponding with a log-linear model. The resulting posterior samples can be used to estimate the coefficients of the regressions in $\mathcal{S}(c)$. The coefficients for the Bayesian model averaging regression $\Pr(Y = y|(n))$ from Equation (3.18) are estimated by sampling from the joint posterior of the coefficients for each individual regression $m_j = [Y|X_{A_j}]$, $j \in B$, for a number of iterations proportional with $\Pr(m_j|(n))$.

3.6 Simulated Examples

3.6.1 First Simulated Example

We apply the MOSS algorithm to perform model selection when the set of potential predictors exhibit strong pairwise correlations. This is a variation of the example suggested by Nott and Green (2004). As in George and McCulloch (1997), we generate $Z_1, \ldots, Z_{15}, Z \sim N_{300}(0, I)$, where $N_{300}(0, I)$ is the 300-dimensional normal distribution with zero mean and identity covariance matrix. Let $X_i = Z_i + 2Z$, $i = 1, 3, 5, 8, 9, 10, 12, 13, 14, 15$ and also $X_2 = X_1 + 0.15Z_2$, $X_4 = X_3 + 0.15Z_4$, $X_6 = X_5 + 0.15Z_6$, $X_7 = X_8 + X_9 - X_{10} + 0.15Z_7$ and $X_{11} = X_{14} + X_{15} - X_{12} - X_{13} + 0.15Z_{11}$. George and McCulloch (1997) point out that this design matrix leads to correlations of about 0.998 between X_i and X_{i+1} for $i = 1, 3, 5$. There are also strong linear associations between (X_7, X_8, X_9, X_{10}) and $(X_{11}, X_{12}, X_{13}, X_{14}, X_{15})$. We let $\tilde{X} = [X^{(1)} X^{(2)}]$ be a 300×30 design matrix obtained by independently simulating two instances $X^{(1)}$ and $X^{(2)}$ of the 300×15 design matrix X.

Consider the 30-dimensional vector of regression coefficients β defined by

$$
\beta_j = \begin{cases} 1.5, & \text{if } j = 1, 3, 5, 7, 11, 12, 13, \\ -1.5, & \text{if } j = 8, \\ 0, & \text{otherwise.} \end{cases}
$$

We generate $Y = \tilde{X}\beta$. The binary response $\tilde{Y}$ is obtained as $\tilde{Y}_j = 1$ if $Y_j \geq 0$ and $\tilde{Y}_j = 0$ if $Y_j < 0$.

We would like to study whether predictors that belong to $X^{(2)}$ are not selected by MOSS. Note that some predictors in $X^{(1)}$ might still be selected even if their regression coefficients are zero because of the complex correlation structure that exists in this block.

We employed MOSS with $c = 0.333$, $c' = 0.0001$, a pruning probability of 0.25 and five different starting points. We allowed MOSS to explore regressions containing at most five predictors. In order to reduce the sample variability, we report the results we obtained by averaging across 100 replicates of this experiment. The percentage of selected predictors that belong to the block $X^{(1)}$ is 100% with a 95% confidence interval of $[50\%, 100\%]$. This implies that MOSS almost never selects predictors from $X^{(2)}$. The mean number of predictors selected is 2 with a 95% confidence interval of $[1, 9]$. The mean number of regressions evaluated by each instance of MOSS was 1403 with a 95% confidence interval of $[957.2, 2615.2]$. The number of possible regressions with 5 predictors or less is 174437 which is indicative of how fast our proposed stochastic search method can move towards models with high posterior probability.

We employed Bayesian model averaging to construct classifiers for $\tilde{Y}$ based on the regressions in the sets $\mathcal{S}(0.333)$. The mean area below the corresponding ROC curves was 0.996 with a 95% confidence interval of $[0.979, 1]$. Therefore MOSS can construct excellent classifiers by identifying predictors that are actually related to the binary response.

3.6.2 Second Simulated Example

The second example was suggested by George and McCulloch (1993). We generate $Z_1, \ldots, Z_{60}, Z \sim N_{120}(0, I)$ and construct the 120×60 design matrix $X = (X_1, \ldots, X_{60})$ with $X_i = Z_i + Z$ for $i = 1, \ldots, 60$. The average pairwise correlations in X is about 0.5 which makes variable selection difficult. Consider the 60-dimensional vector of regression coefficients β given by $\beta_j = 0$ if $j = 1, \ldots, 15$, $\beta_j = 1$ if $j = 16, \ldots, 30$, $\beta_j = 2$ if $j = 31, \ldots, 45$ and $\beta_j = 3$ if $j = 46, \ldots, 60$. We take $Y = \tilde{X}\beta$. The binary response $\tilde{Y}$ is obtained as $\tilde{Y}_j = 1$ if $Y_j \geq 0$ and $\tilde{Y}_j = 0$ if $Y_j < 0$. Here the goal is to see if we do not select the first 15 variables.

We run MOSS with the same choice of parameters as in the first simulated example and replicate the experiment 100 times to reduce the sample variability of the results. The mean percentage of predictors that are among the first 15 variables is 20% with a 95% confidence interval of $[0\%, 40\%]$. This is very close to 25% that represents the actual percentage of "unwanted" predictors in X. We can explain these results by the large correlations among the candidate predictors and by the loss of information that occurs during dichotomization of the response variable and of the predictors themselves. The mean number of predictors selected is 4 with a 95% confidence interval of $[1, 20]$. The mean number of regressions evaluated by each instance of MOSS was 3681 with a 95% confidence interval of $[1841, 7370]$. This is much smaller than 5985198 – the number of all possible regressions with 5 variables or less. Again, this indicates the computational efficiency achieved by MOSS. The Bayesian model averaging classifiers for $\tilde{Y}$ have excellent predictive accuracy. The mean area below the corresponding ROC curves was 1 with a 95% confidence interval of $[0.997, 1]$.

3.7   Real Examples: Gene Expression

3.7.1   Breast Cancer Data

We analyze the breast cancer prognosis dataset from van't Veer et al. (2002). Here the goal is to develop a gene expression classifier to predict which patients are likely to develop metastases within 5 years. Yeung et al. (2005) identified 4919 significantly regulated genes in the training set of 76 samples. The test set comprises 19 samples. van't Veer et al. (2002) select 70 genes based on their high correlation with the response and report that only two samples in the test set were incorrectly classified based on the expression levels of these genes. Yeung et al. (2005) used Bayesian model averaging to produce a classifier that involves only 6 genes. Their predictive model gives 3 classification errors on the test set.

We employed MOSS with $c = 0.5$, $c' = 0.0001$, a pruning probability of 0.1 and five different starting points. We searched for regression models containing at most 3 predictors. The number of models evaluated by MOSS in each of the five instances were: 368779, 368781, 486787, 565463, and 206521. These counts are very small compared to the total number of possible regressions 19837151360.

Using the ordering of the genes of Yeung et al. (2005), we identify seven regressions in the set $\mathcal{S}(0.5)$. These regressions involve 11 genes as follows: TSPYL5 (1), NM_021199 (0.88), Contig31010_RC (0.44), Contig16367_RC (0.22), AA555029.RC (0.1), IFIT3 (0.07), PDIA4 (0.07), Contig1829 (0.06), CASC3 (0.06), GDS1048 (0.05), NUP210 (0.05). The numbers of parentheses represent the posterior inclusion probabilities of each gene. Remark that only gene TSPYL5 was also selected by van't Veer et al. (2002) and Yeung et al. (2005) and is the gene with the highest posterior probability in our ranking. The remaining 10 genes do not appear in the list of Yeung et al. (2005). We were not able to determine the actual overlap between our list and the candidate genes of van't Veer et al. (2002). It is likely that this overlap is empty because van't Veer et al. (2002) select genes based on a univariate dependency measure (correlation), while we take into account combinations of at most three genes and the response. Yeung et al. (2005) consider similar combinations in their stochastic search algorithm.

Figure 3.1 shows the performance of the classifier we produced by employing MOSS for hierarchical log-linear models to determine the most relevant models for the 2^4 contingency tables associated with the binary response and the dichotomized expression levels of the genes in each regression. We chose $c = 0.01$, $c' = 0.0001$, a pruning probability of 0.1 and five starting points. We generated 10000 samples from the posterior distributions of the induced logistic regression parameters using Bayesian IPF. These samples are necessary for parameter estimation and to quantify prediction uncertainty. Only two

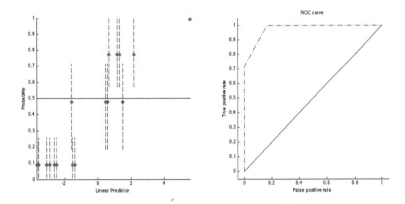

FIGURE 3.1: The left panel shows the performance of the 11 gene classifier of disease-free patients after at least 5 years versus patients with metastases within 5 years. The horizontal lines represent 80% confidence intervals. The right panel shows the ROC curve associated with this classifier.

test samples are misclassified in Figure 3.1. The area below the ROC curve in the right panel of Figure 3.1 is 0.98 which indicates an excellent predictive performance. The Brier score is 1.73 with a standard deviation of 0.19. For comparison, the Brier Score reported by Yeung et al. (2005) is 2.04. Smaller values of the Brier score indicate a better performance.

3.7.2 Leukemia Data

The leukemia dataset of Golub et al. (1999) comprise samples from patients with acute lymphoblastic leukemia (ALL) or acute myeloid leukemia (AML). The initial pre-processing leaves the expression levels of 3051 genes on 38 training samples and 34 test samples – see Yeung et al. (2005). We employed MOSS with $c = 0.5$, $c' = 0.0001$, a pruning probability of 0.1 and five different starting points. We searched for regression models containing at most 2 predictors which gives a total number of possible regressions of 4655827. The number of models evaluated by MOSS in each of the five instances were considerably smaller: 115902, 323301, 189102, 176901, and 79302. The six regressions in the resulting $\mathcal{S}(0.5)$ involve eight genes: MGST1 (0.67), APLP2 (0.33), CCND3 (0.17), TRAC (0.17), NCAPD2 (0.17), ACADM (0.17), MAX (0.17) and PSME1 (0.16). As before, the posterior inclusion probabilities are indicated in parentheses.

We used MOSS to identify relevant hierarchical log-linear models for the corresponding 2^3 contingency tables. We chose $c = 0.01$, $c' = 0.0001$, a pruning probability of 0.1 and five starting points. We generated 10000 samples from the posterior distributions of the induced logistic regression parameters

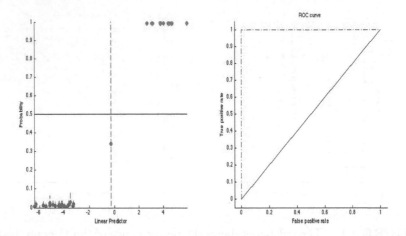

FIGURE 3.2:   The left panel shows the performance of the 8 gene classifier for the 34 test samples in leukemia data. The horizontal lines represent 80% confidence intervals. The right panel shows the ROC curve associated with this classifier.

using Bayesian IPF. Figure 3.2 shows the prediction results for the test data of the classifier obtained by Bayesian model averaging of the six logistic regressions. Only one sample is incorrectly predicted. The Brier score is 0.7 with a standard error of 0.48. The area below the ROC curve is almost 1. Yeung et al. (2005) reports two classification erros based on 20 selected genes and a Brier score of 1.5. Lee et al. (2003) misclassifies one sample based on the expression levels of five genes, while Nguyen and Rocke (2002) reports $1 - 3$ misclassified samples based on $50 - 1500$ genes.

3.7.3   Lymph Node Data

We predict lymph node positivity status in human breast cancer based on the expression levels of 4512 genes – see Hans et al. (2007); Pittman et al. (2004). There are 100 low-risk (node-negative) samples and 48 high-risk (high node-positive). There are two additional predictors: estimated tumor size (in centimeters) and estrogen receptor status (binary variable determined by protein assays).

We employed MOSS with $c = 0.01$, $c' = 0.0001$, a pruning probability of 0.1 and five different starting points. We searched for regression models containing at most 5 predictors due to the larger number of available samples. This choice corresponds with a mean number of samples per cell in the corresponding 2^6 contingency tables of 2.31. The number of models evaluated by MOSS in each

of the five instances (2896078, 1109722, 2057022 and 721769) are significantly
smaller than the total number of possible regressions $1.56 \cdot 10^{16}$. MOSS identifies 49 regressions in $\mathcal{S}(0.01)$. These regressions involve 69 predictors, but
only eight have posterior inclusion probabilities greater than 0.1: ENTPD4
(0.89), tumor size (0.62), ADD3 (0.34), MGLL (0.32), ST6GALNAC2 (0.32),
HEXA (0.18), MEGF9 (0.11) and FAM38A (0.1). Remark that tumor size is
identified as the second most important predictor. The tree models of Pittman
et al. (2004) also found tumor size to be one of the most relevant predictors,
in contrast with the logistic regressions of Hans et al. (2007) that place a
different set of eight genes at the top of their list.

We used MOSS to identify relevant hierarchical log-linear models for the 2^6
contingency tables corresponding with the 49 regressions. We chose $c = 0.01$,
$c' = 0.0001$, a pruning probability of 0.1 and five starting points. We generated 10000 samples from the posterior distributions of the induced logistic
regression parameters using Bayesian IPF. Figures 3.3 and 3.4 show the excellent predictive performance of our classifier. For the fitted values, the area
below the ROC curve is almost one, while the Brier score is 11.68 with a standard deviation of 2.27. Five samples were incorrectly predicted: 97% of the
positives are above 0.5, and 95.8% of the negatives are below 0.5. For comparison, Hans et al. (2007) evaluate the fit of their model to 96% of the positives
above 0.5 and 89% of the negatives below 0.5. We also performed a five-fold
cross-validation check of our models. The area below the ROC curve is 0.98,
the Brier score increases slightly to 16.59 with a standard deviation of 2.64.

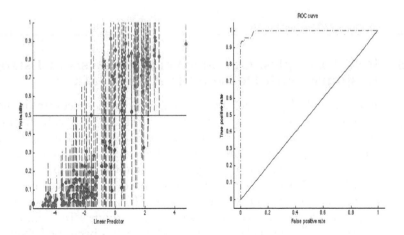

FIGURE 3.3:   Fitted prediction probabilities for the lymph node data. The
horizontal lines represent 80% confidence intervals. The right panel shows the
ROC curve associated with this classifier.

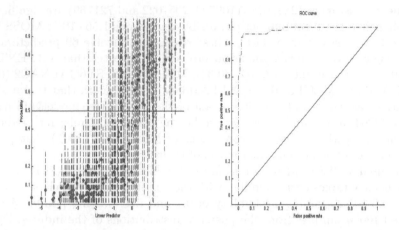

FIGURE 3.4: Fivefold cross-validation prediction probabilities for the lymph node data. The horizontal lines represent 80% confidence intervals. The right panel shows the ROC curve associated with this classifier.

In this case six samples were incorrectly predicted which gives a prediction accuracy of almost 96%.

3.8   Real Examples: Genome-Wide Analysis of Estrogen Response with Dense SNP Array Data

A long-term goal of pharmacogenetics research is the accurate prediction of patient response to drugs, as it would facilitate the individualization of patient treatment. Such an approach is particularly needed in cancer therapy, where currently used agents are ineffective in many patients, and side effects are common. The recent development of genome-wide approaches such as high-density SNP arrays enables the simultaneous measurement of thousands of genes in a single experiment and raises the possibility of an unbiased genome-wide approach to the genetic basis of drug response. To facilitate pharmacogenetics research, a panel of 60 cancer cell lines has been used extensively by the Development Therapeutics Program of the National Cancer Institute (http://dtp.nci.nih.gov). These cell lines have been analyzed for their sensitivity and resistance to a broad range of chemical compounds and thus offer an extensive source of information for testing prediction models of chemotherapy response.

Jarjanazi et al. (2008) focused on the chemosensitivity and resistance to estrogen response (one of the several compounds), which is extremely important in the treatment of breast cancer by hormonal therapy. Determining a genetic signature of estrogen response could help to tailor treatment programs based on patient's genetic background and thus reduce considerably the chance of treatment failure. Each cell line was exposed to estrogen for 48 hours, and the concentration of estrogen required for 50% growth inhibition (GI50 henceforth) was scored. The $\log_{10}$ of the GI50 values were retrieved and normalized to obtain a mean of zero and standard deviation of one across the different cell lines. Using a density estimation of the GI50 values (Wand and Jones (1995)), we labeled 25 cell lines as resistant and 17 cell lines as sensitive. Drug sensitivity tests for estrogen were performed. Genotypes of SNPs in these 42 cell lines were obtained from the Affymetrix 125K chip data – see Garraway et al. (2005) for more details. We retained only 25530 SNPs that were genotyped in at least 90% of the cell lines and had a minimum allele frequency (MAF) of at least 10%.

We employed MOSS with $c = 0.5$, $c' = 0.0001$, a pruning probability of 0.1 and a maximum number of predictors equal to three. We run MOSS one hundred times starting from random regressions. The mean number of models evaluated by MOSS was 2407299, while the minimum and maximum number of models evaluated were 255288 and 9853813, respectively. These counts are very small compared to $2.77 \cdot 10^{12}$ – the total number of regressions with at most three variables. Seven regressions were identified in $\mathcal{S}(0.5)$. These regressions involve 17 SNPs as follows: 1596 (0.39), 8823 (0.33), 5545 (0.33), 7357 (0.33), 14912 (0.25), 14303 (0.25), 7459 (0.18), 11684 (0.13), 23629 (0.12), 4062 (0.12), 9571 (0.12), 394 (0.11), 19083 (0.11), 19132 (0.06), 11103 (0.06), 17403 (0.06), 11775 (0.06). The first SNP is indexed with 1, while the last SNP is indexed with 25530.

Figure 3.5 shows the fitted values of the model-averaged classifier. The area below the ROC curve is one, the Brier score is 0.993 with a standard error of 0.35. No samples were incorrectly predicted. Figure 3.6 shows the two-fold cross-validation results of this classifier. The Brier score is 2.41 with a standard error of 1.67 and all samples are correctly predicted.

The description of the interesting SNPs found by MOSS and their physical distance to known genes are summarized in Table 3.1. Three of these SNPs are located within a known gene, whereas nine were very close to a known gene locus (<150Kb). We also studied the extent of linkage disequilibrium (LD) for the three SNPs that were located within less than 30kb from a known candidate gene region using the software Haploview – see Barrett et al. (2005). In these three cases, we found evidence for LD, suggesting that the associated gene could be implicated in estrogen response. Details about the LD results are given in Tables 3.2, 3.3, and 3.4.

Among the genes identified by MOSS, **Thrombospondin-1 (THBS1)** (also referred to as THBS, TSP, TSP1) has been the most studied in terms of function and association to disease. **THBS1** is a glycoprotein that is in-

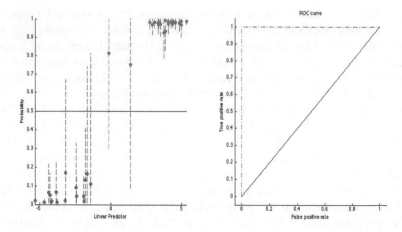

FIGURE 3.5: Fitted prediction probabilities for the SNP data when regressions with at most three variables are considered. The horizontal lines represent 80% confidence intervals. The right panel shows the ROC curve associated with this classifier.

volved in various biological processes including the cell motility, cell adhesion, inflammatory response and multicellular organismal development (Go ontology). **THBS1** regulates the activity of the estrogen in the cell – see Sarkar et al. (2007); Sengupta et al. (2004); Slater et al. (1995). **THBS1** is also extensively implicated in cancer and metastasis in several reviews. Although functionally characterized, the association of the two other candidate genes (**BLNK** and **NSF**) with estrogen response is not well studied. The B cell linker protein **BLNK** at 10q24.1 is a SH2/SH3 containing adaptor protein which serves as a scaffold to assemble downstream targets of antigen activation. It is involved in the regulation of B cell receptor signaling during inflammatory and humoral immune response of the cell. It has been shown to be associated with leukemia, lymphoma and solid tumors. **N-ethylmaleimide-sensitive factor** (**NSF**) at 17q21 is required for vesicle-mediated transport, from the endoplasmic reticulum to the Golgi stack. It is also involved in proteolysis. The functions of the transmembrane protein 26 (TMEM26) at 10q21.2 and other candidate genes in our study are not well studied.

While the identification of **THBS1** has demonstrated that our method is capable of identifying a gene that is extensively involved in estrogen response and cancer development, the remaining genes with limited knowledge represent novel candidates to be further studied. Several important remarks about the application of MOSS to this study are worth noting. First, **THBS1**, the most functionally relevant SNP identified here, would not have been detected

TABLE 3.1: Important SNPs Associated with Estrogen Response from Resistant and Sensitive NIC60 Cell Lines

Variable #	SNP ID	Chr #	Location	Allele1	Allele2	MAF[1]	p-value[2]	FDR_BH[3]	Rank[4] (p-value)	Rank[5] (MOSS)	Related gene[6]	Distance[7] from closest gene
8823	rs199449	17	45283725	A	G	0.23	0.0238	0.5058	1201	2	NSF	0 kb
11684	rs1915437	10	62530796	G	A	0.24	0.0005	0.3299	34	8	TMEM26	0 kb
17403	rs7077062	10	97646812	T	C	0.65	0.0002	0.3020	17	16	BLNK	0 kb
19083	rs1037182	9	110087746	C	T	0.47	0.0015	0.3668	101	13	UGCG	<20 kb
19132	rs7861999	9	110502843	A	G	0.30	0.0047	0.4037	289	14	HSDL2	<20 kb
7459	rs4923835	15	37542125	G	A	0.37	0.0009	0.3448	66	7	THSB1	<30 kb
14912	rs4703882	5	81799438	T	A	0.13	0.6648	0.9308	18211	5	FLJ41309	35 kb
11103	rs1370230	5	58973526	A	G	0.52	0.0576	0.5924	2478	15	MIRN582	40 kb
23629	rs1985065	4	165780046	G	A	0.18	0.9513	0.9919	24471	9	ANP32C	70 kb
11775	rs953654	1	62919199	A	G	0.42	0.2111	0.7609	7073	17	loc199897	95 kb
394	rs2279768	10	2933651	G	C	0.00	0.0004	0.3299	28	12	PFKP	120 kb
9571	rs1830876	6	51112593	C	T	0.21	0.0064	0.4210	385	11	FTHP1	150 kb
1596	rs200032	5	8744094	G	C	0.29	0.0001	0.2686	8	1	SEMA5A	400 kb
14303	—	16	78592091	T	C	0.17	0.0064	0.4210	383	6	WWOX	400 kb
4062	rs1577053	13	21136254	T	C	0.07	0.0000	0.2673	2	10	FTHL7	500 kb
5545	rs1474519	21	28715801	G	A	0.26	0.0071	0.4210	422	3	none	–
7357	rs1401154	8	37083828	T	C	0.13	0.0020	0.3716	139	4	none	–

1 Minor allele frequency
2 P-value from single marker test using Cochran-Armitage trend test
3 False discovery rate from Benjamini and Hochberg (1995)
4 Rank based on single-marker p-value
5 Rank based on MOSS posterior probability for each single marker
6 Closest gene based on physical distance
7 Distance in kilo bases (Kb)

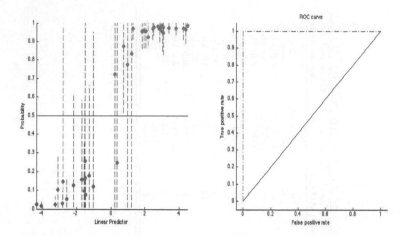

FIGURE 3.6: Two-fold cross-validation prediction probabilities for the SNP data when regressions with at most three variables are considered. The horizontal lines represent 80% confidence intervals. The right panel shows the ROC curve associated with this classifier.

TABLE 3.2:   Linkage Analysis of rs4923835 (marker 932558) with THBS1 and FSIP1

Significant Marker	Markers within Genes	Marker Location	D'*	LOD*	r^{2*}	CI†	Distance (Kb)
rs4923835	rs2228263	exon 18 - THBS1	1	15.31	0.92	(0.84, 1)	59232
rs4923835	rs1051442	3'-UTR - THBS1	1	16.27	0.925	(0.85, 1)	61580
rs4923835	rs17633107	3'-UTR - THBS1	1	16.27	0.925	(0.85, 1)	62212
rs4923835	rs6492905	3'-UTR - FSIP1	1	16.06	0.924	(0.85, 1)	66333
rs4923835	rs17633210	intron 1 - FSIP1	1	15.61	0.92	(0.84, 1)	74497
rs4923835	rs17705806	intron 1 - FSIP1	1	16.27	0.925	(0.85, 1)	74402
rs4923835	rs17706083	intron 1 - FSIP1	1	16.27	0.925	(0.85, 1)	85005

* Measures of linkage disequilibrium (Devlin and Risch, 1995)
† Confidence interval of r^2
Note: LD was mapped using HapMap Caucasian Data (HapMap Data Release 22, NCBI B36 assembly, dbSNP build 126). The LD of rs4923835 was constructed with 7 markers within the THBS1 and FSIP1. All the markers have shown strong LD ($r^2 \geq 0.9$) indicating the linkage of the significant marker to both THBs1 and FSIP1. The distance between the rs4923835 and the other SNPs ranged between 59-85Kb.

TABLE 3.3:   Linkage Analysis of rs7861199 with SNPs in HSDL2, EPF5, and ROD1

Significant Marker	Markers within Genes	Marker Location	D'*	LOD*	r^{2*}	CI†	Distance (Kb)
rs7861999	rs3813855	5' UTR - HSDL2	0.918	13.05	0.844	(0.75, 0.98)	18973
rs7861999	rs7852741	intron 1 - HSDL2	1	12.72	0.79	(0.81, 1)	33384
rs7861999	rs10759544	intron 6 - ROD1/EPF5	0.771	5.9	0.534	(0.51, 0.91)	100518

* Measures of linkage disequilibrium (Devlin and Risch, 1995)
† Confidence interval of r^2

TABLE 3.4: Linkage Analysis of rs1037182 with UGCG (linked to 5′ sequences)

Significant Marker	Markers within Genes	Marker Location	D'^*	LOD*	r^{2*}	CI†	Distance (Kb)
rs1037182	rs10817244	∼10Kb 5′-end of UGCG	1	18.75	0.718	(0.88, 1)	1022
rs1037182	rs7028129	∼20Kb 5′-end of UGCG	1	19.1	0.799	(0.88, 1)	286

* Measures of linkage disequilibrium (Devlin and Risch, 1995)
† Confidence interval of r^2

using an exhaustive testing of all single markers. Indeed, the associated p-value after correction for multiple testing is clearly non significant (p=0.35). Second, the proportion of markers that are biological relevant (i.e. within or close to known genes) is relatively large. This is somewhat unusual compared to other GWAS that have found a substantial fraction of associations in regions that do not contain annotated genes or that are outside of transcriptional units of the genome — see Altshuler and Daly (2007). Third, the predictive ability of the selected subset of SNPs on estrogen response is almost perfect, which implies very promising applications of MOSS in predictive genetic testing. Finally, MOSS provides a unique and comprehensive statistical framework for analyzing GWAS, filling an important gap in the search for combinations of SNPs that have high predictive values in high-dimensional data problems.

3.9 Discussion

We summarize the key steps of our proposed methodology. The use of the conjugate priors for log-linear parameters of Massam et al. (2009) make the variable selection step and the log-linear model selection step coherent. The variable selection step is needed to allow us to focus on the most important covariates, while the model selection step is crucial to determine the most parsimonious representation of the underlying interactions. The efficiency of MOSS in identifying regions of high-posterior probability in the models space allow our method to scale to genomewide studies with data on ≥ 500K assayed SNPs as well as gene expression and clinical information. Bayesian model averaging is crucial to produce classifiers based on the most relevant small subsets of regressors. These subsets embody complex interactions among covariates of any type.

We allowed any combination of covariates as a possible candidate in our variable selection step. The accuracy of our method can be improved by incorporating prior knowledge related to spatial dependencies among markers or genes. Our methodology does not allow for the presence of missing data, hence the individuals for whom complete information is not available would have to be discarded. This could potentially lead to a significant decrease in the total sample size, hence our ability to detect higher-order associations or more complex combinations of covariates might be severely diminished. One alternative would be to use one time imputation methods such as the PHASE software of Stephens et al. (2001); Stephens and Donnelly (2003) for missing genotype/haplotype data. Another alternative that also takes into account the uncertainty related to missing genotype data is the Bayesian model selection and model averaging method proposed by Lunn et al. (2006). Unfortunately their MCMC approach does not seem to scale to whole genome scans.

C++ and MATLAB® code implementing the methods described in this chapter can be downloaded from

http://www.stat.washington.edu/adobra/software/largetables/

Acknowledgments

The authors would like to thank Ka Yee Yeung who provided us with the data for the breast cancer and leukemia examples and Chris Hans who provided the data for the lymph node status example.

References

Altshuler, D. and M. Daly (2007). Guilt beyond reasonable doubt. *Nature Genetics 39*(7), 813–4.

Barrett, J. C., B. Fry, J. Maller, and M. J. Daly (2005). Haploview: analysis and visualization of ld and haplotype maps. *Bioinformatics 21*(2), 263–265.

Benjamini, Y. and Y. Hochberg (1995). Controlling the false discovery rate: a practical and powerful approach to multiple testing. *Journal of the Royal Statistical Society: Series B 57*, 289–300.

Carlin, B. P. and S. Chib (1995). Bayesian model choice via Markov chain Monte Carlo. *Journal of the Royal Statistical Society, Series B 57*, 473–484.

Chipman, H., E. I. George, and R. E. McCullogh (2001). The practical implementation of Bayesian model selection (with discussion). In P. Lahiri (Ed.), *Model Selection*, pp. 66–134. IMS: Beachwood, OH.

Christensen, K. and J. C. Murray (2007). What genome-wide association studies can do for medicine. *New England Journal of Medicine 356*, 1094–1097.

Clark, T. G., M. De Iorio, and R. C. Griffiths (2007). Bayesian logistic regression using a perfect phylogeny. *Biostatistics 8*, 32–52.

Clyde, M. and E. I. George (2004). Model uncertainty. *Statistical Science 19*, 81–94.

Cordell, H. J. and D. G. Clayton (2002). A unified stepwise regression procedure for evaluating the relative effects of polymorphisms within a gene using case/control or family data: application to HLA in type 1 diabetes. *American Journal of Human Genetics 70*, 124–141.

Darroch, J. N. and T. P. Speed (1983). Additive and multiplicative models and interaction. *The Annals of Statistics 11*, 724–738.

Dawid, A. P. and S. L. Lauritzen (1993). Hyper Markov laws in the statistical analysis of decomposable graphical models. *The Annals of Statistics 21*, 1272–1317.

Dellaportas, P. and J. J. Forster (1999). Markov chain Monte Carlo model determination for hierarchical and graphical log-linear models. *Biometrika 86*, 615–633.

Devlin, B. and N. Risch (1995). A comparison of linkage disequilibrium measures for fine-scale mapping. *Genomics 29*(2), 311–22.

Diaconis, P. and D. Ylvisaker (1979). Conjugate priors for exponential families. *The Annals of Statistics 7*, 269–281.

Dobra, A., C. Hans, B. Jones, J. R. Nevins, G. Yao, and M. West (2004). Sparse graphical models for exploring gene expression data. *Journal of Multivariate Analysis 90*, 196–212.

Dobra, A. and H. Massam (2010). The mode oriented stochastic search (MOSS) for log-linear models with conjugate priors. *Statistical Methodology*. To appear DOI:10.1016/j.stamet.2009.04.002.

Dudoit, S., J. Fridlyand, and T. P. Speed (2002). Comparison of discrimination methods for the classification of tumors using gene expression data. *Journal of the American Statistical Association 97*, 77–87.

Edwards, D. E. and T. Havranek (1985). A fast procedure for model search in multidimensional contingency tables. *Biometrika 72*, 339–351.

Efron, B. and R. Tibshirani (2002). Empirical Bayes methods and false discovery rates for microarrays. *Genetics Epidemiology 23*, 70–86.

Fernández, C., E. Ley, and M. F. Steel (2003). Benchmark priors for Bayesian model averaging. *Journal of Econometrics 75*, 317–343.

Furnival, G. M. and R. W. Wilson (1974). Regression by leaps and bounds. *Technometrics 16*, 499–511.

Garraway, L. A., H. R. Widlund, M. A. Rubin, G. Getz, A. J. Berger, S. Ramaswamy, R. Beroukhim, D. A. Milner, S. R. Granter, J. Du, C. Lee, S. N. Wagner, C. Li, T. R. Golub, D. L. Rimm, M. L. Meyerson, D. E. Fisher, and W. R. Sellers (2005). Integrative genomic analyses identify MTF as a lineage survival oncogene amplified in malignant melanoma. *Nature 436*, 117–122.

Geiger, D. and D. Heckerman (2002). Parameter priors for directed acyclic graphical models and the characterization of several probability distributions. *The Annals of Statistics 30*, 1412–1440.

George, E. I. and R. E. McCulloch (1993). Variable selection via Gibbs sampling. *Journal of the American Statistical Association 88*, 881–889.

George, E. I. and R. E. McCulloch (1997). Approaches for Bayesian variable selection. *Statistica Sinica 7*, 339–373.

Golub, T. R., D. K. Slonim, P. Tamayo, C. Huard, M. Gaasenbeek, J. P. Mesirov, H. Coller, M. L. Loh, J. R. Downing, M. A. Caligiuri, C. D. Bloomfield, and E. S. Lander (1999). Molecular classification of cancer: Class discovery and class prediction by gene expression monitoring. *Sci-*

ence *286*, 531–537.

Green, P. J. (1995). Reversible jump Markov chain Monte Carlo computation and Bayesian model determination. *Biometrika 82*, 711–732.

Hans, C., A. Dobra, and M. West (2007). Shotgun stochastic search for "large p" regression. *Journal of the American Statistical Association 102*, 507–516.

Hoggart, C. J., T. G. Clark, M. D. De Iorio, J. C. Whittaker, and D. J. Balding (2008). Genome-wide significance for dense SNP and resequencing data. *Genetic Epidemiology 32*, 179–185.

Hu, J., A. Joshi, and V. E. Johnson (2009). Log-linear models for gene association. *Journal of the American Statistical Association 104*, 597–607.

Jarjanazi, H., J. Kiefer, S. Savas, L. Briollais, S. Tuzmen, N. Pabalan, I. Ibrahim-Zada, S. Mousses, and H. Ozcelik (2008). Discovery of genetic profiles impacting response to chemotherapy: Application to gemcitabine. *Human Mutations 29*, 461–467.

Jones, B., C. Carvalho, A. Dobra, C. Hans, C. Carter, and M. West (2005). Experiments in stochastic computation for high-dimensional graphical models. *Statistical Science 20*, 388–400.

Kass, R. and A. E. Raftery (1995). Bayes factors. *Journal of the American Statistical Association 90*, 773–795.

Lauritzen, S. L. (1996). *Graphical Models*. Clarendon Press, Oxford.

Lee, K. E., N. Sha, E. R. Dougherty, M. Vanucci, and B. K. Mallick (2003). Gene selection: a Bayesian variable selection approach. *Bioinformatics 19*, 90 97.

Liang, F., R. Paulo, G. Molina, M. Clyde, and J. O. Berger (2008). Mixtures of g-priors for Bayesian variable selection. *Journal of the American Statistical Association 103*, 410–423.

Lunn, D. J., J. C. Whittaker, and N. Best (2006). A Bayesian toolkit for genetic association studies. *Genetic Epidemiology 30*, 231–247.

Madigan, D. and A. E. Raftery (1994). Model selection and accounting for model uncertainty in graphical models using Occam's window. *Journal of the American Statistical Association 89*, 1535–1546.

Madigan, D. and J. York (1995). Bayesian graphical models for discrete data. *International Statistical Review 63*, 215–232.

Madigan, D. and J. York (1997). Bayesian methods for estimation of the size of a closed population. *Biometrika 84*, 19–31.

Marchini, J., P. Donnelly, and L. R. Cardon (2005). Genome-wide strategies for detecting multiple loci that influence complex diseases. *Nature Genet-*

ics 37, 413–417.

Massam, H., J. Liu, and A. Dobra (2009). A conjugate prior for discrete hierarchical log-linear models. *The Annals of Statistics 37*, 3431–3467.

Nguyen, D. V. and D. M. Rocke (2002). Tumor classification by partial least squares using microarray gene expression data. *Bioinformatics 18*, 39–50.

Nott, D. J. and P. J. Green (2004). Bayesian variable selection and the Swendsen-Wang algorithm. *Journal of Computational and Graphical Statistics 13*, 1–17.

Pittman, J., E. Huang, H. Dressman, C. F. Horng, S. H. Cheng, M. H. Tsou, C. M. Chen, A. Bild, E. S. Iversen, A. T. Huang, J. R. Nevins, and M. West (2004). Integrated modeling of clinical and gene expression information for personalized prediction of disease outcomes. *Proceedings of the National Academy of Sciences 101*, 8431–8436.

Raftery, A. E., D. Madigan, and J. Hoeting (1997). Bayesian model averaging for linear regression models. *Journal of the American Statistical Association 92*, 1197–1208.

Ruczinski, I., C. Kooperberg, and M. LeBlanc (2003). Logic regression. *Journal of Computational and Graphical Statistics 12*, 475–511.

Sarkar, A. J., K. Chaturvedi, C. P. Chen, and D. K. Sarkar (2007). Changes in thrombospondin-1 levels in the endothelial cells of the anterior pituitary during estrogen-induced prolactin-secreting pituitary tumors. *Journal of Endocrinology 192*, 395–403.

Schaid, D. (2004). Evaluating associations of haplotypes with traits. *Genetic Epidemiology 27*, 348–364.

Sengupta, K., S. Banerjee, N. K. Saxena, and S. K. Banerjee (2004). Thombospondin-1 disrupts estrogen-induced endothelial cell proliferation and migration and its expression is suppressed by estradiol. *Molecular Cancer Research 2*, 150–8.

Slater, M., J. Patava, and R. S. Mason (1995). Thrombospondin co-localises with TGF beta and IGF-I in the extracellular matrix of human osteoblast-like cells and is modulated by 17 beta estradiol. *Experientia 51* (3), 235–44.

Stephens, M. and P. Donnelly (2003). A comparison of Bayesian methods for haplotype reconstruction from population genotype data. *American Journal of Human Genetics 73*, 1162:1169.

Stephens, M., N. J. Smith, and P. Donnelly (2001). A new statistical method for haplotype reconstruction from population data. *American Journal of Human Genetics 68*, 978:989.

Storey, J. D. and R. Tibshirani (2003). Statistical significance for genomewide studies. *Proceedings of the National Academy of Sciences 100*, 9440–9445.

Swartz, M. D., D. C. Duncan, and E. W. Daw (2007). Model selection and Bayesian methods in statistical genetics: Summary of Group 11 contributions to Genetic Analysis Workshop 15. *Genetics Epidemiology 31* (Supplement 1), S96–S102. on behalf of Group 11.

Thomas, D. C., R. W. Haile, and D. Duggan (2005). Recent developments in genomewide association scans: a workshop summary and review. *American Journal of Human Genetics 77*, 337–345.

Tusher, V. G., R. Tibshirani, and G. Chu (2001). Significance analysis of microarrays applied to the ionizing radiation response. *Proceedings of the National Academy of Sciences 98*, 5116–5121.

van't Veer, L. J., D. Hongyue, M. J. van de Vijver, Y. D. He, A. A. M. Hart, M. Mao, H. L. Peterse, K. van der Kooy, M. J. Marton, A. T. Witteveen, G. J. Schreiber, R. M. Kerkhoven, C. Roberts, P. S. Linsley, R. Bernards, and S. H. Friend (2002). Gene expression profiling predicts clinical outcome of breast cancer. *Nature 415*, 530–536.

Verzilli, C. J., N. Stallard, and J. C. Whittaker (2006). Bayesian graphical models for genomewide association studies. *American Journal of Human Genetics 79*, 100–112.

Wand, M. P. and M. C. Jones (1995). *Kernel Smoothing.* Chapman and Hall.

Wang, K. and D. Abbott (2008). A principal components regression approach to multilocus genetic association studies. *Genetic Epidemiology 32*, 108–118.

Yeung, K., R. E. Bumgarner, and A. E. Raftery (2005). Bayesian model averaging: Development of an improved multi-class, gene selection and classification tool for microarray data. *Bioinformatics 21*, 2394–2402.

Zhang, Y. and J. S. Liu (2007). Bayesian inference of epistatic interactions in case-control studies. *Nature Genetics 39*, 1167–73.

Chapter 4

Nonparametric Bayesian Bioinformatics

David B. Dunson
Department of Statistical Science, Duke University

4.1 Introduction

Modern biomedical studies routinely collect complex high-dimensional data, so it is no longer possible for an applied biostatistician to avoid encountering bioinformatics problems. In some ways, bioinformatics is a sub-field of biostatistics. However, due to the inherent challenges involved in analyzing and searching for patterns in massive dimensional data sets, it is often the case that statistical rigor is put aside in favor of practically-motivated algorithmic-based approaches. This has naturally led to an increasing number of computer scientists and machine learning experts with interests in bioinformatics. Algorithmic-based methods have certainly led to valuable insights. However, there are clear advantages to the use of coherent statistical model-based approaches that attempt to flexibly, yet sparsely characterize high dimensional data. Such methods can be used to discover sparse latent structure, while accounting for uncertainty. Given the enormous dimensional model spaces that are routinely encountered in bioinformatics applications, it is crucial to account for uncertainty in model selection in conducting inferences and predictions. Optimization-based strategies that search for the best model based on some criterion have the intrinsic disadvantage that the selected model is almost surely not the true model given that the sample size of the available data is dwarfed by the size of the model space. Bayesian methods represent a natural approach for dealing with uncertainty in model selection, while also incorporating available prior information, which is crucial in addressing the large p small n problems in bioinformatics.

This chapter provides a review and commentary on the use of Bayesian nonparametrics in bioinformatics, while also suggesting a number of promising areas for new research. By "Bayesian nonparametrics", I am referring to approaches utilizing random probability measures, such as the Dirichlet process (Ferguson, 1973, 1974). Section 1.2 provides a brief review of Bayesian nonparametrics, focusing on Dirichlet process mixture models and related approaches. Section 1.3 provides an overview of work on Bayesian nonparametric approaches for multiple testing and high-dimensional regression. Section 1.4 considers clustering and functional data analysis applications. Section 1.5 provides an overview of recent work on using innovative nonparametric priors in population genetics, EST library analyses and other areas. Section 1.6 contains a discussion and commentary.

4.2   Dirichlet Process Mixture Models

A comprehensive review of Bayesian nonparametric methodology is well beyond the scope of this chapter. For more comprehensive reviews of methods for Bayesian nonparametric data analysis, refer to Müller and Quintana (2004) and Dunson (2009a). Here, I focus narrowly on Dirichlet process mixture (DPM) models (Lo, 1984; Escobar and West, 1995), since DPMs have been widely used in biostatistics and bioinformatics and provide a starting point for much of the recent work proposing richer modeling frameworks. For a comprehensive review of properties of the Dirichlet process, refer to Ghosal (2008).

In introducing Dirichlet process mixture models, I focus initially on the simple case in which we have observations $y_1, \ldots, y_n$ for different subjects and want to estimate the density nonparametrically using hierarchical model

$$y_i = \mu_i + \epsilon_i, \quad \epsilon_i \sim N(0, \sigma_i^2),$$
$$(\mu_i, \sigma_i^2) \sim P, \tag{4.1}$$

where $N(\mu, \sigma^2)$ denotes the normal density with mean μ and variance σ^2, and P is a mixture distribution. In particular, instead of characterizing y_i as normally distributed, model (4.1) induces the following form for the density of y_i:

$$f(y) = \int N(y; \mu, \sigma^2) dP(\mu, \sigma^2), \tag{4.2}$$

which is a location-scale mixture of normals. A location mixture of normals could instead be obtained by including a fixed, but unknown variance σ^2 in the normal kernel, and letting P correspond to a mixture distribution for the mean. It is well known that mixtures of normals can approximate any smooth density, but this flexibility relies on a flexible model for P.

Hence, the main question that arises is how to specify P. From a Bayesian perspective, we can allow P to be an unknown distribution by choosing $P \sim \mathcal{P}$, where $\mathcal{P}$ is an appropriate prior. For illustration, consider the location mixture case in which we have $\mu_i \sim P$, with $P \sim \mathcal{P}$ an unknown mixture distribution. Then, $\mathcal{P}$ should be chosen as a distribution on the space of distributions over the real line ($\Re$), so that each realization from $\mathcal{P}$ will correspond to a different distribution over $\Re$. The question is then how do we choose a distribution on the space of distributions? One obvious strategy is to assume that P belongs to some parametric family, meaning that P can be parameterized by finitely many parameters. Then, by specifying a prior distribution for these parameters (e.g., a normal inverse-gamma distribution for the mean and variance of a Gaussian P), one can induce a prior $\mathcal{P}$ for the unknown distribution. Draws from $\mathcal{P}$ are obtained by drawing from the prior distribution for the parameters characterizing P, and then calculating the corresponding P.

Unfortunately, this type of approach induces a prior $\mathcal{P}$ that has support on an extremely small subset of the space of distributions over $\Re$, and the resulting approach for density estimation will do a poor job of approximating densities that are not close to normal. To motivate alternatives, let $P^{(0)}$ denote an unknown true distribution, which can be any distribution over $\Re$. Then, it is appealing to choose a prior with *large support*, meaning that samples from $\mathcal{P}$ have a positive probability of being within an arbitrarily small neighborhood of $P^{(0)}$, as measured by a distance metric $d(P, P^{(0)})$. Large support is a necessary condition for posterior consistency, which means that the posterior distribution of P given the available data concentrates increasingly in an arbitrarily small neighborhood of $P^{(0)}$ as $n \to \infty$.

To simplify the problem of specifying $\mathcal{P}$, suppose that we divide the real line into mutually exclusive subsets, $B_1, \ldots, B_k$, so that $B_h \cap B_{h'} = \emptyset$ for all h, h' and $\Re = \cup_{h=1}^k B_h$. Then, let $\pi_h = \Pr(\mu_i \in B_h)$, for $h = 1, \ldots, k$, denote the probabilities assigned to each of the subsets. Conditionally on P, these probabilities are fixed as $\pi = \{P(B_1), \ldots, P(B_k)\}$, where $P(B)$ denotes the probability assigned to $\mu_i \in B$ by the distribution P. As P is unknown, the probability vector π is a random variable with support on the k-dimensional simplex. The conjugate prior for π is the Dirichlet distribution. Letting P_0 denote an initial guess for the distribution P and α correspond to a precision parameter controlling certainty in this guess, one could choose the prior, $\pi \sim$ Dirichlet$(\alpha P_0(B_1), \ldots, \alpha P_0(B_k))$. Conditionally on realizations $\mu_i \sim P$, for $i = 1, \ldots, n$, we then obtain the posterior,

$$(\pi \mid \mu) \sim \text{Dirichlet}\left(\alpha P_0(B_1) + \sum_{i=1}^n 1(\mu_i \in B_1), \ldots, \alpha P_0(B_k) + \sum_{i=1}^n 1(\mu_i \in B_k)\right),$$

where $1(\mu \in B) = 1$ if $\mu \in B$ and 0 otherwise. It follows that the updated precision parameter is $\alpha + n$, while the updated probability assigned to set B_h is $\mathrm{E}(\pi_h|\mu) = \{P_0(B_h) + \sum_i 1(\mu_i \in B_h)\}/(\alpha + n)$.

The Dirichlet prior is convenient in terms of ease of interpretation, prior elicitation and the conjugacy property. However, instead of choosing a prior for P, we have only chosen a prior for the probabilities that P allocates to the finite collection of subsets $(B_1, \ldots, B_k)$. This may be used as an approximation to P if the number of subsets k is large and the size of the partition sets is sufficiently small. However, it is appealing to avoid sensitivity to the choice of partition. With this goal in mind, Ferguson (1973; 1974) proposed to induce a Dirichlet process (DP) prior for P by simply specifying a finite Dirichlet prior for *all possible partitions*,

$$\{P(B_1), \ldots, P(B_k)\} \sim \mathrm{Diri}\big(\alpha P_0(B_1), \ldots, \alpha P_0(B_k)\big), \quad \text{for all } B_1, \ldots, B_k,$$

where the "for all" statement refers also to any possible k. Ferguson proved that a P exists, which satisfies these distributional assumptions. Denoting the induced prior by $P \sim DP(\alpha P_0)$, he also showed that the conjugacy property is maintained with

$$(P \mid \mu) \sim \mathrm{DP}\left(\alpha P_0 + \sum_{i=1}^{n} \delta_{\mu_i}\right),$$

where δ_μ denotes a degenerate distribution with unit probability mass at μ.

The Ferguson Dirichlet process prior has become routinely used in recent years for density estimation, Bayesian clustering, mixture modeling and other problems. In addition to appealing theoretical properties, such as large support and posterior consistency for a number of model classes, the DP prior is very convenient computationally. Computational efficiency is a crucial issue in nonparametric Bayes models, since it is clear that the random P must be specified in terms of infinitely-many parameters in order to satisfy the large support condition. In fact, the incorporation of infinitely-many parameters is a distinguishing feature of the nonparametric approach. However, this would seem to present insurmountable problems in conducting posterior computation, as we certainly cannot implement Markov chain Monte Carlo (MCMC) update steps for infinitely many unknowns. Indeed, in bioinformatics applications, we are particularly motivated to consider approaches that scale well in sample size and number of observations, so it would seem at first consideration that nonparametric Bayes approaches may be impractical due to computational considerations. The reason that this is very clearly not the case is due to some amazing properties of Dirichlet processes, which also carry over to generalizations and related nonparametric priors.

The first remarkable property is the Polya urn scheme of Blackwell and Mac-Queen (1973). In particular, let $\mu_i \sim P$, for $i = 1, \ldots, n$, with $P \sim DP(\alpha P_0)$. Then, in order to bypass the need for computation of the infinitely-many parameters characterizing P, one can potentially marginalize out P to obtain a prior for $(\mu_1, \ldots, \mu_n)'$, the finitely-many realizations from P represented in the sample. Blackwell and MacQueen showed that such a marginalization leads to

$$\left(\mu_n \mid \mu_1, \ldots, \mu_{n-1}\right) \sim \left(\frac{\alpha}{\alpha + n - 1}\right) P_0 + \left(\frac{1}{\alpha + n - 1}\right) \sum_{j=1}^{n-1} \delta_{\mu_j}, \quad (4.3)$$

so that μ_n is either set equal to one of the previous subject's values or is allocated to a new value drawn from P_0. It is clear from this specification that the DP prior for P induces ties and hence clustering in the observations. Let $\mu^{(i)} = \{\mu_j, j \neq i\}$ denote the set of μ values excluding the value for the ith subject, let $\mu^{*(i)} = \{\mu_h^{*(i)}\}_{h=1}^{k^{(i)}}$ denote the $k^{(i)}$ unique values in the vector $\mu^{(i)}$ and let $S_j^{(i)} = h$ denote that $\mu_j = \mu_h^{*(i)}$, with $S^{(i)} = \{S_j^{(i)}, j \neq i\}$. Then, it follows from (4.3) and exchangeability of the subjects, that

$$\left(\mu_i \mid \mu^{*(i)}, S^{(i)}\right) \sim \left(\frac{\alpha}{\alpha + n - 1}\right) P_0 + \sum_{h=1}^{k^{(i)}} \left(\frac{n_h^{(i)}}{\alpha + n - 1}\right) \delta_{\mu_h^{*(i)}}, \quad (4.4)$$

where $n_h^{(i)} = \sum_{j \neq i} 1(S_j^{(i)} = h)$ is the number of subjects with $\mu_j = \mu_h^{*(i)}$ excluding subject i. The clustering properties of the DP are clear from (4.4). In particular, as subjects are added, the probability of allocating a subject to an existing cluster is proportional to the number of individuals in that cluster, while the probability of allocating the subject to a new cluster is proportional to the DP precision parameter α. The DP prior automatically allows the number of clusters represented in a sample of n subjects to grow at a rate proportional to $\alpha \log n$.

The conditional prior distribution in (4.4) can be used to obtain a simple Gibbs sampling algorithm for posterior computation in DP mixture (DPM) models as proposed by Bush and MacEachern (1996). In particular, consider the local mixture of normals model for Bayesian density estimation, with $y_i \sim N(\mu_i, \sigma^2)$, $\mu_i \sim P$, $P \sim DP(\alpha P_0)$, $\sigma^{-2} \sim \text{gamma}(a, b)$, and P_0 corresponding to a normal distribution with mean μ_0 and precision (1/variance) τ_0. Then, the Gibbs sampler proceeds by cycling through the following sampling steps,

1. Update μ_i from the conditional posterior distribution given $\mu^{*(i)}$, $S^{(i)}$ and y_i by first sampling the cluster allocation $S_i \in \{1, \ldots, k^{(i)}, k^{(i)} + 1\}$ for subject i,

$$\Pr(S_i = h \mid \mu^{*(i)}, S^{(i)}, y_i) \propto \begin{cases} n_h^{(i)} N(y_i; \mu_h^{(i)}, \sigma^2), & h = 1, \ldots, k^{(i)}, \\ \alpha \int N(y_i; \mu, \sigma^2) dP_0(\mu), & h = k^{(i)} + 1 \end{cases}$$

Then, if $S_i = k^{(i)} + 1$ so that subject i is allocated to a new cluster, generate the μ value specific to this new cluster from $N(\widehat{\mu}_i, \widehat{\sigma}_\mu^2)$ with

$$\widehat{\mu}_i = \frac{\sigma^{-2} y_i + \tau_0 \mu_0}{\sigma^{-2} + \tau} \quad \text{and} \quad \widehat{\sigma}_\mu^2 = \frac{1}{\sigma^{-2} + \tau}.$$

Also, $\int N(y_i; \mu, \sigma^2) dP_0(\mu) = N(y_i; 0, \sigma^2) N(\mu_0; 0, \tau_0^{-1}) / N(0; \widehat{\mu}_i, \widehat{\sigma}_\mu^2)$.

2. Update the residual variance by sampling from the gamma conditional,

$$(\sigma^{-2} \mid S, \mu^*, y) \sim \text{gamma}\left(a + \frac{n}{2}, b + \frac{1}{2}\sum_{i=1}^{n}(y_i - \mu_i)^2\right),$$

These steps are straightforward to implement, as they involve sampling from a multinomial with probabilities that can easily be calculated analytically, as well as sampling from normal and gamma distributions. The initial samples should be discarded as a burn-in to allow time for convergence. After convergence, the Gibbs sampler produces autocorrelated draws from the joint posterior distribution of the subject-specific parameters $\mu = \{\mu_i\}_{i=1}^n$, cluster allocation indicators $S = \{S_i\}_{i=1}^n$, cluster-specific parameters $\mu^* = \{\mu_h\}_{h=1}^k$, with k denoting the number of unique values of μ, and residual variance σ^2.

Conducting inferences and predictions based on the Gibbs sampling output involves a number of subtleties, which are worth discussing in detail. First, suppose that interest focuses on density estimation. Then, note that conditionally on $(y_1, \ldots, y_n)'$, $S = (S_1, \ldots, S_n)'$, $\mu^* = (\mu_1^*, \ldots, \mu_k^*)'$ and σ^2 the predictive density of y_{n+1} for a new subject is $f(y_{n+1}) =$

$$\sum_{h=1}^{k}\left(\frac{n_h}{\alpha + n}\right)N(y_{n+1}; \mu_h^*, \sigma^2) + \frac{N(\mu_0; 0, \tau_0^{-1})}{(\alpha + n)N(0; \widehat{\mu}_{n+1}, \widehat{\sigma}_\mu^2)}N(y_{n+1}; 0, \sigma^2).$$

To obtain posterior summaries of $f(y_{n+1})$, including the posterior mean and 95% pointwise credible intervals, one can pre-specify a dense grid of different possible y_{n+1} values and then save $f(y_{n+1})$ at each Gibbs sampling iteration after the burn-in for each y_{n+1} value in the grid. By calculating the posterior mean for each grid value and connecting the points with lines, one obtains a Bayes estimate of the density. In my experience, this Bayes density estimator performs better than typical frequentist estimators, such as the kernel estimator with default bandwidth value used in R and other software packages. Even better performance can be obtained by using a location-scale instead of a location mixture. The above Gibbs sampling algorithm is straightforward to modify to accommodate location-scale mixtures of Gaussians.

The allocation $S = (S_1, \ldots, S_n)'$ of the n subjects to k clusters varies across the Gibbs iterations, with varying number of clusters k and sizes of clusters $n_1, \ldots, n_k$. This is appealing in fully accommodating uncertainty in the clustering process in obtaining estimates of the density. In contrast, commonly used approaches for model-based clustering and density estimation based on finite mixture models fitted with the EM algorithm condition on an estimate of the number of clusters, $\widehat{k}$. Such an estimate is typically obtained by fitting the model for varying values of k and then using the BIC for selection, though the development of appropriate model selection criteria in mixture models remains an active research topic. For finite mixture models, one can instead specify a prior for k and use a reversible jump Markov chain Monte Carlo (RJMCMC) algorithm for posterior computation, as described in Richardson and Green (1997). The DPM model bypasses the need for RJMCMC and also avoids selecting a single k that holds regardless of sample size. To obtain inferences on the number of clusters represented in the sample, one can estimate $\Pr(k = h|y)$ as simply the proportion of Gibbs draws such that $k = h$.

Although allowing uncertainty in k and S is appealing, challenges arise when interest focuses on estimating the cluster-specific parameters based on the Gibbs sampling draws. The main issue is the *label switching* problem (Stephens, 2000; Jasra et al., 2005). In particular, there is nothing in the prior or the model distinguishing mixture component h from component h'. Hence, the marginal posterior distributions for μ_h^* and $\mu_{h'}^*$ should be identical, with both being multimodal, and it is meaningless to estimate μ_h^* or $\mu_{h'}^*$ based on simple averages of the Gibbs samples of these parameters. In fact, unless the mixture components are well separated and there is slow mixing between the components, one commonly observes label switching behavior in trace plots showing the Gibbs samples of μ_h^* for each iteration. For example, if the truth is that $f(y)$ is a two component mixture, with one component centered at zero and one component centered at one, then the μ_1^* samples may initially oscillate around one for hundreds or thousands of iterations and then suddenly switch to oscillating around zero. If one averaged these samples, the estimated values of μ_1^* and μ_2^* may both be similar, having values between zero and one. To solve this problem, Stephens (2000) proposed a useful post-processing approach, which can be implemented by running the Gibbs sampler described above without considering the label switching problem and then re-labeling the cluster-specific parameters prior to calculating posterior summaries. Jasra et al. (2005) provide a more recent review of approaches for dealing with the label switching problem.

Although the Polya urn scheme is useful in allowing straightforward posterior computation through marginalizing out the mixture distribution P before conducting posterior computation, the disadvantage is that one looses the ability to conduct inferences on P. In addition, the Ferguson (1973; 1974) definition of the DP is quite indirect, so that it becomes difficult to obtain a intuition for what realizations from $P \sim DP(\alpha P_0)$ look like. A commonly used alternative to the DP is to rely on a finite mixture model, which specifies P as

$$P(\cdot) = \sum_{h=1}^{k} \pi_h \delta_{\theta_h}(\cdot), \tag{4.5}$$

where $(\pi_1, \ldots, \pi_k)'$ are probabilities summing to one, and δ_θ is a degenerate distribution with all its mass at θ. The prior $\mathcal{P}$ is then induced through priors for the mixture weights, $\pi = (\pi_1, \ldots, \pi_k)'$, and atoms, $\theta = (\theta_1, \ldots, \theta_k)'$. Note that using (4.5) results in the finite mixture of normals, $f(y) = \sum_{h=1}^{k} \pi_h N(y; \theta_h)$, where $\theta_h = (\mu_h, \sigma_h^2)$, for $h = 1, \ldots, k$.

Ishwaran and Zarepour (2002) show that (4.5) provides an approximation to $P \sim DP(\alpha P_0)$ for large k in the special case in which $\pi \sim$

Dirichlet$(\alpha/k, \ldots, \alpha/k)$ and $\theta_h \sim P_0$ independently. Hence, as a finite approximation to the Dirichlet process mixture model, one could use (4.5). The Gibbs sampler described above can be easily modified in this case to alternate between allocating subjects to one of the k components through sampling from a multinomial full conditional, updating the component-specific parameters, updating the weights from a conjugate Dirichlet distribution, and updating the residual precision from an inverse-gamma.

In a ground-breaking article, Sethuraman (1994) showed that $P \sim DP(\alpha P_0)$ can be equivalently expressed as

$$P(\cdot) = \sum_{h=1}^{\infty} \pi_h \delta_{\theta_h}(\cdot), \quad \theta_h \sim P_0, \tag{4.6}$$

so that the DP prior generalizes the discrete finite mixture model in (4.5) to incorporate infinitely-many components. If these components were treated as exchangeable, then this result would not be as useful practically. However, Sethuraman introduced a *stick-breaking* prior for the probabilities $\{\pi_h\}_{h=1}^{\infty}$, with this prior ensuring that π_h is stochastically larger than $\pi_{h'}$, for $h < h'$. Under the stick-breaking prior, the probability weight on the first component is $\pi_1 = V_1 \sim \text{beta}(1, \alpha)$. Removing a V_1 length piece from a unit probability stick, $1 - V_1$ is left available to allocate to the remaining components. The second component is then assigned $\pi_2 = V_2(1 - V_1)$, with $V_2 \sim \text{beta}(1, \alpha)$, leaving $(1 - V_1)(1 - V_2)$ to allocate to components $h = \{3, 4, \ldots, \infty\}$. This process is continued so that $\pi_h = V_h \prod_{l<h}(1 - V_l)$, with $V_h \sim \text{beta}(1, \alpha)$ and $h = 1, \ldots, \infty$. Because the expectation of V_h is $1/(1 + \alpha)$, the DP precision parameter controls how rapidly the probability weights π_h decrease towards close to zero as h increases.

Figure 4.1 shows realizations from (4.6) for varying values of α. In generating this plot, I only plot the first N components, with N chosen so that $1 - \prod_{h=1}^{N}(1 - V_h)$, the total probability allocated to components $N+1, \ldots, \infty$, is less than 0.0001. It is clear that for small α there tends to be a few dominant components having probability weights much higher than the other components. Hence, even though there are infinitely-many components that can be used, it tends to be the case that only a small number of these (relative to the sample size n) will be occupied by subjects in the sample. Therefore, the effective number of parameters to be estimated is often quite small even though the number of parameters in the model is infinite. A key to the practical performance of nonparametric Bayes methods is the ability to use only those parameters that are needed and grow the effective model dimension adaptively to accommodate lack of fit in lower dimensional models. Due to the Bayes intrinsic penalty on model complexity, this does not lead to overfitting and the approaches tend to have good out of sample predictive performance.

In practice, one can assign a flat prior for α, since the data tend to be highly informative. For example, suppose that y_i is approximately normally distributed. Then, the tendency will be to update the hyperprior for α to favor

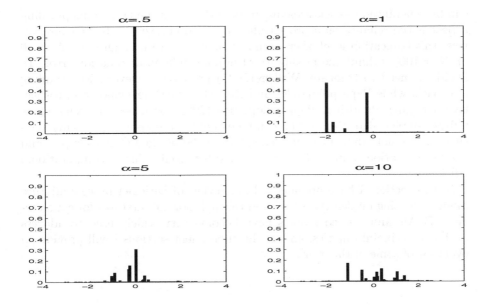

FIGURE 4.1: Realizations from a Dirichlet process prior for different choices of the DP precision parameter α.

small values, so that the posterior for $\pi_1 = V_1$ is concentrated on values close to one, subjects are allocated to a single component with high probability and the posterior mean estimate for $f(y)$ is close to normal. In contrast, if the true density is heavy tailed and one allows unknown mean, variance and degrees of freedom in P_0, then the tendency will be to favor high values of α *a posteriori*, allocate individuals to many clusters, and obtain a posterior mean estimate for $f(y)$ that resembles a t-distribution. If the true density is skewed or multi-modal, then one will instead favor moderate values for α, allocate subjects to a few dominate components, and obtain an accurate estimate of $f(y)$. These observations are based on my own experience in using DPMs for density estimation in a variety of simulated data settings.

Truncations of stick-breaking priors were originally proposed in Muliere and Tardella (1998) and were further considered by Ishwaran and James (2001) who proposed a blocked Gibbs sampling algorithm for posterior computation. The blocked Gibbs sampler relies on truncating the stick-breaking representation in (4.6) by letting $V_N = 1$ and hence $\pi_h = 0$ for $h = N + 1, \ldots, \infty$. The algorithm proceeds by updating the allocation to components for each subject by sampling from multinomial conditional posteriors, updating the stick-breaking parameters $V_1, \ldots, V_{N-1}$ from beta conditional posteriors, updating the component-specific parameters, and updating the residual precision from a gamma distribution. Although this algorithm is easy to implement and can be generalized to stick-breaking priors other than the DP, an unappealing aspect is the reliance on a pre-specified truncation level N. To be conservative, this truncation level often needs to be chosen to be high (e.g., $N = 50$ or $N = 100$) to limit the possibility of non-neglible approximation error. To avoid the need to truncate, Walker (2007) proposed a clever slice sampling approach, while Papaspiliopoulos and Roberts (2008) developed a retrospective sampling algorithm. Papaspiliopoulos (2008) proposed an exact block Gibbs sampler algorithm, which combines both these approaches. This algorithm is no more difficult to implement than the blocked Gibbs sampler and avoids truncation, potentially leading to substantial gains in computational efficiency.

In this section, I have provided a brief review of Dirichlet process mixture models focusing on density estimation through normal mixtures for concreteness. DPMs and related models have become very widely used to address problems in bioinformatics, and in the subsequent sections I will provide an overview of some of this work.

4.3 Multiple Testing and High-Dimensional Regression

Multiple testing issues are one of the greatest challenges faced in bioinformatics. To illustrate the usefulness of the Dirichlet process in this setting, we start by considering an application to assessing equivalence in p treatment groups. In particular, let y_{ij} denote the response for the j subject in treatment group i, for $j = 1, \ldots, n_i$ and $i = 1, \ldots, p$, and suppose that $y_{ij} \sim N(\mu_i, \sigma_i^2)$, so that the observations within a group are normally distributed but the mean and variance can vary across groups. Then, following the approach proposed by Gopalan and Berry (1998), let $\phi_i = (\mu_i, \sigma_i^2) \sim P$, with $P \sim DP(\alpha P_0)$. From (4.3) it is then clear that this approach induces a prior over the model space corresponding to all possible groupings of equivalent treatment groups. For example, the global null hypothesis that all groups are equivalent, $H_0 : \phi_1 = \cdots = \phi_p$, is assigned probability $\Pr(H_0) = \prod_{i=1}^{p-1}(\alpha + i)^{-1}$, while the probability that treatment groups i and i' are equivalent is $1/(1 + \alpha)$. Applying simple to implement Gibbs sampling algorithms, such as that proposed by Bush and MacEachern (1996), one can estimate posterior probabilities of equivalence in any two groups or sets of groups by simply averaging hypothesis indicators across the MCMC iterations after discarding a burn-in to allow convergence.

Dahl and Newton (2007) recently proposed a closely related idea as an independent development that overlooked the article by Gopalan and Berry (1998). As one application, they considered gene expression data, letting d_{gtr} denote a transformed expression level for gene g replicate r and treatment condition t, with the gene-specific mean in an arbitrarily chosen reference group subtracted to bypass modeling of an intercept. They then let $d_{gtr} \sim N(\tau_{gt}, \lambda_g)$, with τ_{gt} a treatment effect, λ_g a gene-specific intercept, $\tau_g = \{\tau_{gt}\}$, and $(\tau_g, \lambda_g) \sim P$, with P assigned a Dirichlet process prior. Unlike Gopalan and Berry (1998), who used the DP to cluster treatment groups directly, the Dahl and Newton (2007) approach borrows information across the high-dimensional vector of genes by using the DP to cluster genes. This is a very effective strategy for flexibly reducing dimensionality through only requiring estimation of a relatively small number of unique treatment effects and variances. Because clustering is soft and the allocation to clusters varies across MCMC iterations, one still obtains gene-specific treatment effects and variances.

Do, Müller and Tang (2005) proposed an alternative approach for modeling of differential expression across two experimental conditions. The first component corresponds to non-differentially expressed genes, while the second component corresponds to differentially expressed genes. Hence, posterior probabilities of allocation to the second component can be used as a weight of evidence that a gene is differentially expressed. There is a rich literature on using two component mixtures in this setting, with Efron et al. (2001) proposing an empirical Bayes approach that avoids parametric assumptions on the distributions within the two components. Do, Müller and Tang (2005) instead used a fully Bayes approach, which used Dirichlet process mixtures of normals to characterize the distributions in each of the components, with the base measure in the DP carefully chosen to avoid ambiguity in the two components.

MacLehose et al. (2006) instead addressed the problem of modeling of the effects of a high dimensional vector of correlated predictors, $\mathbf{x}_i = (x_{i1}, \ldots, x_{ip})'$, on a binary health outcome, y_i. In particular, they considered the logistic regression model

$$\text{logitPr}(y_i = 1 \mid x_{i1}, \ldots, x_{ip}) = \beta_0 + \sum_{j=1}^{p} \beta_j x_{ij}, \qquad (4.7)$$

where β_0 is an intercept and $\beta = (\beta_1, \ldots, \beta_p)'$ is a high-dimensional vector of regression coefficients. In large p settings, maximum likelihood estimation is unreliable, and a common strategy is to use a shrinkage estimator, which penalizes large coefficients. Shrinkage estimators typically have a Bayesian interpretation, with a prior for the coefficients inducing the penalty. In particular, for the logistic regression model (4.7), one could let $\beta_j \sim P$, for $j = 1, \ldots, p$. The prior P is often chosen to be centered on zero with a heavy tail. For example, the Lasso procedure (Tibshirani, 1996; Genkin, Lewis and Madigan, 2007) corresponds to a double exponential prior, while the relevance vector machine (Tipping, 2001) results from a t prior. Mixture priors with one component concentrated at zero and another more diffuse are also very commonly used.

MacLehose et al. (2006) noted that it seems overly restrictive to use a prior that is unimodal and centered on zero. Instead, it would be appealing to allow the prior P to be unknown in order to allow more flexible shrinkage. To accomplish this, they assigned P a prior consisting of a mixture of a point mass at zero, corresponding to null predictors, and a Dirichlet process, allowing clustering of the coefficients for non-null predictors. Predictors were assumed to have the same measurement scale, which can be ensured by normalization. Based on simulation studies, MacLehose et al. (2006) demonstrated that the approach resulted in better performance in terms of hypothesis testing and estimation compared to standard mixture priors. The idea of obtaining improved performance through multiple shrinkage is not new, dating back to George (1986), though earlier work did not consider nonparametric Bayes approaches or variable selection.

A potential criticism of the approaches of Gopalan and Berry (1998), Dahl and Newton (2007) and MacLehose et al. (2006) is the restriction that coefficients within a cluster are exactly identical. For example, in the gene expression application, it seems unlikely that treatment effects are exactly identical for different genes, though this may provide a reasonable approximation. One side effect of assuming exact equivalence within a cluster is that the number of clusters tends to be larger than if one allowed deviations within a cluster. As a more flexible approach motivated by applications to high dimensional SNP data, MacLehose and Dunson (2007) proposed a Bayesian multiple Lasso procedure, relying on a DP mixture of double exponential priors for P.

Motivated by case-control studies with massive numbers of candidate predictors, Pittman et al. (2004) proposed a Bayesian approach to prediction using a binary tree model with Dirichlet process priors on the distributions of meta genes underlying high-dimensional gene expression data. Using model averaging over the trees, the approach was applied to predict breast tumor status.

A limitation of all of the above approaches is the assumption of exchangeability of genes or predictors. In many bioinformatics applications, all predictors are not equally likely to be important *a priori*, so it may be unappealing to treat them exchangeably (Wacholder et al., 2004). For example, in attempting to identify single nucleotide polymorphisms predictive of a disease phenotype from among a large number of candidates, one may expect polymorphisms in pathways involved in the development of disease to be more likely candidates. The incorporation of gene functional annotation and other information into the analysis can potentially decrease false discovery rates substantially. In large p, small n settings it is particularly important to utilize all available prior information. In addition to gene function, the location of the SNP may be informative.

One possibility in terms of relaxing the exchangeability assumptions is to group predictors *a priori* and borrow strength more strongly within a group than across groups. For example, in applications with SNP predictors, a group may correspond to SNPs within a gene or genes within a pathway. In such settings, the group Lasso for logistic regression approach of Meier, van de Geer and Buehlmann (2008) can potentially be used. However, as discussed above, the Lasso approach of shrinking all the coefficients towards zero instead of borrowing information across the non-null predictors may not be ideal in certain settings. Dunson, Herring and Engel (2008) proposed an alternative semiparametric Bayes approach that uses a hierarchical logistic regression model, with zero-inflated Dirichlet process priors for the distributions of group- and predictor-specific coefficients to borrow information within and across groups in variable selection and clustering. This approach was particularly motivated by applications to SNPs in functionally related genes.

It is appealing to consider more general approaches for accommodating multiple discrete and continuous features of each predictor. In particular, if we have a vector of predictors, $x_i = (x_{i1}, \ldots, x_{ip})'$ for subject i, then the jth predictor could have an associated feature vector, $w_j = (w_{j1}, \ldots, w_{jp})'$. The feature vector w_j may consist of a single group index or it may include other information, such as location of the SNP within a chromosome or multiple group indices. Focusing on the logistic regression case in (4.7), we can generalize the approach of MacLehose et al. (2007) through use of a hierarchical structure of the form

$$\beta_j \sim P_j = \sum_{h=1}^{\infty} \pi_h(w_j)\delta_{\theta_h}, \quad j = 1, \ldots, p, \tag{4.8}$$

where $\theta_h \sim P_0$, for $h = 1, \ldots, \infty$, and $\pi_h(w)$ is a probability weight on component h, which is defined to be a function of features w. Note that this formulation incorporates a common set of atoms, $\theta = \{\theta_h\}_{h=1}^{\infty}$, while allowing the weights on these atoms to vary with predictors. The base measure, P_0, can be chosen to correspond to a mixture distribution with a mass at zero to allow $\beta_j = 0$ with positive probability.

To allow the weight functions to be unknown, we can use the kernel stick-breaking prior (KSBP) of Dunson and Park (2008). In particular, under the KSBP, we have

$$\pi_h(w) = V_h K(w, \Gamma_h) \prod_{l<h} \{1 - V_l K(w, \Gamma_l)\}, \tag{4.9}$$

where $V_h \sim \text{beta}(1, \alpha)$ as in the stick-breaking representation of the DP in (??), $K(\cdot)$ is a kernel bounded above by one, and $\Gamma_h \sim H$, for $h = 1, \ldots, \infty$, are random kernel locations. This form assigns the atom, θ_h, to a location, Γ_h, in the feature space. The weight $\pi_h(w)$ is then a function of the distance from w to Γ_h, with the distance metric depending on the choice of kernel, $K(\cdot)$. In this manner, one obtains a higher probability of clustering β_j and $\beta_{j'}$ if w_j and $w_{j'}$ are close together. For example, suppose that w is a scalar location of SNP j on the chromosome and we choose a Gaussian kernel, $K(w, \Gamma) = \exp\{-\psi(w - \Gamma)^2\}$, with ψ unknown. Then, in the limiting case as $\psi \to 0$, we obtain $\beta_j \sim P$, with $P \sim DP(\alpha P_0)$, and the location of the SNP does not play a role in clustering. However, as ψ increases, the location is increasingly important in forming the clusters, with only SNPs in small local regions having identical coefficients if ψ is large. For large number of SNPs, computation can be made more efficient by choosing H to be a discrete grid of locations, so that Gibbs sampling can be used (e.g., via a slice sampler).

4.4   Clustering and Functional Data Analysis

In additional to multiple testing and problems in variable selection for massive dimensional regression models, bioinformatics research often involves searching for low dimensional structure and patterns in complex and high dimensional data. Simple methods for exploratory data analysis used in lower dimensional problems do not necessarily scale up to high dimensions, so a number of innovative strategies have been developed. One tool for exploratory analysis of high-dimensional data, which has been widely exploited, is clustering. For example, it is commonly of interest to identify genes with similar expression patterns. Bayesian methods for model-based clustering typically rely on mixture models, which assume that data are drawn from one of k components, with individuals drawn from the same component belonging to the same cluster (refer, for example, to Yeung et al., 2001).

One issue that is unavoidable in clustering problems is how to choose an appropriate number of clusters, k. This is the same issue that was used to motivate the generalization from the finite mixture model in (1.3) to the Dirichlet process-based infinite mixture model in (1.4). As further motivation for use of an infinite mixture model instead of a finite mixture model in the gene expression application, consider the following argument. Suppose that we fit a finite mixture model to gene expression data for an initial set of p_0 genes, and based on the BIC criteria estimate that there are $\hat{k} = 5$ clusters in these genes. Then, suppose we are interested in incorporating additional gene expression data containing the expression measurements for the original p_0 genes and an additional p_1 genes. Typical use of the finite mixture model would require the p_1 new genes to be allocated to one of the $\hat{k}$ clusters estimated from the initial p_0 genes. This seems clearly inappropriate, since some of the p_1 genes may have very different expression profiles. It would seem most realistic biologically to allow clusters to be added as additional genes are added, but at a slow rate so that we can obtain relative few clusters and hence sparser and more interpretable results.

Motivated by these types of issues, Medvedovic and Sivaganesan (2002) proposed an approach for identifying genes with similar expression patterns through a Dirichlet process mixture model. In addition, to solve the label switching problem in which the number and meaning of the clusters varies across the MCMC iterations, they proposed a simple and effective method for identifying a single clustering of genes based on thresholding of the pairwise posterior probabilities that two genes are grouped together. This approach has been widely used in the literature, with Dahl (2006) and Lau and Green (2007) providing alternative methods to estimate a single optimal clustering based on the MCMC output. Xiang, Qin and He (2007) recently developed an easy to use web server for clustering of gene expression data using a Dirichlet process mixture model.

Note that many of the methods for multiple testing and massive dimensional regression described in Section 1.3 also relied on a Dirichlet process mixture model, with borrowing of information and equalities in genes, treatment groups and regression coefficients obtained through flexible model-based clustering. The approach described in expressions (1.7) and (1.8) to generalize DP clustering to include predictors through a kernel stick-breaking process can also be used in settings in which the primary focus is not on hypothesis testing or gene selection but instead on the clustering itself. For example, we can include pathway information as gene-specific features to increase the probability that genes within a pathway are clustered together, without restricting this to be the case. Even when it is reasonable to treat genes exchangeably *a priori*, it is important to keep in mind that the Dirichlet process is not the only nonparametric Bayes option available. The DP introduces clusters at a rate proportional to $\alpha \log n$, which may be overly-restrictive in certain applications. Hence, it is worth exploring more flexible alternatives instead of always using the DP as the default. For example, Lijoi, Mena and Prünster (2007b) recently developed nonparametric Bayes methodology for clustering based on generalized gamma (GG) process priors. The GG-based clustering is more flexible than the DP, introducing clusters at power law rate.

There are some issues that arise in basing inferences on an estimated clustering that do not arise in using clustering as a flexible tool for borrowing of information and dimensionality reduction. First, it is important to keep in mind that the space of possible clusterings of genes is incredibly enormous for large p. Due to this fact it is very hard to adequately explore the space of clusterings given current algorithms and computing resources. Hence, the optimal clustering may never even be visited by an MCMC, simulated annealing or population Monte Carlo algorithm. In addition, even if one could identify the optimal clustering under some criteria, it is likely that there are a very large number of clusterings that have essentially identical performance. For example, in ranking clusterings based on posterior probabilities when p is large, the probability assigned to any particular clustering is extremely small and there may be thousands of clusterings within ϵ of the highest posterior probability clustering. Hence, one should try to avoid over-interpreting results of any clustering analysis, and should try to focus inferences on quantities that can be better estimated. For example, the marginal posterior probability that two genes are clustered together accounting for uncertainty in the other genes in the cluster provides a useful basis for inferences, which can be used to bypass focusing on a single estimated clustering in the guess. Optimization-based approaches to cluster analysis can produce misleading results, as one typically does not have a measure of uncertainty in the estimate so may be likely to over-interpret the results.

Until this point, I have focused on data consisting of a high-dimensional vector of responses or predictors. However, there are increasing numbers of applications that collect high-dimensional functional data. One example that has received abundant focus in recent years in the statistical literature is mass spectrometry data, which are often used as a tool in proteomics for identifying proteins and searching for biomarkers. Ghosh et al. (2008) recently proposed an innovative Bayesian semiparametric approach for identifying biomarkers from mass spectrometry data motivated by an application to distinguishing urinary metabolic profiles between trauma patients and healthy controls. The Ghosh et al. (2008) approach used a Dirichlet mixture of beta distributions as a prior for a density on [0,1], with this density used to construct a functional profile from multiple measurements on a subject. Ghosh and Dey (2008) later extended the methodology to incorporate covariates.

Another area that has received increasing attention is gene copy number data collected in comparative genomic hybridization (CGH) microarrays. Broet and Richardson (2006) proposed a mixture model, with states corresponding to unmodified, delete and amplified sequences. A spatial model was then used to allow allocation to the three mixture components to depend on location along the chromosome. Pique-Regi et al. (2008) proposed an alternative approach based on starting with a high-dimensional piecewise constant model, with a sparse Bayesian learning algorithm used to estimate the knot locations. Potentially, one can modify these approaches by using a nonparametric Bayes approach in which the copy number variability along the chromosome within an individual is viewed as functional data. Using a piecewise constant basis representation, one can express the measured copy number at location j within subject i as

$$y_{ij} = \sum_{l=1}^{p} \theta_{il} b_l(t_{ij}) + \epsilon_{ij}, \quad \epsilon_{ij} \sim N(0, \sigma^2)$$
$$\theta_i \sim P, \tag{4.10}$$

where $b = \{b_l\}_{l=1}^{p}$ correspond to a large number of piecewise constant basis functions, t_{ij} is the location of the jth measurement for individual i, $\theta_i = (\theta_{i1}, \ldots, \theta_{ip})'$ are individual-specific basis coefficients (random effects), and P is a random effects distribution characterizing variability among individuals in the copy number profile. One possibility for allowing flexible modeling and borrowing of information across individuals, while accommodating differences in the knot locations, would be to let $P \sim DP(\alpha P_0)$, with P_0 corresponding to a mixture of a point mass at zero, allowing knots to drop out, and a continuous probability measure, such as a Gaussian. Note that this approach would allow the knots to be different for each component of the DP.

A disadvantage of using a Dirichlet process prior for P is that the DP induces global functional clustering, so that two individuals allocated to the same component by the DP will be required to have the same basis coefficient vectors. This certainly seems overly-restrictive in the array CGH application, since there may be particular regions of the chromosome for which multiple subjects have similar profiles, but the entire profile is unlikely to be identical everywhere for any two subjects. Hoff (2006) proposed a subspace clustering approach, which used a DP to cluster individuals in terms of the locations and magnitude of the deviation from a mean vector. He applied this method to array CGH data. His approach does not entirely solve the problem of local borrowing of information across individuals, allowing certain regions to be more likely to have deleted or amplified genes. Potentially this can be addressed using recently proposed local generalizations of the Dirichlet process, such as the matrix stick-breaking process (Dunson, Xue and Carin, 2008), the hybrid functional Dirichlet process (Petrone, Guindani and Gelfand, 2008) or the local partition process (LPP) (Dunson, 2009b).

Focusing on the local partition process, we let $\theta_i \sim P$, with the random effects distribution assigned the prior $P \sim \text{LPP}(\alpha, \beta, P_0)$. The LPP is a generalization of the DP that allows dependent local clustering. In particular, local clustering allows $\theta_{ij} = \theta_{i'j}$ for some $j \in \{1, \ldots, p\}$, while $\theta_{ij} \neq \theta_{i'j}$ for other js. In the array CGH application, identical basis coefficients for two subjects imply the same level of deletion or amplification in copy number in a particular location of the genome. The LPP has the appealing properties of being fully flexible in modeling of $\Pr(\theta_{ij} = \theta_{i'j})$ and the conditional probability $\Pr(\theta_{ij} = \theta_{i'j} \mid \theta_{ij'} = \theta_{i'j'})$. By allowing dependence in local clustering, the approach allows learning of similarities between two subjects increasingly as locations are added. In addition, choosing P_0 to have a mixture structure with a mass at zero allows a different selection of knots for each subject, while borrowing of information since each subject's function is formed from local selection of basis coefficients from the same basis coefficient vectors. This structure tends to lead to a very sparse representation of complex functional data, exploiting underlying commonalities while allowing each function to differ.

4.5 Additional Topics

There are a number of additional application areas that fall under the broad umbrella of bioinformatics for which Bayesian nonparametric methods can be usefully applied. One area is analysis of expressed sequence tags (ESTs), which provides a technology for gene discovery. In collecting EST data from a cDNA library, the rate of discovery of new genes is a function of redundancy in the library. Hence, in deciding whether it is worthwhile to continue sequencing the library and, if so, how many samples to take, one would like to estimate the number of new genes in the library and the rate of discovery in a new sample based on previous data obtained for the library. Lijoi et al. (2007a) proposed an approach for addressing this problem based on the theoretical framework developed in Lijoi et al. (2007c) for nonparametric Bayesian estimation of the probability of developing a new species.

The Lijoi et al. (2007c) approach is based on a class of species sampling priors, which generalize the Dirichlet process to a broader class of priors, which induce Gibbs-type random partition structure. In addressing the problem of estimating a new species, it is particularly important to be flexible in characterizing the prior on the partition of samples into clusters. The DP induces a prior, which implies that species are discovered at a rate proportional to the natural logarithm of the number of samples. In applications, there is no guarantee that this rate is appropriate, so it is appealing to use a more flexible prior that allows power law rate of discovery, with the power estimated based on an initial sample from the EST library under consideration. The two parameter Poisson-Dirichlet process provides a generalization of the DP that has this property and is convenient in terms of mathematical and computational tractability. The Bayesian nonparametric approach has clear advantages over frequentist estimators of species discovery rate, which tend to become unstable as the number of new samples to be drawn increases.

Nonparametric Bayes methods have also been utilized increasingly in population genetics. In Section 4.3 I mentioned some approaches for incorporating high-dimensional SNP data as predictors in a regression analysis. There has been an increasing focus on collecting very high-dimensional SNP data spanning the entire genome, with such data often referred to as whole genome scans. Note that a single SNP corresponds to a pair of amino acids (nucleotides) at a single location (loci), with one on the chromosome inherited from the father and one on the chromosome inherited from the mother. Since the chromosomes are inherited instead of the individual SNPs, there are actually two *haplotypes*, corresponding to the sequences of amino acids on each chromosome, underlying the observed SNP sequence. It is well known that the number of haplotypes occurring in a population tends to be extremely small relative to the number possible, so one can potentially obtain a reduction in dimensionality in using haplotypes as the basis for analysis instead of SNPs. However, this reduction in dimensionality is counter-balanced by the fact that current widely used sequencing technology is *un-phased*, so only provides SNPs, with no information on chromosome of origin. For this reason, haplotype-based analyses involve a missing data problem, with widely-used approaches for inferring the unknown haplotypes recently shown to be inaccurate (Andrés et al., 2007).

In addition to the inherent uncertainty involved in imputation of missing data, one of the reasons for this inaccuracy may be use of an overly-simplified model. Xing, Jordan and Sharan (2007) recently proposed a nonparametric

Bayes approach, which improved upon standard finite mixture model-based analyses by allowing an unknown number of haplotypes in the population through a Dirichlet process mixture model. Often interest focuses on problems of inferring population substructure and regions along the chromosome with high rates of recombination (hotspots) on the basis of SNP data collected for different individuals. Inferences of this type require a biologically-based model of genetic recombinations among large numbers of founders followed by a coalescence with mutation process applied to the genealogies across the generations leading up to the current population. It is crucial that the model characterizing this complex process is biologically realistic, as the inferences require extrapolating backwards in time on the basis of modern day data, with limited ability to check modeling assumptions given the lack of genetic information for past generations. Given these inherent limitations, it is important that researchers are modest in making conclusions about population structure and migration patterns in the distant past on the basis of statistical methods applied to modern sequence data.

Xing and Sohn (2006) proposed a nonparametric Bayes model, referred to as the hidden Markov Dirichlet process (HMDP), for flexible genetic inference. The proposed model generates current haplotypes through a sequence of recombination events that proceeds via a first order Markov process starting with an unlimited number of founder haplotypes, with mutation also allowed. For a related approach, which was independently developed, refer to Sun et al. (2007). Xing et al. (2006) proposed an alternative approach that allows dependence within and between populations through a hierarchical Dirichlet process (HDP) mixture model (Teh et al., 2006). The HDP allows for a common set of haplotypes across the different populations, but with different frequencies in the individual populations.

An additional area, which is understudied statistically, is molecular epidemiology studies using single cell gel electrophoresis (comet assay) to measure the frequency of DNA strand breaks on the level of individual cells. In these studies, immortalized cell lines are first established for different individuals. Samples of cells are then drawn and subjected to a variety of experimental conditions, with the comet assay used to measure damage. For example, in studies of polymorphisms predictive of DNA damage and repair rates, a sample of 100 cells could be drawn at baseline, with additional samples of 100 cells drawn immediately after exposure to a known genotoxic agent and after some time has been allowed for repair. In addition, SNP data may be available for each individual, so that the focus is on identifying genotypes predictive of rates of baseline damage, sensitivity to induced damage, and repair. This is a challenging problem, because the density of comet assay measures of DNA damage across cells in a sample tends to be non-Gaussian, with the density changing in shape across experimental conditions.

Dunson (2006) proposed a dynamic mixture of Dirichlet processes (DMDP) framework, which allowed the distribution of latent DNA damage to change nonparametrically across dose groups, while flexibly borrowing information.

Pennell and Dunson (2008) adapted this framework to allow nonparametric testing of equalities in the distributions between dose groups. In order to allow heterogeneity in cells drawn from cell lines established for different individuals due to measured genotypes and unknown factors, it is appealing to consider general hierarchical regression models in which the natural response for an individual is a collection of unknown distributions. This problem is conceptually related to functional data analysis, though functional data methods are not directly applicable to this setting. De Iorio et al. (2004) proposed an ANOVA framework for dependent collections of densities. This approach can accommodate comet assay data from a single cell line with multiple categorical experimental conditions, such as dose level and repair time, but does not accommodate variability among individuals and predictors impacting the profile of DNA damage within an individual across experimental conditions.

To address these problems, Rodriguez, Dunson and Taylor (2009) proposed a general hierarchically weighted finite mixture (HWFM) model framework. Their proposed approach involves incorporation of a finite number of parametric basis distributions, consisting of Gaussian densities with unknown means and variances, with the distribution specific to an individual under a particular condition modeled as a mixture of these bases. The mixture weights are then characterized using a hierarchical probit model for computational convenience. A variable selection component can be incorporated in the mixture weight model to allow selection of genotypes predictive of baseline damage, sensitivity to exposure and repair rates. In future work, it will be interesting to generalize these methods to infinite mixtures. A promising idea in this regard is to rely on a probit stick-breaking process, which replaces the beta stick-breaking weights in the Dirichlet process with an unknown Gaussian framework that leads to more tractable generalizations to hierarchical modeling, variable selection and other complications. Using such an approach, one can select predictors having any impact on the distribution of a response variable. Such an approach can detect predictors that have no effect on the mean, but impact distributional shape (Chung and Dunson, 2008).

4.6 Discussion

This chapter has provided a brief overview of recent work using nonparametric Bayes approaches in bioinformatics, with the term "nonparametric Bayes" used to refer to models incorporating random probability measures, such as the Dirichlet process. The "nonparametric" terminology is somewhat of a misnomer in that nonparametric Bayes methods are actually infinitely parametric. The flexibility comes in through defining a prior with large support, which is flexible enough to characterize a very broad class of distributions and data generating mechanisms. The excellent performance often observed by nonparametric Bayes methods in practice arises through the combination of an infinite-dimensional and extremely flexible model with the intrinsic Bayes penalty for model complexity. The result is that a very sparse structure tends to be favored with additional components incorporated as the evidence builds up in the data that these components are really needed. Hence, there does not tend to be any problem with over-fitting. In bioinformatics, nonparametric Bayes methods provide an appealing black box for sparsely yet flexibly characterizing very high-dimensional data, while allowing variable selection and shrinkage through a full probabilistic framework for accommodating uncertainty in a principal manner.

However, there are a number of challenges involved in the implementation and interpretation of results from nonparametric Bayes analyses that motivate additional research. The first issue is the prior specification. For example, specification of a Dirichlet process prior involves choice of a total mass parameter, α, and a base probability measure, P_0. The total mass parameter is important in determining the posterior on the number of mixture components, so it is appealing for flexibility to choose a hyperprior, such as a gamma. As the data are highly informative in most cases, a flat prior can even be chosen by letting the gamma parameters be close to zero. The choice of P_0 is more challenging. Suppose a particular parametric choice is made, such as the Gaussian or double-exponential. Then, the issue is how to choose the parameters characterizing P_0. For robustness, one can choose a hyperprior, such as a normal inverse-gamma for the mean and variance of the Gaussian. However, it is still necessary to choose the hyperparameters in the hyperprior. It is important to keep in mind that it is not appropriate to choose a high variance prior, because the posterior distribution on the partition of subjects into clusters depends on a marginal likelihood. The tendency is to allocate all the subjects into a single cluster with high probability if the variance of P_0 is extremely high. Hence, it is important to be careful to choose P_0 to have an appropriate scale. This can be done based on the application. For example, if a DPM is being used for coefficients on binary genotype predictors in a logistic regression model, choosing P_0 to correspond to a $N(0, 2)$ distribution is reasonable as a default, since log odds ratios are quite unlikely to be very far from 0. In the absence of such prior information, one can instead use an empirical Bayes-type approach, which either standardizes the data prior to analysis and chooses unit scale in the prior or uses the data in choosing P_0 (McAuliffe, Blei and Jordan, 2006).

A second issue is the computational implementation. For Dirichlet process mixture models and other nonparametric Bayes models, there is a rich literature on simple and computationally efficient methods for posterior computation. MCMC analyses of a wide variety of nonparametric Bayes models can be implemented in the R package, DPpackage (Jara, 2007), and WinBUGS can also be used for many models (refer to Ghosh, Ghosh and Tiwari, 2008 for a recent example). For massive dimensional problems, MCMC algorithms may not scale appropriately, motivating a literature on fast approximations. In the machine learning literature, variational Bayes methods have been increasingly widely used to approximate the posterior in nonparametric Bayes models (refer to Blei and Jordan, 2006, and Kurihara, Welling and Teh (2007) for methodology articles and Qi, Paisley and Carin, 2007 for a recent application). These methods can be implemented with a simple EM algorithm, though it is important to use multiple starting points, since the posterior is typically multimodal in mixture models and the EM will converge to a local mode. Intelligent default choice of these starting points based on preliminary analyses of the data is a good idea, since the number of modes may be large.

A third challenge in nonparametric Bayes inference in bioinformatics is interpretability. For example, suppose that a highly flexible mixture model is defined that allows the distribution of a health response to change nonparametrically with a high-dimensional vector of candidate predictors, while also allowing interactions. If predictions are the focus, then the nonparametric Bayes mixture model may provide a useful blackbox that allows for a sparse structure underlying the high-dimensional data without imposing restrictive assumptions, such as normality and linearity. However, now suppose that interest is in inferences on the impact of a particular predictor on the health response. This tends to be more challenging, since the number of mixture components and role of the predictor of interest can change across the MCMC iterations, making it difficult to obtain a clearly interpretable summary of the effect of the predictor. My general recommendation in this regard is to try to avoid making cluster-specific inferences due to the problem of label switching and to the inherent uncertainty in clustering. In most cases, the clustering process in the mixture model can be viewed as a tool for building a flexible model, with over-interpretation of the clusters as a dangerous and potentially misleading exercise. Ideally, one can instead base inferences on a well-defined quantity, obtained in averaging over the posterior distribution on the clusters. For example, in assessing predictors effects, one can use the marginal posterior probability of inclusion and summaries of the marginal impact of the predictor on different summaries of the response density.

References

Andrés, A.M., Clark, A.G., Boerwinkle, E., Sing, C.F. and Hixson, J.E. (2007). Assessing the accuracy of statistical haplotype inference with sequence data of known phase. *Genetic Epidemiology*, 31, 659–671.

Blei, D.M. and Jordan, M. (2006). Variational inference for Dirichlet process mixtures. *Bayesian Analysis*, 1, 121–144.

Broet, P. and Richardson, S. (2006). Detection of gene copy number changes in CGH microarrays using a spatially correlated mixture model. *Bioinformatics*, 22, 911–918.

Bush, C.A. and MacEachern, S.N. (1996). A semiparametric Bayesian model for randomised block designs. *Biometrika*, 83, 275–285.

Chung, Y. and Dunson, D.B. (2008). Nonparametric Bayes conditional distribution modeling with variable selection. *Discussion Paper*, Department of Statistical Science, Duke University, Durham, NC, USA.

Dahl, D.B. (2006). Model-based clustering for expression data via a Dirichlet process mixture model. In *Bayesian Inference for Gene Expression and Proteomics*, Kim-Anh Do, Peter Müller, Marina Vannucci (eds), Cambridge University Press.

Dahl, D.B. and Newton, M.A. (2007). Multiple hypothesis testing by clustering treatment effects. *Journal of the American Statistical Association*, 102, 517–526.

De Iorio, M., Müller, P., Rosner, G.L. and MacEachern, S.N. (2004). An ANOVA model for dependent random measures. *Journal of the American Statistical Association*, 99, 205–215.

Do, K.A., Müller, P. and Tang, F. (2005). A Bayesian mixture model for differential gene expression. *Applied Statistics*, 54, 627–644.

Dunson, D.B. (2006). Bayesian dynamic modeling of latent trait distributions. *Biostatistics*, 7, 551–568.

Dunson, D.B. (2009a). Nonparametric Bayes applications to biostatistics. *Nonparametric Bayesian Modeling*. Cambridge University Press, to appear.

Dunson, D.B. (2009b). Nonparametric Bayes local partition models for random effects. *Biometrika*, 96, 249–262.

Dunson, D.B., Herring, A.H. and Engel, S.M. (2008). Bayesian selection and clustering of polymorphisms in functionally related genes. *Journal of the American Statistical Association*, 103, 534–546.

Dunson, D.B. and Park, J-H. (2008). Kernel stick-breaking processes. *Biometrika*, 95, 307–323.

Dunson, D.B., Xue, Y. and Carin, L. (2008). The matrix stick-breaking process: Flexible Bayes meta analysis. *Journal of the American Statistical Association*, 103, 317–327.

Efron, B., Tibshirani, R., Storey, J.D. and Tusher, V. (2001). Empirical Bayes analysis of a microarray experiment. *Journal of the American Statistical Association*, 96, 1151–1160.

Escobar, M.D. and West, M. (1995). Bayesian density estimation and inference using mixtures. *Journal of the American Statistical Association*, 90, 577–588.

Ferguson, T.S. (1973). A Bayesian analysis of some nonparametric problems. *Annals of Statistics*, 1, 209–230.

Ferguson, T.S. (1974). Prior distributions on spaces of probability measures. *Annals of Statistics*, 2, 615–629.

Genkin, A., Lewis, D.D. and Madigan, D. (2007). Large-scale Bayesian logistic regression for text categorization. *Technometrics*, 49, 291–304.

George, E.I. (1986). Minimax multiple shrinkage estimation. *Annals of Statistics*, 14, 188–205.

Ghosh, S. and Dey, D.K. (2008). An unified modeling framework for metabonomic profile development and covariate selection for acute trauma subjects. *Statistics in Medicine*, to appear.

Ghosh, K., Ghosh, P. and Tiwari, R.C. (2008). Commentary on "The nested Dirichlet process", *Journal of the American Statistical Association*, to appear.

Ghosh, S., Grant, D.F., Dey, D.K. and Hill, D.W. (2008). A semiparametric modeling framework for potential biomarker discovery and the development of metabonomic profiles. *BMC Bioinformatics*, 9, 38.

Ghosal, S. (2008). Dirichlet process, related priors and posterior asymptotics. *Nonparametric Bayesian Modeling*. Cambridge University Press, to appear.

Gopalan, R. and Berry, D.A. (1998). Bayesian multiple comparisons using Dirichlet process priors. *Journal of the American Statistical Association*, 93, 1130–1139.

Hoff, P.D. (2006). Model-based subspace clustering. *Bayesian Analysis*, 1, 321–344.

Ishwaran, H. and James, L.F. (2001). Gibbs sampling methods for stick-breaking priors. *Journal of the American Statistical Association*, 96, 161–173.

Ishwaran, H. and Zarepour, M. (2002). Exact and approximate representations for the sum Dirichlet process. *Canadian Journal of Statistics*, 30, 269–283.

Jara, A. (2007). Applied Bayesian non- and semi-parametric inference using DPpackage. *Rnews*, 7, 17–26.

Jasra, A., Holmes, C.C. and Stephens, D.A. (2005). Markov chain Monte Carlo methods and the label switching problem in Bayesian mixture modeling. *Statistical Science*, 20, 50–67.

Kurihara, K., Welling, M. and Teh, Y-W. (2007). Collapsed variational Dirichlet process mixture models. *The Twentieth International Joint Conference on Artificial Intelligence* (IJCAI, 2007).

Lau, J.W. and Green, P.J. (2007). Bayesian model-based clustering procedures. *Journal of Computational and Graphical Statistics*, 16, 526–558.

Lijoi, A., Mena, R.H. and Prünster, I. (2007a). A Bayesian nonparametric method for prediction in EST analysis. *BMC Bioinformatics*, 8, 339.

Lijoi, A., Mena, R.H. and Prünster, I. (2007a). Bayesian nonparametric estimation of the probability of discovering new species. *Biometrika*, 94, 769–786.

Lijoi, A., Mena, R.H. and Prünster, I. (2007b). Controlling the reinforcement in Bayesian non-parametric mixture models. *Journal of the Royal Statistical Society* B, 69, 715–740.

Lo, A.Y. (1984). On a class of Bayesian nonparametric estimates. 1. Density estimates. *Annals of Statistics*, 12, 351–357.

MacLehose, R.F. and Dunson, D.B. (2007). Bayesian semi-parametric multiple shrinkage. *Discussion Paper*, 2007–22, Department of Statistical Science, Duke University.

MacLehose, R.F., Dunson, D.B., Herring, A.H. and Hoppin, J.A. (2007). Bayesian methods for highly correlated exposure data. *Epidemiology*, 18, 199–207.

McAuliffe, J.D., Blei, D.M. and Jordan, M.I. (2006). Nonparametric empirical Bayes for the Dirichlet process mixture model. *Statistics and Computing*, 16, 5–14.

Medvedovic, M. and Sivaganesan, S. (2002). Bayesian infinite mixture model based clustering of gene expression profiles. *Bioinformatics*, 18, 1194–1206.

Meier, L., van de Geer, S.V. and Buehlmann, P. (2008). The group lasso for logistic regression. *Journal of the Royal Statistical Society* B, 70, 53–71.

Muliere, P. and Tardella, L. (1998). Approximating distributions of random functionals of Ferguson-Dirichlet priors. *Canadian Journal of Statistics*, 26, 283–297.

Müller, P. and Quintana, F.A. (2004). Nonparametric Bayesian data analysis. *Statistical Science*, 19, 95–110.

Papaspiliopoulos, O. (2008). A note on posterior sampling from Dirichlet process mixture models. *Working Paper*, 08–20, Centre for Research in Statistical Methodology, University Warwick, Coventry, UK.

Papaspiliopoulos, O. and Roberts, G.O. (2008). Retrospective Markov chain Monte Carlo methods for Dirichlet process hierarchical models. *Biometrika*, 95, 169–186.

Pennell, M.L. and Dunson, D.B. (2008). Nonparametric Bayes testing of changes in a response distribution with an ordinal predictor. *Biometrics*, OnlineEarly.

Petrone, S., Guindani, M. and Gelfand, A.E. (2008). Hybrid Dirichlet processes for functional data. *Submitted*.

Pique-Regi, R., Monso-Varona, J., Ortega, A., Seeger, R.C., Triche, T.J. and Asgharzadeh, S. (2008). Sparse representation and Bayesian detection of genome copy number alterations from microarray data. *Bioinformatics*, 24, 309–318.

Pittman, J., Huang, E., Nevins, J., Wang, Q.L. and West, M. (2004). Bayesian analysis of binary prediction tree models for retrospectively sampled outcomes. *Biostatistics*, 5, 587–601.

Qi, Y.T., Paisley, J.W. and Carin, L. (2007). Music analysis using hidden Markov mixture models. *IEEE Transactions on Signal Processing*, 55, 5209–5224.

Rodriguez, A., Dunson, D.B. and Taylor, J. (2009). Bayesian hierarchically weighted finite mixture models for samples of distributions. *Biostatistics*, 10, 155–171.

Richardson, S. and Green, P.J. (1997). On Bayesian analysis of mixtures with an unknown number of components. *Journal of the Royal Statistical Society* B, 59, 731–758.

Stephens, M. (2000). Dealing with label switching in mixture models. *Journal of the Royal Statistical Society* B, 62, 795–809.

Sun, S., Greenwood, C.M.T. and Neal, R.M. (2007). Haplotype inference using a Bayesian hidden Markov model. *Genetic Epidemiology*, 31, 937–948.

Teh, Y.W., Jordan, M.I., Beal, M.J. and Blei, D.M. (2006). Hierarchical Dirichlet processes. *Journal of the American Statistical Association*, 101, 1566–1581.

Tibshirani, R. (1996). Regression shrinkage and selection via the Lasso. *Journal of the Royal Statistical Society* B, 58, 267–288.

Tipping, M.E. (2001). Sparse Bayesian learning and the relevance vector machine. *Journal of Machine Learning Research*, 1, 211–244.

Wacholder, S., Chanock, S., Garcia-Closas, M., El Ghormli, L. and Rothman, N. (2004). Assessing the probability that a positive report is false: An approach for molecular epidemiology studies. *Journal of the National Cancer Institute*, 96, 434–442.

Walker, S.G. (2007). Sampling the Dirichlet mixture model with slices. (2007). *Communications in Statistics - Simulation and Computation*, 36, 45–54.

Xiang, Z.S., Qin, Z.H.S. and He, Y.Q. (2007). CRCView: a web server for analyzing and visualizing microarray gene expression data using model-based clustering. *Bioinformatics*, 23, 1843–1845.

Xing, E.P., Jordan, M.I. and Sharan, R. (2007). Bayesian haplotype inference via the Dirichlet process. *Journal of Computational Biology*, 14, 267–284.

Xing, E.P. and Sohn, K. (2007). Hidden Markov Dirichlet process: Modeling genetic recombination in open ancestral space. *Bayesian Analysis*, 2.

Xing, E.P., Sohn, K-A., Jordan, M.I. and Teh, Y-W. (2006). Bayesian multi-population haplotype inference via a hierarchical Dirichlet process mixture. *Proceeding of the 23rd International Conference on Machine Learning* (ICML 2006).

Yeung, K.Y., Fraley, C., Murua, A., Raftery, A.E. and Ruzzo, W.L. (2001). Model-based clustering and data transformations for gene expression data. *Bioinformatics*, 17, 977–987.

Symbol Description

bvcn   Intentionally Blank.

Chapter 5

Measurement Error and Survival Model for cDNA Microarrays

Jonathan A. L. Gelfond[1] and **Joseph G. Ibrahim**[2]

[1] *University of Texas Health Science Center San Antonio and* [2] *University of North Carolina at Chapel Hill*

5.1   Introduction

The discovery of associations between gene expression and patient outcome relies on statistical models which typically use the expression measurement directly as a predictor. A more accurate approach would assume that the expression measurement is merely an indication of the underlying true expression level. The proposed model recognizes that gene expression has measurement error. Experiments involving many tumors will often perform only one microarray assay on each tumor because of cost limitations. Thus, the distributions of measurement errors are not directly identifiable. This model extends the additive error-in-variable survival model for Affymetrix data of Tadesse et al. (2005) to two color microarrays with correlated multiplicative errors. The error model is included within the framework of a piecewise exponential survival model. Robustness analyses are performed, and the model is applied to a breast cancer study dataset.

DNA microarrays are assays that simultaneously measure the expression of thousands of genes. Characterizing the association between time-to-event data and gene expression is similar to the differential expression problem because event data constitutes a biological state, although the state is complex in that the state space is censored and infinite. The broader problem of differential expression requires that the gene expression for a particular gene on an array is measured or computed. The aforementioned value of a gene's expression is often referred to as the gene expression index (GEI) (Li and Wong, 2001). The GEI is computed in numerous ways depending on the type of array and the model used. The probes or spots on an individual array have complementary subsequences highly specific to the corresponding gene's RNA in the samples. The two major types of arrays in most common use are Affymetrix and cDNA arrays. In cDNA two-color arrays, there is typically only one probe per gene, and two samples that are dyed red and green are applied to each array. The first is often a standardized reference used in some sense to calibrate the sensitivity of the probe and to represent a standard level of gene expression. The second sample is from the biological state of interest. Affymetrix arrays have a very different design (Li and Wong, 2001) with sets of mismatch and match pairs for each gene. The GEI is typically calculated as a function of the raw array data whose parameters are derived from a model fit. Examples of this for Affymetrix data are the Li and Wong model (Li and Wong, 2001) and Robust Microarray Analysis (Irizarry et al., 2003a). For cDNA two color arrays with a standardized reference, the log-ratio of the sample intensity over the reference intensity is used. However, performance of model based estimates of GEI have been shown superior to simpler ones at least for the case of Affymetrix data (Irizarry et al., 2003b). Using the log-ratio for cDNA arrays is also problematic in that information about the magnitude of the two channels is lost by taking ratios. The magnitude of such measurements contains information regarding reliability, and this reliability should be considered in the data analysis. Further, the log-ratio has a dependence on the reference channel used which can complicate comparisons between two experiments using different references.

Ideally, one would like to consider the variability due to microarray measurement error in inferences. The microarrays are accepted as having significant amounts of assay noise (Yang et al., 2002). The presence of noise is obvious in the case of assay replication, but assay noise can be confounded with biological variation depending on experimental design. In the case where the biological state is finite (i.e. treatment and control), there is often biological replication of the states. Here, the association between biological state and a gene's expression may be done strictly on a gene by gene basis or may pool information across genes. The assay noise in the presence of biological or technical replication is accounted for by estimating the variance within replicates in the manner of t-statistics (Dudoit, Yang, Callow, and Speed, 2002). There are parametric models that relate the variance of measurements to their mean which have been proposed by Ideker, Thorsson, Siegel, and Hood (2000) and Rocke and Durbin (2001). There are also some models of microarray noise that use semiparametric methods such as the mixture distribution method of Newton et al. (2004) and a quasi-likelihood approach of Strimmer (2003). These semiparametric models were designed to identify differential expression between a finite number of biological states, not for the type of biological variation and lack of replication in tumor sample experiments with time-to-event being the biological state. Cho and Lee (2004) developed a Bayesian model for oligonucleotide arrays which can apply a measurement error model to only one replication of a particular biological state, but this is developed in a very limited context of differential expression, and it is not clear how to apply this to tumor samples with time-to-event outcome. No two tumors constitute the same biological state. Unless the same tumor is assayed more than once, the assay noise will be confounded with biological variation between tumors. The analysis of noise in the absence of either biological or technical replication is not straightforward, but it is of interest to account for the effects of assay noise when dealing with time-to-event data. It is a well known phenomena that failure to account for measurement error in covariates results in bias of the estimated effect toward the null (Nakamura, 1992). This gives us motivation to develop a model that includes assay noise and avoids biased inference. Tadesse et al. (2005) have recently shown how inferences concerning microarrays and survival can be affected by not accounting for measurement error in Affymetrix microarrays, and we would like to build a similar model for cDNA microarray data.

The aim is to construct a model that accounts for the effects measurement error in cDNA microarray experiments on the assessment of associations between gene expression and time-to-event data. We present a Bayesian hierarchical latent variable model linked to a piecewise constant proportional hazards model for the time-to-event data. The latent variable corresponds to the gene expression index, and the hazard function is conditional upon this latent GEI. The model is shown to have favorable properties such as robustness to misspecification and GEIs that do not explicitly depend on platform specific parameters. Platform specific parameters include the sensitivity of the red probe compared to the green probe and the reference sample. We apply the model to a particular breast cancer experiment that previously demonstrated novel subtypes of breast cancer based on gene expression profiles. The time-to-event of interest is time-to-death due to disease. It is important to understand the influence of a particular gene's influence on time-to-death for the purposes of specifying treatment and identifying potential drug targets. Further, more complex models for the prediction of survival involve a preliminary selection of genes that are individually associated with survival (Bair and Tibshirani, 2004).

Gene expression profiling of breast cancer has and will play an important role in the understanding and treatment of the disease. Microarrays from this dataset have already proven their usefulness in categorizing breast cancer into novel, biologically meaningful subtypes predictive of prognosis as in Sorlie et al. (2001) and Perou et al. (2000). The subtypes of breast tumors such as basal and luminal were confirmed to predict outcome by Sotiriou et al. (2003). Van't Veer et al. (2002) demonstrated that a group of 70 genes expression values from cDNA microarrays can be used to construct a classifier that outperformed clinical and histopathological data in predicting metastasis. A follow-up study done by Vijver et al. (2002) showed that the same 70 genes were an effective predictor of outcome in 295 patients. In the Netherlands, a pilot study is underway in which the doctor and patient are given the tumor's profile of the 70 gene predictors (Tuma, 2004). This study's goal is to characterize what influence the microarrays will have on therapeutic decisions. A prospective randomized trial called MINDACT (Microarray for Node Negative Disease may Avoid Chemotherapy Trial) will test if using microarrays to affect treatment decisions improves outcome (Hampton, 2004).

The paper is organized as follows. Section 5.2 outlines the data structure and some preliminary analyses. In Section 5.3, the general expression and survival models are presented. The specification of priors is given in Section 5.4. Section 5.5 describes the two stage model fit procedure. The application of the proposed model to the breast cancer dataset and comparisons with the standard Cox proportional hazards model are given in Section 5.6. Operating characteristics and a robustness analysis are given in Section 5.7. Section 5.8 summarizes the article with a brief discussion.

5.2   The Data Structure

The data analyzed were obtained from experiments performed on breast cancer samples with similar types of cDNA microarrays. There were a total of 85 microarrays of tumor (78), normal tissue (4), and other tissue (3) from 84 individuals, but clinical information was available from only 77 of the individuals who corresponded to tumors. Of these 77 individuals, 75 had time-to-event data available. This subset of the data is the focus of the paper. There were six batches of microarrays, some arrays having 24k probes and others having 8k probes. A common subset of 7,938 probes were selected. In the green channel, one of three batches of standardized reference were used. The red channel for each array consisted of the 75 tumor samples. It has been reported that the differences in the array type and the batch effect due to reference add some variability to the analysis (Sorlie et al., 2001), but this noise is not considered here. The dataset is available from the Stanford Microarray Database (http://genome-www.stanford.edu/microarray) which is described in Gollub et al. (2003).

The endpoint studied was time to death due to disease in months. Survival times were between 0 and 100 months (mean 35.43, median 30.0). Twenty-six of the 75 patients experienced the event after time 0. A Kaplan–Meier curve of the 75 patients is shown in Figure 5.1.

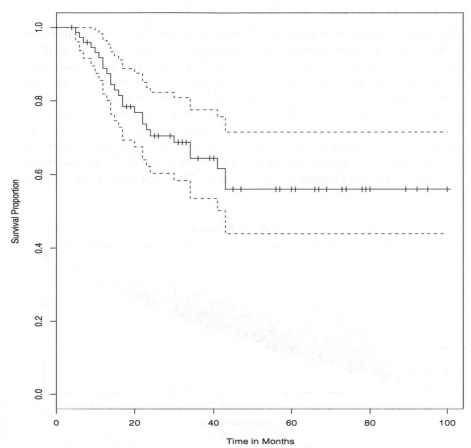

FIGURE 5.1: Survival curve for breast cancer dataset.

5.3 The General Model

The goal is to characterize the association between gene expression of particular genes and time-to-event data. The model proposed will integrate both survival times and a measurement error model. The data is inherently trivariate in that the red and green channels of any probe are potentially correlated with survival and jointly modeled. Some notation will be introduced for this two color data. Each spot on the array will be described as P_{gir} which are vectors of length two whose indices g, i and r refer to the gth gene and the ith individual at the rth replicate respectively. The elements of P_{gir} are

R_{gir} and G_{gir} which are the red and green fluorescent measurements of the spot. P_{gir} may be written as $Probe_{gir} \equiv P_{gir} = \begin{bmatrix} R_{gir} \\ G_{gir} \end{bmatrix}$.

The measurement error model is adapted from one proposed by Ideker et al. (2000). Ideker's model consists of a bivariate normal error with an additive component and a multiplicative component. The multiplicative component will be called the spot effect ($spot \equiv S$). The spot effect is the motivation for taking the ratio of R_{gir}/G_{gir}. By dividing R by G, the general assumption is that the multiplicative error will cancel. The additive component is related to the background effect (B). An examination of the data reveals the relationship between the mean probe intensity and the variance of the probe. Figure 5.2 shows the log of the sample variance plotted against log of the sample mean in the green and red channels of our dataset. There appears to be a strong

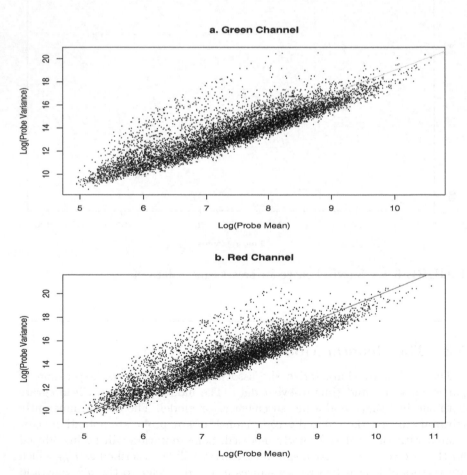

FIGURE 5.2:　Variance vs. mean relationship with model fit lines.

linear relationship between log(probe mean) and log(probe variance). Stating
the model in equation form we have

$$P_{gir} = M_{gi}S_{gir} + B_{gir}, \qquad (5.1)$$

where $S_{gir} \sim N_2 \left(\begin{bmatrix} 1 \\ 1 \end{bmatrix}, \begin{bmatrix} \sigma_{mR}^2 & \rho_m \sigma_{mR} \sigma_{mG} \\ \rho_m \sigma_{mR} \sigma_{mG} & \sigma_{mG}^2 \end{bmatrix} \right)$,

$$B_{gir} \sim N_2 \left(\begin{bmatrix} 0 \\ 0 \end{bmatrix}, \begin{bmatrix} \sigma_{aR}^2 & \rho_a \sigma_{aR} \sigma_{aG} \\ \rho_a \sigma_{aR} \sigma_{aG} & \sigma_{aG}^2 \end{bmatrix} \right),$$

and $M_{gi} = \begin{bmatrix} \mu_{Rgi} & 0 \\ 0 & \mu_{Ggi} \end{bmatrix}$. The diagonal elements of M_{gi} are interpreted as
the mean intensities for gene g and state i since $E[P_{gir}] = [\mu_{Rgi} \quad \mu_{Ggi}]'$, and
this is what motivates the mean vector of $\begin{bmatrix} 1 \\ 1 \end{bmatrix}$ for S_{gir}. The covariance parameters for the multiplicative error are σ_{mR}^2, σ_{mG}^2, and ρ_m which represent
the variability due to a multiplicative effect in assay replication of biologi-
cally identical samples. Similarly, the covariance parameters for the additive
variability due to replication are σ_{aR}^2, σ_{aG}^2, and ρ_a.

There are other models for cDNA data with both additive and multiplicative components. Rocke and Durbin (2001) suggest a log-normal multiplicative
error with a normal additive error. This model presents major computational
challenges because P_{gir} does not have a standard distribution, and the likelihood cannot be written in closed form. Rocke and Durbin (2001) suggest an
iterative fitting procedure on different subsets of genes for the additive and
the multiplicative components separately. However, we do not choose this
model for three reasons. First, analysis of the residuals of the log transformed
data suggest that the log-transformation over-corrects for the relatively small
amount of skewness in the data. Second, the difficulty of dealing with a nonstandard distribution adds to an already heavy computational burden. Third,
we show in Section 1.8 that the estimation of survival parameters and GEI's
with a model based on a normality assumption are robust to this type of misspecification of the multiplicative error distribution. However, even the Ideker
model is not identifiable unless there is technical replication in both the red
and green channels. In the dataset considered in this paper, we do not have
such replication except in a small number of duplicate probes on each array
(about 180). An analysis of these probe measurements was performed on the
green and the red channels separately, and the estimates of σ_{mR}^2 and σ_{mG}^2 were
found to be approximately equal. With this justification, we set the constraint
$\sigma_{mR}^2 = \sigma_{mG}^2 = \sigma_m^2$ for the purpose of model identifiability. Also, the variance
parameters due to the additive components (σ_{aR}^2, σ_{aG}^2, and ρ_a) were found
to be very small relative to the multiplicative error, so we set them to zero.
The parameters μ_{Rgi} and μ_{Ggi} are the means within a biological state. When
a common reference is used, μ_{Ggi} becomes μ_{Gg} and it represents the mean

intensity of the reference channel, and μ_{Rgi} is the mean of the sample channel. In experiments with biological replication within a channel, the means of the intensities measured are often considered to be derived from the same underlying population, so that $\mu_{Rgi} = \mu_{Rg}$ for replicates within a biological state. We must account for the biological variability in tumor samples, and thus we consider an additional hierarchical component to the model and take $\mu_{Rgi} = \mu_{Rg}(1 + \beta_{gi})$. The parameter β_{gi} is the latent GEI and represents the ith tumor's and the gth gene's deviation from mean of that gene (μ_{Rg}). β_{gi} is taken to be a truncated normal variable with $\beta_{gi} > -1$ because $\beta_{gi} + 1$ is considered to be proportional to a concentration, and therefore, $\beta_{gi} + 1$ must be positive. The method of identifying the GEI as a latent variable is novel. It is well suited for tumor samples because it gives a structure to the variation in a gene's expression. The structure of the truncated normal distribution acts to resist outlying measurements so that the GEI's have a regression to the mean. The model can be restated in another equivalent form.

$$R_{gi} = \mu_{Rg}(1 + \beta_{gi})\epsilon_{Rgi} \tag{5.2}$$

$$G_{gi} = \mu_{Gg}\epsilon_{Ggi} \tag{5.3}$$

where

$$S_{gi} \equiv \begin{bmatrix} \epsilon_{Rgi} \\ \epsilon_{Ggi} \end{bmatrix} \sim \left(\begin{bmatrix} 1 \\ 1 \end{bmatrix}, \begin{bmatrix} \sigma_m^2 & \rho_m\sigma_m^2 \\ \rho_m\sigma_m^2 & \sigma_m^2 \end{bmatrix} \right)$$

and

$$\beta_{gi} \sim N_{\{\beta_{gi} > -1\}}(0, \sigma_{bio}^2).$$

We have a simple physical model that assumes that the intensity of a probe (P) is roughly proportional to product of the concentration ([mRNA]) of the target mRNA in the sample and the sensitivity of the probe (ϕ). In equation form we have $P \approx [\text{mRNA}] \times \phi$. The physical model is motivated in part because of the Li and Wong model for Affymetrix data which takes the following form for a single gene:

$$P_{ij} = \nu_j + \theta_i\phi_j + \epsilon_{ij} \tag{5.4}$$

Here, P_{ij} represents the ith measurement j^{th} probe with sensitivity ϕ_j and background ν_j. θ_i is the gene expression index and ϵ_{ij} is a normally distributed error term. The difference between our model and models like that of Li and Wong is that the GEI of individual i is not a random effect. That is, in the Li and Wong model, the biological variation of GEI's is not modeled explicitly. We extend the form of the Li and Wong model to cDNA data here for the case of a standard reference in the green channel by taking

$$P_{gir} = \Theta_{gi}\Phi_g S_{gir}, \tag{5.5}$$

where $\Phi_g = \begin{bmatrix} \phi_{Rg} & 0 \\ 0 & \phi_{Gg} \end{bmatrix}$ and $\Theta_{gir} = \begin{bmatrix} [red]_{gi} & 0 \\ 0 & [green]_g \end{bmatrix}$.

The parameters ϕ_{Rg} and ϕ_{Gg} are the platform specific sensitivities of the red and green channels respectively. The Θ_{gi} denotes a matrix whose diagonal elements $[red]_{gi}$ and $[green]_g$ are the concentrations of RNA on the specified array. This model statement is consistent with (5.1) if we let $\Theta_{gi}\Phi_g = M_{gi}$ and set $B_{gi} = 0$. The problem of gene by dye interaction occurs for some genes when the intensity of the red channel and the green channel respond differently to the same concentration gradient. Using the language of this model, gene by dye interaction can be stated as $\phi_{Rg} \neq \phi_{Gg}$. The connection with this model and the log-ratio can be seen by considering the special case that $\rho_m = 1$. The log-ratio is given by

$$\psi_{gir} \equiv \log(R_{gir}/G_{gir}) = \log\left((\phi_{Rg}/\phi_{Gg})([red]_{gi}/[green]_g)\right). \tag{5.6}$$

The three deficiencies of ψ_{gir} can be noticed. First, if the values in the red or green channel are negative, then the log-ratio cannot be computed, and this generates missing data despite the clear informativeness of low values. Second, the platform specific parameters of ϕ_{Rg} and ϕ_{Gg} are contained in the GEI. Third, the reference specific parameter μ_{Gg} is also affecting the GEI, and these two problems complicate the interpretation and the cross platform comparisons of the log-ratio. Now, consider the parameter β_{gi}. The parameter can be stated in terms of the ratio of intensity parameters as $\beta_{gi} = (\mu_{Rgi}/\mu_{Rg}) - 1$. According model (5.9), $\mu_{Rgi} = \phi_{Rg}[red]_{gi}$ and $\mu_{Rg} = \phi_{Rg}[red_g]$ then,

$$\beta_{gi} = (\mu_{Rgi}/\mu_{Rg}) - 1 = ([red_{gi}]/[red]_g) - 1. \tag{5.7}$$

Thus, β_{gi} does not explicitly depend on platform or reference specific parameters for the reason that it is a function of the ratio of the mean intensities, and that ratio is not dependent on the probe sensitivity or the reference channel.

The parameter β_{gi} will also be linked to the following piecewise constant hazards survival model. This model divides the survival time axis into J adjacent disjoint intervals $(s_0, s_1], (s_1, s_2], ..., (s_{J-1}, s_J]$ where $0 = s_0 < s_j < s_{j'}$ if $(0 < j < j')$ and $j = 1, \dots J$. Within each interval is a constant baseline hazard $h_0(y) = \lambda_j$ when $y \in (s_{j-1}, s_j]$. We let $\nu_i = 1$ be the failure indicator for the ith individual ($\nu_i = 0$ otherwise), and let $\delta_{ij} = 1$ if the ith individual was either censored or failed in the jth interval ($\delta_{ij} = 0$ otherwise). The survival component contribution of the likelihood for the ith individual becomes

$$f(y_i|\beta_{gi}, \gamma_c) = \prod_{j=1}^{J} (\lambda_j \exp(\eta_i))^{\delta_{ij}\nu_i} \exp\Bigg\{ - \delta_{ij}$$
$$\times \left[\lambda_j(y_i - s_{j-1}) + \sum_{k=1}^{j-1} \lambda_k(s_k - s_{k-1})\right] \exp(\eta_i)\Bigg\} \tag{5.8}$$

where $\eta_i = \log(\beta_{gi} + 1)\gamma_g + Z_i'\gamma_c$ is the linear predictor. Z_i is the $p \times 1$ vector of clinical covariates for the ith individual, and γ_c is the corresponding $p \times 1$

vector of coefficients. Note that β_{gi} has been log transformed for comparisons with the log-ratio models.

In this paper, we consider only one gene's ($g = g'$) association with survival at a time so $\beta_{g'}$ refers to the vector of latent GEI's for the $g'^{(th)}$ gene, but P refers to all probe data, that is all of the red and green channel measurements. The model parameters are $\Omega = \{\lambda_j, \beta_{g'i}, \gamma'_g, \gamma_{ck}, \mu_{Rg}, \mu_{Gg}, \sigma_m, \rho_m, \sigma_{Bg}\}$.

The dataset consists of $D = \{P_{gi}, Y, \nu_i, \delta_{ij}, \}$. The full likelihood function is the given by

$$L(\Omega|D) \propto$$

$$\prod_{i=1}^{n} \prod_{g=1}^{G} \phi_2 \left(P_{gi}; \begin{bmatrix} \mu_{Rg}(1 + \beta_{gi}) \\ \mu_{Gg} \end{bmatrix}, \begin{bmatrix} \mu_{Rgi}^2 \sigma_m^2 & \rho_m \sigma_m^2 \mu_{Rgi} \mu_{Gg} \\ \rho_m \sigma_m^2 \mu_{Rgi} \mu_{Gg} & \mu_{Gg}^2 \sigma_m^2 \end{bmatrix} \right)$$

$$\times \phi_{\{\beta_{gi} > -1\}}(\beta_{gi}; 0, \sigma_{Bg}^2) \left[f(y_i | \beta_{gi}, \gamma_c) \right]^{I[g=g']}. \tag{5.9}$$

where $\phi_2()$ is the bivariate normal density, and $\phi_{\{\beta_{gi} > -1\}}()$ is the left truncated normal density. Again, $\eta_i = \log(\beta_{g'i} + 1)\gamma_{g'} + Z_i \gamma_c$ is the linear predictor for survival involving only one gene (g'). Also, $\mu_{gRi} = \mu_{gR}(1 + \beta_{gi})$ for convenience.

The likelihood has two parts. The first part will pertain to the measurement error model, and the second part is the survival model. This dichotomy of the likelihood motivates the two stage fitting procedure described in Section 5.5.

5.4   Priors

Bayesian models involve the specification of priors as well as the likelihood, therefore the specification of priors will complete our model. We do not have information about parameters from previous studies, and therefore we choose priors that are relatively non-informative or vague. We use the following priors for the parameters

$$\mu_{Rg}^{-1} | \mu_i, \sigma_{\mu_i}^2 \sim N_{\{\mu_{Rg} > 0\}}(\mu_i m_{Rg}^{-1}, \sigma_{\mu_i}^2 m_{Rg}^{-2}) \quad [m_{Rg} = \frac{1}{n_g} \sum_{i=1}^{n_g} R_{gi}] \tag{5.10}$$

$$\mu_{Gg}^{-1} | \mu_i, \sigma_{\mu_i}^2 \sim N_{\{\mu_{Gg} > 0\}}(\mu_i m_{Gg}^{-1}, \sigma_{\mu_i}^2 m_{Gg}^{-2}) \quad [m_{Gg} = \frac{1}{n_g} \sum_{i=1}^{n_g} G_{gi}] \tag{5.11}$$

$$\sigma_m^{-2} | \alpha_m, \omega_m \sim \text{gamma}(\alpha_m, \omega_m) \tag{5.12}$$

$$\rho_m \sim \text{Unif}(0,1) \tag{5.13}$$

$$\sigma_{Bg}|\alpha_B, \omega_B \sim \text{gamma}(\alpha_B, \omega_B) \tag{5.14}$$

$$\lambda_j|\alpha_0, \omega_0 \sim \text{gamma}(\alpha_0, \omega_0/\lambda_{j-1})(\lambda_0 = 1) \tag{5.15}$$

$$\gamma_{g'}|\sigma_g^2 \sim N(0, \sigma_{gene}^2) \tag{5.16}$$

The gamma priors on the λ_j's are chosen because they are strictly positive, conjugate, and they induce correlation between adjacent $\lambda's$. Such correlated priors create smoothness in the baseline hazard and were introduced by Arjas and Gasbarra (1994). Such correlated priors are also discussed in Ibrahim, Chen and Sinha (2001). The prior for σ_{Bg} was chosen to be a vague gamma prior; the prior was taken for on σ_{Bg} instead of the precision parameter σ_{Bg}^{-2} because the former is more easily interpreted, and the precision parameter of a truncated normal does not have a conjugate gamma prior. The prior for σ_m^{-2} is a vague gamma prior because this is the conjugate form. A vague normal prior was selected for the survival coefficients γ_g and to let the likelihood drive the inference and make the survival parameters comparable to the Cox model for comparison. The μ parameters in both models had priors that cover the range of the measurements, and a vague prior is placed on μ_{Gg}^{-1} and μ_{Rg}^{-1} instead of the reciprocal to take advantage of the log-concave posterior which facilitates a more efficient Gibbs sampling scheme. See the appendix for computational details. The array data is scaled to avoid numerical problems. This scaling by m_{Gg} and m_{Rg} results in the choice of $\mu_i = 1$.

5.5 Model Fit

Our goal is to fit the model (5.9) on a gene by gene basis in a computationally efficient manner, and the parameter of interest is $\gamma_{g'}$ because $\gamma_{g'}$ determines the association between gene expression and time-to-event. We could fit the full model likelihood for each gene, but doing so would be computationally expensive because parameters such as $(\beta_{gi}, \mu_{gR}, \text{and}\mu_{gG})$ relating to other genes would then be estimated as well. The number of these nuisance parameters is on the order of $n * G \approx 100,000$. To facilitate a more feasible fitting scheme, the model was fit using an MCMC method in two stages. These two stages correspond to the two parts of the likelihood. In the full likelihood, the first part contains information about the measurement error parameters $(\sigma_m, \rho_m, \sigma_{Bg})$ for all genes, and the second part contains the parameters of the survival model. One may notice that the measurement error parameters are shared across all genes and that one individual gene's contribution to the likelihood should be relatively small. Further, our analysis has shown that these parameters can be estimated to a reasonably high precision by using a large number of genes (≥ 500). Thus, in the first stage of the model fitting, we will estimate the measurement error parameters with using likelihood

$$L(\Omega|D) = \prod_{i=1}^{n}\prod_{g=1}^{G}\phi_2\left(P_{gi};\begin{bmatrix}\mu_{Rg}(1+\beta_{gi})\\ \mu_{gG}\end{bmatrix},\begin{bmatrix}\mu_{Rgi}^2\sigma_m^2 & \rho_m\sigma_m^2\mu_{Rgi}\mu_{Gg}\\ \rho_m\sigma_m^2\mu_{Rgi}\mu_{Gg} & \mu_{Gg}^2\sigma_m^2\end{bmatrix}\right)$$
$$\times \phi_{\{\beta_{gi}>-1\}}(\beta_{gi};0,\sigma_B^2) \tag{5.17}$$

The biological variance parameter σ_B is chosen in this stage to be the same for each gene for computational convenience and to borrow strength across genes. Alternatively, one could select a subset of housekeeping genes thought to have the same low biological variability, and use only these genes to estimate the measurement error parameters. From this model fit, we will use the estimates of the measurement error parameters $\hat{\sigma}_m$ and $\hat{\rho}_m$ and substitute them into (5.9) and this will constitute the second stage of the model fit:

$$L(\Omega|D) \propto$$
$$\prod_{i=1}^{n}\phi_2\left(P_{gi};\begin{bmatrix}\mu_{Rg}(1+\beta_{gi})\\ \mu_{g'G}\end{bmatrix},\begin{bmatrix}\mu_{Rgi}^2\hat{\sigma}_m^2 & \hat{\rho}_m\hat{\sigma}_m^2\mu_{Rgi}\mu_{Gg}\\ \hat{\rho}_m\hat{\sigma}_m^2\mu_{Rgi}\mu_{Gg} & \mu_{Gg}^2\hat{\sigma}_m^2\end{bmatrix}\right)$$
$$\times \phi_{\{\beta_{g'i}>-1\}}(\beta_{g'i};0,\sigma_{Bg'}^2)$$
$$\times \prod_{j=1}^{J}(\lambda_j\exp(\eta_i))^{\delta_{ij}\nu_i}\exp\left\{-\delta_{ij}\left[\lambda_j(y_i-s_{j-1})+\sum_{k=1}^{j-1}\lambda_k(s_k-s_{k-1})\right]\right.$$
$$\times \exp(\eta_i)\}. \tag{5.18}$$

The second stage will be applied to each gene, and the parameters associated with the measurement error $(\hat{\sigma}_m, \hat{\rho}_m)$ remain fixed. Further, we found that the model is weakly identifiable when σ_B becomes large ($\sigma_B > 2$). For large σ_B, the parameters σ_B and μ_{Rg} become confounded. So, for the second stage of the analysis, we fixed $\mu_{Rg} = \frac{1}{n}\sum_{i=1}^{n}R_{gi}$. We found that this constraint only had slight influence on the inferences regarding the parameter of interest (γ_g). In order to classify the genes as either significantly associated with a survival or not, we will use the highest posterior density (HPD) intervals for the γ_g parameter. If and only if the interval does not contain 0, then the gene will be included in the list of genes associated with survival. We will examine both the 95% and the 99% HPD intervals.

5.6 Case Study in Breast Cancer

We use the model in the previous section to examine the breast cancer data described in Section 5.2. As mentioned above, the model was in two stages, measurement error parameter estimation and survival analysis.

5.6.1   Estimating the Measurement Error Parameters

We normalized the microarrays before applying our model. There are many normalization procedures available for cDNA (Yang et al., 2002; Quackenbush, 2002). However, most of these methods are applied to the log-ratio as opposed to the red and green channel individually. For our purposes, we jointly model the red and green channel instead of modeling $\log(R/G)$. Moreover, there is no replication of samples that is an important component of many normalization procedures. For normalization, we choose to perform a simple scaling procedure as follows. One array without major problems such as poor green or red dye measurements is chosen as the standard, and the red channel measurements from that array are scaled so that the mean of the red channel is equal to the mean of the green channel. Then, all other arrays are scaled so that the means of each channel's probes are equal to the mean of the green channel of the first array. This method was chosen above quantile normalization because it better preserved the correlation between the red and the green channels across arrays. When we compare our method to one that uses the log-ratio, we used the log-ratio normalization procedure used by Sorlie et al. (2001).

After the arrays were normalized, we estimated the measurement error parameters by sampling 500 probes at random from the original 7,938 probes. Prior parameters were selected as follows: $(\alpha_B, \omega_B) = (2, 0.1)$; $(\alpha_m, \omega_m) = (2, 0.1)$; $(\mu_i, \sigma_{\mu_i}^2) = (1, 100)$. The burn-in period of 10,000 Gibbs sample was used to achieve convergence, and the number of samples used was 50,000. The convergence of the Gibbs sampler was diagnosed with parallel chains by using the Gelman and Rubin $\sqrt{\hat{R}}$ statistic (Gelman and Rubin, 1992). Convergence diagnostics were computed with the R coda package (Plummer, Best, Cowles, and Vines, 2004). See Figure 5.3 for trace plots. There was some autocorrelation in the parameters that slowed convergence, but the effect on parameter estimation was small as the mean of the posterior estimates were within 1% of their final estimates very early in the chain (i.e. after a few hundred iterations). The measurement error model then yielded estimates for these parameters as follows: $\sigma_B = 0.5752(0.0062)$; $\sigma_m = 0.6082(0.0029)$; $\rho_m = 0.9347(0.0021)$.

These parameters suggest a large amount of variation due to assay noise. The coefficient of variation due to the multiplicative technical error is $\hat{\sigma}_m = 0.6082$, and the correlation of the red and green components of this multiplicative effect is $\hat{\rho}_m$ which suggests that the log-ratio has significant error. These parameters will now be considered fixed in the gene by gene survival analysis stage.

5.6.2   Data Preprocessing

Before survival models are fitted, there is some data preprocessing including gene filtering and imputation of missing data. The large number of genes relative to the number of independent observations makes it beneficial to limit the analysis to a subset of probes that meet some threshold of variability

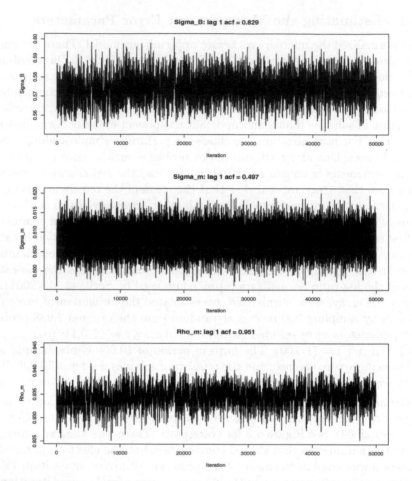

FIGURE 5.3: Trace plots of select parameters (σ_B, σ_m, and ρ_m) of measurement error model with lag 1 autocorrelation.

across samples. We used a similar inclusion threshold to that of the original analysis. We considered only probes that had at least 3 samples that were a 4-fold change from the median log-ratio. From that list, we took a subset of those probes which had missing data in the green or red channels for no more than 10 out of the 75 arrays. This left 991 probes for examination, but there were duplicated gene names in the probe list. All duplicate gene names were removed for the survival analysis which left 942 genes. The missing data in the reduced set was then imputed using the log-ratio. Specifically, imputation was performed using the Statistical Analysis for Microarrays package (Tusher, Tibshirani, and Chu, 2001) with a K-nearest neighbor algorithm in which K=10.

TABLE 5.1:   Comparison of Lists of Significant Genes
for Gene Only Model 99% Cutoff

		Proposed Model		
		Significant	Not Significant	Total
Cox	Significant	65	18	83
Model	Not Significant	13	846	859
	Total	78	864	942

5.6.3   Results: Genes Identified by the Gene Only Model

Our goal is to find a list of genes that are associated with time-to-event in breast cancer. We perform two types of analyses and compare the results with a conventional Cox proportional hazards model. The analysis presented here tests the gene's survival association without additional clinical covariates. For comparison, we fit a Cox proportional hazards model with standard software (R package, 2004) using the log-ratio as the GEI covariate. When constructing gene lists using the Cox model, the p-value of the corresponding regression parameter was used to determine association. Specifically, lists with genes having a p-value cutoff of < 0.05 and < 0.01 for the regression parameter will be compared to lists including genes whose γ_g parameters have 95% and 99% HPDs that do not contain 0. The latent variable and the Cox models were fit to the 942 genes. The prior hyperparameters for the survival model are as follows: $(\alpha_B, \omega_B) = (2, 0.1)$; $(\alpha_0, \omega_0) = (0.01, 0.01)$; $(\mu_i, \sigma^2_{\mu_i}) = (1, 100)$; $\sigma^2_{gene} = 100$.

The Gelman–Rubin statistic was again used to assess convergence, but because of the number of models (942), convergence could not be thoroughly examined except for a few genes. Based on these models, conservative estimates for the number of iterations needed to achieve convergence was used for all genes. A burn-in of 5,000 cycles, and 10,000 samples were used to summarize the posterior estimates. The results compare lists of genes selected to have a significant association with survival by the proposed model and the Cox model given in Table 5.1.

There is significant agreement between the two lists. For the sake of brevity, we will focus on the lists of the 99% cutoff. The intersection of the two lists includes genes which have known associations with breast cancer such as the estrogen receptor (Perou et al., 2000), gamma glutamyl hydrolase (Rhee, Wang, Nair, and Galivan 1993), and the angiotensin receptor 1 (AGTR1) gene (De Paepe, Verstraeten, De Potter, Vakaet, and Bullock, 2001). Of the 13 genes that were detected by the proposed model only, we have found that some of them have associations with breast cancer such as estrogen regulated LIV-1 protein (Dressman et al., 2001) and the 5T4 oncofetal trophoblast glycoprotein gene (Kopreski, Benko, and Gocke, 2001).

5.7   Robustness Analysis and Operating Characteristics

5.7.1   Deviation from Normality in the Data

According to the model, the array data in the green channel for a particular probe is normally distributed about the same mean so that $green_{gi} \sim N(\mu_{gG}, \mu_{gG}^2 \sigma_m^2)$. One may calculate the scaled residuals in a typical manner of subtracting the sample mean and then dividing by the sample standard deviation for each gene. To examine the validity of the distributional assumption, we show a histogram of the scaled residuals in Figure 5.4a. The normal density is overlaid. One can detect that the distribution is skewed to the right with a heavier tail. One could consider a transformation, but transformations dilute the relationship between the mean and the variance. The distribution of the red channel is much more complicated under the model because it is the product of a normal and a truncated normal random variable. The residuals for the red channel are shown in Figure 5.4b.

5.7.2   Simulations Demonstrating Robustness to Nonnormality

In order to characterize the effects of this deviation from normality, we performed a robustness analysis with a simulation. We used the log-normal model of Rocke and Durbin (2001) without the normal additive error to simulate a dataset and applied our two stage model fitting procedure to 200 different datasets with $n = 75$ individuals. The true measurement error parameters of the simulation were $\sigma_m = 0.6$, $\rho_m = 0.9$, and $\sigma_{Bg} = 0.5$. A total of 500 genes were simulated with $\mu_{gR} = |X_{gi}|$ and $\mu_{gG} = \mu_{gR} Y_{gi}$ where $X_{gi} \sim N(10, 000, 3, 000)$ and $Y_{gi} \sim \text{gamma}(2, 2)$. The estimates (and SD's) from the model fit were ($\hat{\sigma}_m = 0.651$ (0.005), $\hat{\rho}_m = 0.985$ (0.001), and $\hat{\sigma}_{Bg} = 0.59$ (0.05)). Then, 200 survival datasets were generated with the survival time y_i being exponentially distributed with rate parameter equal to $\exp[\gamma_g log(\beta_{gi} + 1)]$ with a censoring probability of 0.7. The regression coefficient γ_g was drawn uniformly from the interval $[-2, 2]$, and $\beta_{gi} \sim N_{\{\beta_{gi} > -1\}}(0, 25.0)$. The parameters μ_{Rg} and μ_{Gg} were simulated as above. We are primarily interested in the γ_g parameter, but we also show results of β_{gi}. Figure 5.5 shows the $\hat{\gamma}_g$ plotted against the true values. The bars in the plot indicate the 95% HPD intervals. The 95% and 99% HPD intervals contained the true values of γ_g 92.0% and 98.5% of the time respectively, which indicates that the model is estimating γ_g fairly accurately. Also, the $log(\hat{\beta}_{gi} + 1)$ were highly correlated with the true values, see Figure 5.6. Another test of robustness of the $\hat{\beta}_{gi}$ as GEI's is the correlation that they have with the conventional log-ratio GEI's. The mean and median correlation of $log(\hat{\beta}_{gi} + 1)$ with the log-ratio estimates for each of the 942 genes of interest

a. Scaled Residuals of Green Channel

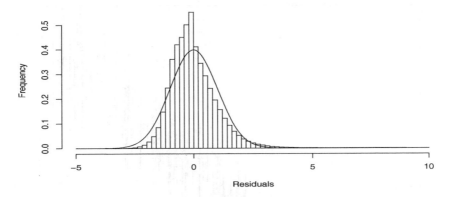

b. Scaled Residuals of Red Channel

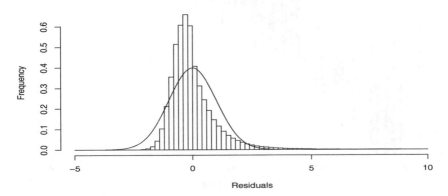

FIGURE 5.4: Scaled residuals with normal density curve.

are 0.90 and 0.97 respectively. These high correlations between the log-ratio GEI and the latent GEI estimates suggests substantial agreement of the two estimates of the biological variability present in the data.

An analysis of the operating characteristics of the model demonstrates that the model has good type I and type II error rate control for inference regarding

TABLE 5.2: Operating Characteristics under True Model

N	γ_g	Estimate	(SD)	$\gamma_g \in$ 95%HPD	$\gamma_g \in$ 99%HPD	$0 \notin$ 95%HPD	$0 \notin$ 99%HPD
950	0	0.0004	(0.18)	0.943	0.99	0.057	0.008
25	1	1.11	(0.24)	0.96	1.0	1.0	1.0
25	-1	-1.01	(0.22)	0.96	1.0	1.0	0.96

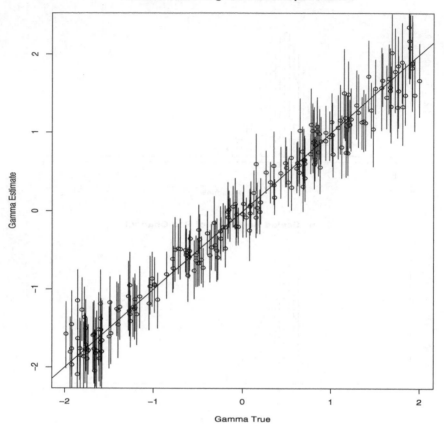

FIGURE 5.5: Estimation of regression parameter with 95% HPD under misspecified error model.

the γ_g parameter. The Ideker model (Ideker et al., 2000) was used to simulate the datasets, and the same measurement error parameters were used as above with these parameters treated as known. A total of 1,000 datasets with $n = 75$ individuals were simulated with 950 genes under the null ($\gamma_g = 0$) and 50 genes under the alternative ($\gamma_g = 1$ and $\gamma_g = -1$, 25 times each). The results for the simulation are given in Table 5.2. Table 5.2 shows that the properly specified model has no strong evidence of type I error rate inflation and has good power for moderate effect size. Also, the HPDs have accurate coverage probabilities, and the estimated coefficients γ_g show no indication of bias. For comparison, we retested the operating characteristics using the log-normal multiplicative error mentioned above and using same simulation parameters as well as the same estimated measurement error parameters. Again, 1,000 datasets were

Model Under Log–Normal Misspecification

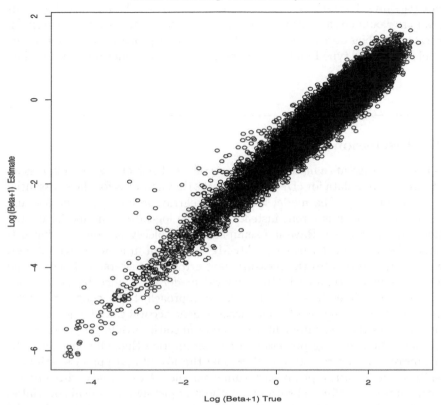

FIGURE 5.6: Estimation of latent expression under misspecified error model.

simulated with $n = 75$, for fitting the survival model. Table 5.3 represents the model fit under a grossly misspecified error structure, and this degree of skewness in the error is greater than that of the observed data. Despite this large deviation from normality, the model is seen to be quite robust to this kind of misspecification. One can see a slight inflation of the type I error rate ($0.05 \rightarrow 0.059$ and $0.01 \rightarrow 0.013$). The power is not seen to decrease, and this

TABLE 5.3: Operating Characteristics under Misspecified Model

N	γ_g	Estimate	(SD)	$\gamma_g \in$ 95%HPD	$\gamma_g \in$ 99%HPD	$0 \notin$ 95%HPD	$0 \notin$ 99%HPD
950	0	0.004	(0.15)	0.941	0.987	0.059	0.013
25	1	0.989	(0.19)	1.0	1.0	1.0	1.0
25	-1	-0.958	(0.16)	0.92	1.0	1.0	1.0

may be surprising. However, one must remember that the measurement error is not the same. The estimates of the γ_g coefficients are slightly biased towards the null as would be the case for models that did not account for measurement error, but this bias does not affect the HPD coverage probabilities. Overall, the model shows good type I and type II error rate control under misspecification.

5.8 Discussion

This paper presents a model to find associations between a gene's expression and time-to-event data for cDNA microarrays that accounts for the substantial measurement error. The model for the microarray probes is parametric and creates a GEI which is latent instead using the log-ratio. The model for the time-to-event data is a Bayesian semiparametric piecewise constant hazards model. We fit the model using an MCMC algorithm in a two stage process. The first stage estimates the measurement error parameters, and the second stage uses these estimates in the survival model on a gene by gene basis. There may be biases in such a two stage approach if the genes are vastly non-homogenous in terms of their measurement accuracy, and a joint model that estimates the measurement error simultaneously would be preferable in that case. The two stage process and the assumption that all genes have the same measurement error variance does offer the advantage of precise estimates despite having no true replication of tumor samples. A case study with a breast cancer dataset is performed with and without adjusting for clinical covariates. The new model is shown to be generally consistent with a conventional model that uses the log-ratios in a Cox proportional hazards model, and potentially important genes selected by the proposed model only are found to have known connections with breast cancer. That is, conventional models that do not account for measurement error may fail to detect these genes' associations between event and gene expression. In addition to detecting associations, the conventional models may underestimate the strength of these associations because models not accounting for measurement error are known to be biased towards the null, and this bias may be avoided in the proposed model. The model was shown to be robust to some parametric assumptions for inference about the parameter of interest, and the new GEI's are found to be highly correlated with the log-ratios. Further, the model is demonstrated to have good operating characteristics concerning type I and type II error rates as well as accurate coverage of the parameter values by the HPDs. However, the issue of False Discovery Rates (FDR) is not addressed here. Conceivably, permutation of the survival times could be applied to the data in order to estimate the false discovery rate. Permutation is regularly applied in the case of the Cox model and in other frequentist approaches in microarray data

(Sorlie et al., 2001), yet such permutations would be not computationally feasible for a Bayesian analysis using this model, and permutation is only valid under exchangability which excludes more complex models with clinical covariates. The problem of estimating FDR for Bayesian models is one of current research (Efron, Tibshirani, Storey, and Tusher, 2001; Ibrahim, Chen, and Gray, 2002; Tadesse, Ibrahim, and Mutter, 2003; Newton, 2004; Tadesse et al., 2005), and the estimation of the FDR can be obtained by using the mean posterior probability. If one is interested in which genes are most likely to be associated with the time-to-event data, an ordering of the genes in terms of association is required. In the frequentist setting, the p-values for the test statistics can generate the ordering. One may easily derive such an ordering from the model presented here by calculating the posterior probability that $\gamma_g = 0$ as in Tadesse et al. (2005).

In our model, we assumed the independence of genes, but expression is often correlated between genes due to involvement within a gene network or metabolic pathway. This correlation between genes is quite difficult to estimate in practice because of the small sample size given the larger of genes. However, the model may be extended to include a structured correlation of the latent gene expression values for a carefully chosen set of genes. For instance, one might assume a compound symmetric covariance for a set of genes that are known or estimated to be co-expressed. This can be implemented by adding a layer to the hierarchical model so that the latent gene expressions of different genes within the cluster are a function of a latent network activity level. One might then link the latent level of network activity to survival probability. Overall, this model has an important advantage over the conventional one in that it accounts for measurement error which is a significant additional source of variation.

References

Arjas, E. and Gasbara, D., (1994), "Nonparametric Bayesian inference from right censored survival data, using the Gibbs sampler," *Statistica Sinica, 4*, 505–524.

Bair, E., and Tibshirani, R.,(2004) "Semi-supervised methods to predict patient survival from gene expression data," *Plos Biology, 2*, 0511–0521.

Cho, H., and Lee, J. K., (2004) "Bayesian hierarchical error model for analysis of gene expression data," *Bioinformatics, 20*, 2016–2025.

De Paepe, B., Verstraeten, V. L., De Potter, C.R., Vakaet, L.A., Bullock, G.R., (2001), "Growth stimulatory angiotensin II type-1 receptor is upregulated in breast hyperplasia and in situ carcinoma but not in invasive carcinoma," *Histochemistry and Cell Biology, 116*, 247–254.

Buchholtz, S., Kwon, I., Ellis, M.J., Polymeropoulos, M.H., (2001), "GrGenes that co-cluster with estrogen receptor alpha in microarray analysis of breast biopsies," *Pharmacogenomics Journal, 1*, 135–141.

Dudoit, S., Yang, Y. H., Callow, M. J., and Speed, T. P.,(2002), "Statistical methods for identifying differentially expressed genes in replicated cDNA microarry experiments," *Statistica Sinica, 12*, 111–139.

Efron, G., Tibshirani, R., Storey, J. D., and Tusher, V.,(2001), "Empirical Bayes analysis of a microarray experiment," *Journal of the American Statistical Association, 96*, 1151–1160.

Gelman, A., Rubin, D. B., (1992), "Inference from iterative simulation using multiple sequences," *Statistical Science, 7*, 457–472.

Gilks, W. R., Best, N. G., and Tan, K. K. C., (1995),"Adaptive rejection Metropolis sampling within Gibbs sampling," *Applied Statistics, 44*, 445–472.

Gilks, W. R. and Wild, P., (1992),"Adaptive rejection sampling for Gibbs sampling," *Applied Statistics, 41*, 337–348.

Gollub, G., Ball, C.A., Binkley, G., Demeter, J., Finkelstein, D.B., Hebert, J.M., Hernandez-Boussard, T., Jin, H., Kaloper, M., Matese, J.C., Schroeder, M., Brown, P.O., Botstein, D., Sherlock, G., (2003), "The Stanford microarray database," *Nucleic Acids Research, 31(1)*, 94–96.

Hampton, T., (2004),"Breast cancer gene chip study under way," *Journal of the American Medical Association, 291*, 2927–2930.

Ibrahim, J. G., Chen, M.-H., and Gray, R. J., (2002), "Bayesian models for gene expression with DNA microarray data," *Journal of the American Statistical Association, 97*, 88–99.

Ibrahim, J. G., Chen, M.-H., and Sinha, D., (2001), *Bayesian Survival Analysis*, New York: Springer Verlag, pp. 106–109.

Ideker, T., Thorsson, V., Siegel, A. F., and Hood, L. E., (2000), "Testing for differentially-expressed genes by maximum-likelihood analysis of microarray data," *Journal of Computational Biology, 7*, 805–817.

Irizarry, R., Hobbs, G., Collin, F., Beazer-Barclay, Y. D., Antonellis K. J., Scher, U., and Speed, T. P. (2003a), "Exploration, normalization, and summaries of high density oligonucleotide array probe level data," *Biostatistics, 4*, 249–264.

Irizarry, R., Bolstad, B. M., Collin, F., Cope, L. M., and Speed, T. P. (2003b), "Summaries of Affymetrix GeneChip probe level data," *Nucleic Acids Research, 31(4)*, e15.

Kopreski, M.S., Benko, F.A., Gocke, C.D., (2001), "Circulating RNA as a tumor marker: detection of 5T4 mRNA in breast and lung cancer patient serum," *Annals of the New York Academy of Science, 945*, 172–178.

Li, D., Wong, W. H., (2001), "Model-based analysis of oligonucleotide arrays: Expression index computation and outlier detection," *Proceedings of the National Academy of Science U S A, 98*, 31–36.

Nakamura, T., (1992), "Proportional hazards model with covariates subject to measurement error," *Biometrics, 48*, 829–838.

Neal, R. M. (1998), "Suppressing random walks in Markov Chain Monte Carlo using ordered overrelaxation." In *Learning in Graphical Models*, Jordan, M. I., ed., Dordrecht: Kluwer Academic Publishers, pp. 205–228.

Newton, M. A., Noueiry, A., Sarkar, D., Ahlquist, P., (2004), "Detecting differential gene expression with a semiparametric hierarchical mixture method," *Biostatistics, 5*, 155–176.

Perou, C. M., Sorlie, T., Elsen, M. B., van de Rijn, M., Jeffrey, S.S., Rees, C.A., Pollack, J.R., Ross, D.T., Johnsen, H., Akslen, L.A., Fluge, O., Pergamenschikov, A., Williams, C., Zhu, S.X., Lonning, P.E., Borresen-Dale, A.L., Brown, P.O., Botstein, D., (2000), "Molecular portraits of human breast tumors," *Nature, 406*, 747–752.

Plummer, M., Best, N., Cowles, K., Vines, K., (2004), *coda: Output analysis and diagnostics for MCMC*. R package version 0.8–3.

Quackenbush, J. (2002), "Microarray normalization and transformation," *Nature Genetics, 32*, 496–501.

R Development Core Team, (2004), *R: A language and environment for statistical computing*, Vienna, Austria: R Foundation for Statistical Computing.

Rhee, M. S., Wang, Y., Nair, M. G., and Galivan, J., (1993), "Acquisition of resistance to antifolates caused by enhanced gamma-glutamyl hydrolase activity," *Cancer Research, 53*, 2227–2230.

Rocke, D. M. and Durbin, B., (2001), "A model for measurement error for gene expression arrays," *Journal of Computational Biology, 8*, 557–569.

Sorlie, T., Perou, C. M., Tibshirani, R., Aas, T., Geisler, S., Johnsen, H., Hastie, T., Eisen, M.B., van de Rijn, M., Jeffrey, S.S., Thorsen, T., Quist, H., Matese, J.C., Brown, P.O., Botstein, D., Eystein Lonning, P., Borresen-Dale, A.L., (2001), "Gene expression patterns of breast carcinomas distinguish tumor subclasses with clinical implications," *Proceedings of the National Academy of Science U S A, 98*, 10869–10874.

Sotiriou, C., Neo, S-Y., McShane, L. M., Korn, E.L., Long, P.M., Jazaeri, A., Martiat, P., Fox, S.B., Harris, A.L., Liu, E.T., (2003), "Breast cancer classification and prognosis based on gene expression profiles from a population based study," *Proceedings of the National Academy of Science U S A, 100*, 10393–10398.

Strimmer, K., (2003), "Modeling gene expression measurement error: a quasi-likelihood approach," *BMC Bioinformatics, 4*, 10.

Tadesse, M. G., Ibrahim, J. G., and Mutter, G., (2003), "Identification of differentially expressed genes in high-density oligonucleotide arrays accounting for the quantification limits of the technology," *Biometrics, 59*, 542–554.

Tadesse, M. G., Ibrahim, J. G., Gentleman, R., Chiaretti, S., Ritz, J. and Foa, R., (2005) "Bayesian error-in-variable survival model for the analysis of genechip arrays," *Biometrics, 61*, 488–497.

Tuma, R. S., (2004), "A big trial for a new technology: TransBIG project takes microarrays into clinical trials," *Journal of the National Cancer Institute, 96*, 648–649.

Tusher, V., Tishirani, R., Chu, C., (2001), "Significance analysis of microarrays applied to ionizing radiation response," *Proceedings of the National Academy of Science U S A, 98*, 5116–5121.

van 't Veer, L.J., Dai, H., van de Vijver, M.J., He, Y.D., Hart, A.A., Mao, M., Peterse, H.L., van der Kooy, K., Marton, M.J., Witteveen, A.T., Schreiber, G.J., Kerkhoven, R.M., Roberts, C., Linsley, P.S., Bernards, R., Friend, S.H., (2002), "A gene expression profiling predicts clinical outcome of breast cancer," *Nature, 415*, 530–536.

Vijver, M. J., He, Y. D., Van't Veer, L. J., Dai, H., Hart, A.A., Voskuil, D.W., Schreiber, G.J., Peterse, J.L., Roberts, C., Marton, M.J., Parrish, M., Atsma, D., Witteveen, A., Glas, A., Delahaye, L., van der Velde, T., Bartelink, H., Rodenhuis, S., Rutgers, E.T., Friend, S.H., Bernards, R., (2002), "A gene-expression signature as a predictor of survival in breast cancer," *New England Journal of Medicine, 347*, 1999–2009.

Yang, Y. H., Dudoit, S., Luu, P., Lin, D.M., Peng, V., Ngai, J., Speed, T.P.,(2002), "Normalization for cDNA microarray data: a robust composite method addressing single and multiple slide systematic variation," *Nucleic Acids Research , 30(4)*, e15.

Appendix: Computational Details

We fit the models using a Gibbs sampling technique in which samples from the joint posterior $(L(D|\Omega)\pi(\Omega))$ are obtained by successively sampling from the full conditionals for a number of iterations after convergence criteria are met. Computation for the Gibbs sampler was performed using the C language. The full conditionals of ρ_m, σ_m, and β_{gi} were sampled using the Adaptive Rejection with Metropolis Sampling (ARMS) algorithm of Gilks, Best and Tan (1995). The μ^{-1} parameters have a log-concave density, and could be sampled directly using Adaptive Rejection Sampling (ARS) (Gilks and Wild, 1992). The parameter σ_m^{-2} has a gamma distribution which could be sampled using standard statistical algorithms. The ordered overrelaxation technique of Neal (1998) was used when sampling from the σ_m^{-2}, μ_{Rg}^{-1} and μ_{Gg}^{-1} full conditionals to reduce autocorrelation of the Gibbs sampler and improve convergence. The γ_g parameters have full conditionals that are log-concave so that the ARS algorithm is potentially applicable; however, numerical imprecision sometimes yielded non-concave log-likelihood functions despite the analytical log-concavity of the conditionals. Since ARMS is a more general sampling method, it was used for these parameters.

Chapter 6

Bayesian Robust Inference for Differential Gene Expression

Raphael Gottardo
Department of Statistics, University of British Columbia

6.1 Introduction

Following the success of genome sequencing, DNA microarray technology has emerged as a valuable tool in the exploration of genome functionality. cDNA microarrays (Schena et al., 1995; Lockhart et al., 1996) consist of thousands of individual DNA sequences printed on a high density array using a robotic arrayer. A microarray works by exploiting the ability of a given labeled cDNA molecule to bind specifically to, or hybridize to, a complementary sequence on the array. By using an array containing many DNA samples, scientists can measure—in a single experiment—the expression levels of thousands of genes within a cell by measuring the amount of labeled cDNA bound to each site on the array. A common task with microarrays is to determine which genes are differentially expressed under two or more experimental conditions.

In recent years the problem of identifying differentially expressed genes has received much attention; an early statistical treatment can be found in Chen et al. (1997). A common approach is to test a hypothesis for each gene using variants of t or F-statistics and then attempting to correct for multiple testing (Tusher et al., 2001; Efron et al., 2001; Dudoit et al., 2002). Due to the small number of replicates, variation in gene expression is often poorly estimated. Tusher et al. (2001) and Baldi and Long (2001) suggested using a modified t statistic where the denominator has been regularized by adding a small constant to the gene specific variance estimate. Similar to an empirical

Bayes approach, this results in shrinkage of the empirical variance estimates towards a common estimate. Lönnstedt and Speed (2002) proposed an empirical Bayes normal mixture model for gene expression data, which was later extended to the two condition case by Gottardo et al. (2003) and to more general linear models by Smyth (2004). Though Smyth (2004) did not use mixture models but simply empirical Bayes normal models for variance regularization. In each case, the authors derived explicit gene specific statistics and did not consider the problem of estimating p, the proportion of differentially expressed genes. Newton et al. (2001) developed a method for detecting changes in gene expression in a single two-channel cDNA slide using a hierarchical gamma-gamma (GG) model. Kendziorski et al. (2003) extended this to replicate chips with multiple conditions and provided the option of using a hierarchical lognormal-normal (LNN) model. More recently, fully Bayesian approaches where "exact" estimation is carried out by Markov chain Monte Carlo (MCMC) have also been proposed (Broët et al., 2004; Tadesse and Ibrahim, 2004; Lewin et al., 2006). These methods allow for more complex and realistic models compared to empirical Bayes methods that often rely on unrealistic assumptions.

In this chapter, we review a fully Bayesian framework named BRIDGE, for Bayesian Robust Inference for Differential Gene Expression, that can be used to test for differentially expressed genes in a robust way (Gottardo et al., 2006b). Robustness is achieved by using a t distribution Besag and Higdon (1999), which is more robust to outliers than the usual Gaussian model because of its heavier tails. The model allows each gene to have a different variance while regularizing these using information from all genes. Inference is based on the posterior probabilities of differential expression estimated using MCMC. The goal of this chapter is to concentrate on the application of BRIDGE, showing novel results.

The chapter is organized as follows: We first introduce the methodology, including the model and the parameter estimation procedure. We then present results from applying BRIDGE to two experimental datasets as well as a comparison of BRIDGE to four other commonly used methods using the same two datasets. Finally, we discuss our results and possible extensions.

6.2 Model and Prior Distributions

In this section, we introduce the Bayesian hierarchical model that BRIDGE uses to test for differentially expressed genes between two experimental conditions. We will denote by y_{gcr} the log expression value from the rth replicate of gene g under experimental condition c $(= 1, 2)$. BRIDGE models the measurements from each gene as the sum of a condition effect and an error term. The condition effects and error variances are assumed to arise from a genomewide distribution with hyperparameters specific to each condition. In other words, the log expression values are modeled as follows,

$$y_{gcr} = \gamma_{gc} + \frac{\epsilon_{gcr}}{\sqrt{w_{gcr}}}, \tag{6.1}$$

$$(\epsilon_{gcr}|\lambda_{\epsilon_{gc}}) \sim N(0, \lambda_{\epsilon_{gc}}^{-1}),$$

$$(w_{gcr}|\nu_{cr}) \sim \mathcal{G}a(\nu_{cr}/2, \nu_{cr}/2),$$

where w_{gcr} and ϵ_{gcr} are independent. It follows that $\epsilon_{gcr}/\sqrt{w_{gcr}}$ has a t-distribution with ν_{cr} degrees of freedom and scale parameter $\lambda_{\epsilon_{gc}}^{-1}$. Note that the degrees of freedom parameter is replicate specific. Assuming that each replicate is done on a different array (in each condition), this would account for the fact that some arrays might contain more outliers than others.

In (6.1), we model γ_{gc}, the effect of experimental condition c on gene g, as a random effect with a mixture of two mutually singular Gaussian distributions, i.e.

$$(\gamma_g|\lambda_\gamma, p) \sim (1-p)N(\gamma_{g1}|0, \lambda_{\gamma_{12}}^{-1})\mathbf{1}(\gamma_{g1} = \gamma_{g2}) \tag{6.2}$$

$$+ pN(\gamma_{g1}|0, \lambda_{\gamma_1}^{-1})N(\gamma_{g2}|0, \lambda_{\gamma_2}^{-1})\mathbf{1}(\gamma_{g1} \neq \gamma_{g2}),$$

where $\gamma_g = (\gamma_{g1}, \gamma_{g2})'$, $\lambda_\gamma = (\lambda_{\gamma_1}, \lambda_{\gamma_2}, \lambda_{\gamma_{12}})$ and $\mathbf{1}(\cdot)$ is an indicator function equal to one if the expression between parentheses is true. The first component corresponds to the genes that are not differentially expressed ($\gamma_{g1} = \gamma_{g2}$), while the second component corresponds to the genes that are differentially expressed ($\gamma_{g1} \neq \gamma_{g2}$). Note that the formulation is not standard as it is not absolutely continuous with respect to the two-dimensional Lebesgue measure. However, it defines a proper distribution with respect to a more general dominating measure, namely the sum of a one-dimensional Lebesgue measure on the line $\gamma_{g1} = \gamma_{g2}$ and the two-dimensional Lebesgue measure (Gottardo and Raftery, 2008).

In terms of priors, we use an exchangeable prior for the precisions, so that information is shared between genes, namely $\lambda_{\epsilon_{gc}} \sim \mathcal{G}a(a_{\epsilon_c}^2/b_{\epsilon_c}, a_{\epsilon_c}/b_{\epsilon_c})$, i.e. a gamma distribution with mean a_{ϵ_c} and variance b_{ϵ_c}. Similarly, we use a vague but proper exponential prior with mean 200 for $\lambda_{\gamma_{12}}, \lambda_{\gamma_1}, \lambda_{\gamma_2}$, namely $\lambda_\gamma \sim \mathcal{G}a(1, 0.005)$. For a_{ϵ_c} and b_{ϵ_c} we use a vague uniform prior between 0 and 1000. The prior for the mixing parameter, p, is uniform over $[0, 1]$. Finally, the prior for the degrees of freedom, ν_{cr}, is uniform on the set $\{1, 2, \ldots, 10, 20, \ldots, 100\}$. By using a prior that allows degrees of freedom between 1 and 100, we allow a wide range of sampling errors from the heavy-tailed Cauchy ($\nu = 1$) to nearly Gaussian ($\nu = 100$).

6.3 Parameter Estimation

Realizations can be generated from the posterior distribution using Markov chain Monte Carlo (MCMC) algorithms Gelfand and Smith (1990). All updates are straightforward except for γ, for which the update is nonstandard since the distribution is formed by two mutually singular components. However, there is a common dominating measure and so the Metropolis-Hastings algorithm can be used Gottardo and Raftery (2008). In our case, the full conditional of γ is available and the Gibbs sampler can be used. It can be shown (see Gottardo and Raftery (2008)) that the full conditional of γ_g is given by

$$(\gamma_g|\ldots) \propto (1-p)k_g \mathrm{N}(\gamma_{g1}|\mu_g^*, \lambda_g^{*-1})\mathbf{1}(\gamma_{g1}=\gamma_{g2}) \tag{6.3}$$
$$+ pk_{g1}k_{g2}\mathrm{N}(\gamma_{g1}|\mu_{g1}^*, \lambda_{g1}^{*-1})\mathrm{N}(\gamma_{g2}|\mu_{g2}^*, \lambda_{g2}^{*-1})\mathbf{1}(\gamma_{g1}\neq\gamma_{g2}),$$

where

$$\lambda_g^* = \sum_{r,c} w_{gcr}\lambda_{\epsilon_{gc}} + \lambda_{\gamma_{12}}, \qquad \mu_g^* = \lambda_g^{*-1}\sum_{r,c} w_{gcr}\lambda_{\epsilon_{gc}}y_{gcr},$$

and

$$\lambda_{gc}^* = \lambda_{\epsilon_{gc}}\sum_r w_{gcr} + \lambda_{\gamma_c}, \qquad \mu_{gc}^* = \lambda_{gc}^{*-1}\sum_r w_{gcr}\lambda_{\epsilon_{gc}}y_{gcr}.$$

The constants k_g, k_{g1} and k_{g2} are given by

$$k_g = \sqrt{\frac{\lambda_{\gamma_{12}}}{\lambda_g^*}}\exp\left\{-0.5\sum_{r,c} w_{gcr}\lambda_{\epsilon_{gc}}y_{gcr}^2 + 0.5\lambda_g^{*-1}\left(\sum_{r,c} w_{gcr}\lambda_{\epsilon_{gc}}y_{gcr}\right)^2\right\},$$

and

$$k_{gc} = \sqrt{\frac{\lambda_{\gamma_c}}{\lambda_{gc}^*}}\exp\left\{-0.5\lambda_{\epsilon_{gc}}\sum_r w_{gr}y_{gcr}^2 + 0.5\lambda_{gc}^{*-1}\left(\sum_r w_{gr}\lambda_{\epsilon_{gc}}y_{gcr}\right)^2\right\}.$$

To update γ, one draws new pairs $(\gamma_{g1}, \gamma_{g2})$ from the first component of the full conditional with probability $(1-p_g^*) \equiv (1-p)k_g/[pk_{g1}k_{g2}+(1-p)k_g]$, or from the other component with probability p_g^*.

For each gene g, inference will be based on the marginal posterior probability of differential expression. This probability, corresponding to the posterior probability that $\gamma_{g1} \neq \gamma_{g2}$ given the data, can easily be estimated from the MCMC output. For a given posterior sample S of size B we estimate the posterior probability by $\frac{1}{B}\sum_{k\in S}\mathbf{1}(\gamma_{g1}^{(k)} \neq \gamma_{g2}^{(k)})$, where $\gamma_{g1}^{(k)}$ and $\gamma_{g2}^{(k)}$ are the values generated at the kth MCMC iteration and $\mathbf{1}(\gamma_{g1}^{(k)} \neq \gamma_{g2}^{(k)})$ is the indicator function equal to one if $\gamma_{g1}^{(k)} \neq \gamma_{g2}^{(k)}$. In our case, when using the Gibbs sampler to update the pair $(\gamma_{g1}, \gamma_{g2})$, Rao-Blackwelization can be used to improve the estimates. One simply needs to replace the indicator function by the conditional probability $P(\gamma_{g1} \neq \gamma_{g2} | \cdots)$ computed at each iteration, i.e. the probability that $\gamma_{g1} \neq \gamma_{g2}$ given all the other parameters. Note this conditional probability is available analytically and is simply given by p_g^*, which is computed at each iteration of the MCMC when updating γ_g. Thus, the Rao-Blackwell estimates can be obtained at no extra cost.

For the datasets explored here, trace plots and autocorrelation plots were used as convergence diagnostic tools. We found that a sample of 50,000 iterations with 1,000 burn-in iterations and storing every 10th iteration was sufficient to produce reliable results. Guided by this, and leaving some margin, we used 100,000 iterations with 5,000 burn-in iterations, and storing every 10th iteration after the burn-in period. This took about 1 hour for the HIV data and 4 hours for the spike-in data on an Intel Xeon processor at 3GHz. An R software package called `bridge` implementing the method is available from Bioconductor at `www.bioconductor.org`.

6.4 Application to Experimental Data

6.4.1 Data Description

To illustrate our methodology we use two publicly available microarray datasets: one cDNA experiment and one Affymetrix spike-in experiment. These two datasets have the advantage that in each case the true state (differentially expressed or not) of all or some of the genes is known.

The HIV-1 data: The expression levels of 4608 cellular RNA transcripts were measured one hour after infection with human immunodeficiency virus type 1 (HIV-1) using four replicates on four different slides. 13 HIV-1 genes have been included in the set of RNA transcripts to serve as positive controls, i.e., genes known in advance to be differentially expressed. Meanwhile, 29 non-human genes have also been included and act as negative controls, i.e., genes known not to be differentially expressed. Another dataset was obtained by repeating the four aforementioned experiments but with an RNA preparation different from that for the first dataset. For easy reference, in this paper we label the two datasets as HIV-1A and HIV-1B respectively. See van't Wout et al. (2003) for more details on the HIV-1 data. For both of these datasets, BRIDGE can be run with the following command:

```
hiv1<-bridge.2samples(C,T,B=60000,min.iter=10000,affy=FALSE)
```

where C corresponds to the matrix of raw intensity from the treatment (HIV infected) condition and T corresponds to the matrix of raw intensity from the control (non HIV infected) condition. Note that by default, BRIDGE takes the logarithm of the input data. The object *hiv1* would then contain the MCMC output, including the posterior probabilities of differential expression. The option *B=60000* sets the number of MCMC iterations to 60,000 while the *min.iter=10000* option sets the number of burn-in iterations to 10,000. Finally, the *affy=FALSE* tells bridge that the data correspond to cDNA microarray data, and thus the dimension of the C and T matrices should be the same.

The spike-in data: This dataset was obtained from a spike-in study done with HGU133A arrays. 42 spiked-in genes were organized in 14 groups, and the concentrations used were 0, 0.125, 0.25, 0.5, 1, 2, 4, 8, 16, 32, 64, 128, 256 and 512 pM. Each concentration was done in three replicates. The data were preprocessed using GCRMA (Irizarry et al., 2003) resulting in a data set of 22300 genes across 42 samples (3 replicates × 14 concentrations). In addition to the original 42 spiked-in genes, we claim that another 20 genes should also be included in the spiked-in gene list as they consistently show significant differential expression across the array groups in the exploratory data analysis. Similar observations have been made by Sheffler et al. (2005). Moreover, the probe sets of three genes contain probe sequences exactly matching those for the spiked-ins. These probes should be hybridized by the spiked-ins as well. As a result, our expanded spiked-in gene list contains 65 entries in total. To evaluate BRIDGE on this data, we will compare the first array group to the other array groups, leading to 13 comparisons with three replicates each. In each of the comparisons, only the 65 spiked-in genes should be differentially expressed. Similar to the HIV data, the following code can be used to run bridge on the spike-in data:

```
spike.in<-bridge.2samples(C,T,B=60000,min.iter=10000,affy=TRUE)
```

where C corresponds to the matrix of raw intensity from the treatment (array group i>1) sample and T corresponds to the matrix of raw intensity from the control (array group 1) sample. The object *spike-in* would then contain the MCMC output, including the posterior probabilities of differential expression. This time, the *affy=TRUE* tells bridge that the data corresponds to Affymetrix (or more generally oligo based) data, and thus the dimensions of the C and T matrices need not be the same.

6.4.2   Results

We fitted model (6.1), described in Section 6.2, to both the HIV data and the 13 comparisons of the spike-in data. The posterior modes of the degrees of freedom of the t-distribution ranged from 2 to 4 for the spike-in data, indicating that the sampling errors are clearly heavier-tailed than the Gaussian distribution. For the HIV data, the degrees of freedom of the t-distribution ranged from 10 to 100, suggesting that the sampling errors are more Gaussian. The proportion of differentially expressed genes is estimated to be 0.003 for both the HIV-1A and HIV-1B data, while it ranges from 0.0025 to 0.004 for the 13 comparisons of the spike-in data. In the case of the spike-in data, the true number of differentially expressed (DE) genes is 65, and thus the proportion of DE genes is 65/22300=0.003. This shows that the proportion of DE genes is well estimated by BRIDGE. Figure 6.1(a) and 6.1(b) show plots of the posterior probabilities against the posterior means of the log-ratios computed from our model for the HIV-1A data and one of the 13 comparisons for the spike-in data. Most of the log-ratios are shrunk close to zero and have very low posterior probabilities of differential expression. To evaluate the effect of the t-distribution, we fitted the model given by (6.1) replacing the t errors with Gaussian errors. In the case of the spike-in data, the t-based model clearly leads to a more powerful procedure than the Gaussian one as the corresponding posterior probabilities are higher (Figure 6.1(b)). The model with the t-distribution detects more spiked-in genes, 60, as against 57 for the Gaussian model, at the 0.5 posterior threshold. In contrast, for the HIV data, the posterior probabilities for the Gaussian and t models are almost identical (Figure 6.1(a)). This is not surprising as the estimated degrees of freedom parameter for the HIV data is quite large. This shows that even if the errors are more Gaussian, fitting a t model with unknown degrees of freedom won't affect the overall performance. The converse is not true, however, as shown by the spike-in data.

We now compare BRIDGE to four other methods: EBarrays, both gamma-gamma (GG) and lognormal-normal (LNN) models (Newton et al., 2001; Kendziorski et al., 2003), the popular Significance Analysis of Microarrays (SAM) (Tusher et al., 2001) and Linear Models for Microarray (LIMMA) data (Smyth, 2004). In addition, we have included the BRIDGE model with Gaussian errors for comparison purposes. The results have been organized in Tables 6.1–6.2.

In the analysis of the HIV-1 data, we obtain the number of genes called differentially expressed (DE) for each method. Among those genes called DE, we look at the number of true positives (TP), i.e., genes known to be DE in advance, and the number of false positives (FP), i.e., genes known to be not DE. Gottardo et al. (2006a) showed that one of the HIV genes, which was expected to be highly differentially expressed, had a very small estimated log ratio and did not properly hybridize in the second experiment (HIV-1B). We removed the corresponding gene from the list of known differentially expressed genes. Thus there are 13 genes known to be DE in the first experiment and 12 in the second. To compare the performance between the five methods, we control the false discovery rate (FDR) at a fixed level of 0.1. In the literature, FDR values between 5% and 10% are commonly used Tusher et al. (2001); Efron (2004). This said, we have also tried other FDR cutoffs and the results were similar (data not shown). Using our BRIDGE posterior probabilities, the FDR cutoffs can be selected using a direct posterior probability calculation

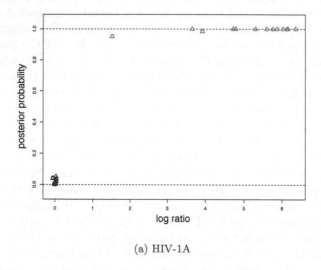

(a) HIV-1A

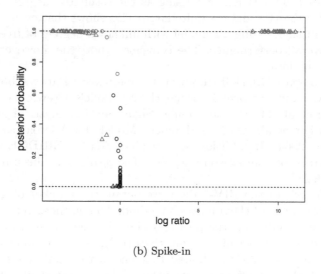

(b) Spike-in

FIGURE 6.1: BRIDGE posterior probabilities with both Gaussian ($\triangle$) and t-errors ($\circ$) plotted against the posterior differences between γ_1 and γ_2 (estimated log-ratios) from the model with t-distribution for the HIV-1A data (a) and the spike-in data (b). Positive controls are shown in red while negative controls are shown in black for the spike-in data and blue for the HIV data. The use of the t-distribution increases the posterior probabilities of expression for several of the known DE genes for the spike-in data whereas it does not change much for the HIV data due to the large value for the estimated degrees of freedom.

TABLE 6.1: Analysis of Differential Expression with the HIV-1 Data

Method	HIV-1A			HIV-1B		
	DE	TP	FP	DE	TP	FP
GG	24	13	0	18	11	1
LNN	18	13	1	18	11	1
LIMMA	13	13	0	11	11	0
SAM	13	13	0	13	11	0
BRIDGE Gaussian	14	13	0	14	11	0
BRIDGE	14	13	0	14	11	0

Note: The FDR is controlled at 0.1 for each method. The numbers of true positives (TP) and false positives (FP) are based on the controls, namely, the 13 (resp. 12 in the second experiment) HIV-1 and the 29 non-human genes of which the states are known in advance, only. They do not represent the true numbers of TP and FP in the entire data.

as described in Newton et al. (2004). For the HIV-1A dataset, when the FDR is controlled at 0.1, all methods can identify the 13 positive controls.

Meanwhile, EBarrays LNN has made one FP. Similar results are observed when the HIV-1B dataset is considered. All methods detect 11 out of the 12 positive controls but both versions of EBarrays (GG and LNN) have made one FP. Recall that both EBarrays model specifications rely on the assumption of a constant coefficient of variation across genes and thus may be more likely to incorrectly classify genes with high variances. This is not the case for the other methods compared. Concluded from the HIV-1 datasets, along with LIMMA and SAM, BRIDGE appears to perform the best as it recognizes the most positive controls and does not find any FP. Finally, as suspected, BRIDGE with Gaussian errors performs just as well because the estimated degrees of freedom for this particular dataset is quite large indicating that the errors might be more Gaussian. This is not the case for the spike-in data, as we will see below.

For the spike-in data, since we know the actual status of each gene, we can check the true FDR of each method against the desired FDR in addition to the number of true and false positives. SAM has considerably more FP cases than the other methods with an FDR over the desired 10%. LIMMA exhibits similar FDR performance with less FPs but misses many spiked-in genes. The FDRs for EBarrays GG and LNN methods are quite high, demonstrating that the respective methods detect too many false positives, as was seen with the HIV data. Finally, BRIDGE with Gaussian errors is clearly inferior to BRIDGE with an underestimated FDR and misses more spiked-in genes. Overall, BRIDGE performs the best with a true FDR of 0.11, closest to the desired 0.10 and detects, on average, 59 of the 65 known DE genes.

TABLE 6.2: Analysis of Differential Expression with the Spike-in Data

Method	DE	TP	FP	FDR
GG	75	59	16	0.21
LNN	75	59	16	0.20
LIMMA	37	25	12	0.16
SAM	73	52	21	0.16
BRIDGE Gaussian	57	54	3	0.03
BRIDGE	67	59	8	0.11

Note: The FDR is controlled at 0.1 for each method. The numbers of true positives (TP) and false positives (FP), and the true false discovery rate (FDR) are based on the 65 differentially expressed genes (spiked-in). Here, the number of differentially expressed (DE) genes is simply TP+FP and FDR=FP/DE. The values of DE, TP, FP, and FDR shown are actually the averages across the 13 comparisons.

6.5   Discussion

We have provided an overview of BRIDGE in the context of differential gene expression with two conditions, but BRIDGE can also be used in situations where there are more than two conditions and differences in expression of the same gene between any two conditions may be of interest. Here the alternative hypothesis is not as simply defined as before, because there are many possible patterns of differential gene expression. In order to account for all possible patterns, the prior for the γ's in (6.3) needs to be modified. An example with three conditions can be found in Gottardo et al. (2006b), and is implemented in the `bridge` package.

In the HIV data, we have assumed that for each array the two experimental conditions were independent. However, with cDNA microarrays, the two RNA samples (corresponding to the two conditions) are usually spotted on the same slide, which could induce some correlation. In this case, it is possible to model such correlation; see Gottardo et al. (2006b) for an attempt. In our experience, the improvement is marginal and this is why we have chosen not to present the correlation model here. In addition, we assumed that normalization was done as a preprocessing step but it is possible to include normalization effects in the model as in Gottardo et al. (2006a).

In this chapter, we have compared our model with four alternatives, but there are many other methods for detecting differentially expressed genes with gene expression data. We chose these four because they are either obvious baseline methods or widely used; they are also representative of other methods. For example, there are several other empirical Bayes methods that we could have used. These include the lognormal-normal models of Lönnstedt and Speed (2002) and Gottardo et al. (2003) and the less parametric approaches of Efron et al. (2001) and Newton et al. (2004). There are also several other fully Bayesian methods we could have used (Broët et al., 2004; Do et al., 2005; Hein and Richardson, 2006; Lewin et al., 2007). In particular, in the Affymetrix dataset it would be interesting to compare BRIDGE to BGX (Hein et al., 2005; Hein and Richardson, 2006), which directly models the available probe intensities representing the gene in a condition. However, given the large number of samples, such a comparison would be computationally intensive and beyond the scope of this chapter. More comparisons between statistical tests can be found in Cui and Churchill (2003), Gottardo et al. (2006b) and Lo and Gottardo (2007).

Acknowledgments

The author thanks Adrian Raftery, Ka Yee Yeung and Roger Bumgarner with whom the BRIDGE work originated, Kenneth Lo for help with some of the results presented, and Luke Bornn for careful proofreading.

References

Baldi, P. and A. Long (2001). A Bayesian framework for the analysis of microarray expression data: Regularized t-test and statistical inferences of gene changes. *Bioinformatics 17*, 509–519.

Besag, J. E. and D. M. Higdon (1999). Bayesian analysis of agricultural field experiments (with discussion). *Journal of the Royal Statistical Society, Series B 61*, 691–746.

Broët, P., A. Lewin, S. Richardson, C. Dalmasso, and H. Magdelenat (2004). A mixture model-based strategy for selecting sets of genes in multiclass response microarray experiments. *Bioinformatics 20*(16), 2562–71.

Chen, Y., E. R. Dougherty, and M. L. Bittner (1997). Ratio-based decisions and the quantitative analysis of cDNA microarray images. *Journal of Biomedical Optics 2*, 364–374.

Cui, X. and G. Churchill (2003). Statistical tests for differential expression in cDNA microarray experiments. *Genome Biology 4*, 210.

Do, K., P. Muller, and F. Tang (2005). A Bayesian mixture model for differential gene expression. *Journal of the Royal Statistical Society Series C 54*, 627–644.

Dudoit, S., Y. H. Yang, M. J. Callow, and T. P. Speed (2002). Statistical methods for identifying differentially expressed genes in replicated cDNA microarray experiments. *Statistica Sinica 12*, 111–139.

Efron, B. (2004). Large-scale simultaneous hypothesis testing: The choice of a null hypothesis. *Journal of the American Statistical Association 99*, 96–104.

Efron, B., R. Tibshirani, J. D. Storey, and V. Tusher (2001). Empirical Bayes analysis of a microarray experiment. *Journal of the American Statistical Association 96*, 1151–1160.

Gelfand, A. E. and A. F. M. Smith (1990). Sampling-based approaches to calculating marginal densities. *Journal of the American Statistical Association 85*, 398–409.

Gottardo, R., J. A. Pannucci, C. R. Kuske, and T. S. Brettin (2003). Statistical analysis of microarray data: a Bayesian approach. *Biostatistics (Oxford, England) 4*(4), 597–620.

Gottardo, R. and A. Raftery (2008). Markov chain Monte Carlo computations with mixture of singular distributions. *J. Comput. Graph. Stat. In Press.*

Gottardo, R., A. Raftery, K. Yeung, and R. Bumgarner (2006a). Quality control and robust estimation for cdna microarrays with replicates. *J Am Stat Assoc 101*, 30–40.

Gottardo, R., A. E. Raftery, K. Y. Yeung, and R. E. Bumgarner (2006b). Bayesian robust inference for differential gene expression in microarrays with multiple samples. *Biometrics 62*(1), 10–18.

Hein, A.-M. K. and S. Richardson (2006). A powerful method for detecting differentially expressed genes from genechip arrays that does not require replicates. *BMC Bioinformatics 7*, 353.

Hein, A.-M. K., S. Richardson, H. C. Causton, G. K. Ambler, and P. J. Green (2005). Bgx: a fully Bayesian integrated approach to the analysis of affymetrix genechip data. *Biostatistics (Oxford, England) 6*(3), 349–73.

Irizarry, R. A., B. M. Bolstad, F. Collin, L. M. Cope, B. Hobbs, and T. P. Speed (2003). Summaries of affymetrix genechip probe level data. *Nucleic Acids Res 31*(4), e15.

Kendziorski, C., M. Newton, H. Lan, and M. N. Gould (2003). On parametric empirical Bayes methods for comparing multiple groups using replicated gene expression profiles. *Statistics in Medicine 22*, 3899–3914.

Lewin, A., N. Bochkina, and S. Richardson (2007). Fully Bayesian mixture model for differential gene expression: simulations and model checks. *Statistical Applications in Genetics and Molecular Biology 6*, Article 36.

Lewin, A., S. Richardson, C. Marshall, A. Glazier, and T. Aitman (2006). Bayesian modeling of differential gene expression. *Biometrics 62*(1), 1–9.

Lo, K. and R. Gottardo (2007). Flexible empirical Bayes models for differential gene expression. *Bioinformatics 23*(3), 328–35.

Lockhart, D. J., H. Dong, M. C. Byrne, M. T. Follettie, M. V. Gallo, M. S. Chee, M. Mittmann, C. Wang, M. Kobayashi, H. Horton, and E. L. Brown (1996). Expression monitoring by hybridization to high-density oligonucleotide arrays. *Nat Biotechnol 14*(13), 1675–80.

Lönnstedt, I. and T. P. Speed (2002). Replicated microarray data. *Statistica Sinica 12*, 31–46.

Newton, M., A. Noueiry, D. Sarkar, and P. Ahlquist (2004). Detecting differential gene expression with a semiparametric hierarchical mixture method. *Biostatistics 5*, 155–176.

Newton, M. C., C. M. Kendziorski, C. S. Richmond, F. R. Blattner, and K. W. Tsui (2001). On differential variability of expression ratios: Improving statistical inference about gene expression changes from microarray data. *Journal of Computational Biology 8*, 37–52.

Schena, M., D. Shalon, R. W. Davis, and P. Brown (1995). Quantitative monitoring of gene-expression patterns with a complementary-DNA microarray. *Science 270*, 467–470.

Sheffler, W., E. Upfal, J. Sedivy, and W. S. Noble (2005). A learned comparative expression measure for affymetrix genechip dna microarrays. *Proceedings / IEEE Computational Systems Bioinformatics Conference, CSB IEEE Computational Systems Bioinformatics Conference*, 144–54.

Smyth, G. K. (2004). Linear models and empirical Bayes methods for assessing differential expression in microarray experiments. *Statistical Applications in Genetics and Molecular Biology 3*, Article 3.

Tadesse, M. G. and J. G. Ibrahim (2004). A Bayesian hierarchical model for the analysis of affymetrix arrays. *Ann N Y Acad Sci 1020*, 41–8.

Tusher, V., R. Tibshirani, and G. Chu (2001). Significance analysis of microarrays applied to the ionizing radiation response. *Proceedings of the National Academy of Sciences 98*, 5116–5121.

van't Wout, A. B., G. K. Lehrma, S. A. Mikheeva, G. C. O'Keeffe, M. G. Katze, R. E. Bumgarner, G. K. Geiss, and J. I. Mullins (2003). Cellular gene expression upon human immunodeficiency virus type 1 infection of $CD4^+ - T -$ Cell lines. *Journal of Virology 77*, 1392–1402.

Symbol Description

L_k	The kth mapped clone or DNA fragment from the p-telomere of a chromosome		tent state j where $j = 1, \ldots, 4$
n	The number of mapped clones on a given chromosome	A	Matrix of stationary transition probabilities
Y_k	Normalized $\log_2$ ratio observed at clone L_k	a_{ij}	Element of matrix A on row i and column j where $i, j = 1, \ldots, 4$
s_k	Copy number state associated with the clone L_k	a_i	Row i of matrix A for $i = 1, \ldots, 4$
μ_1	Expected $\log_2$ ratio of copy number losses	$\mathcal{H}(A)$	Hidden Markov process with parameter A
μ_2	Expected $\log_2$ ratio of the copy neutral state	ϵ	Parameter determining the support of μ_j for $j = 1, \ldots, 4$
μ_3	Expected $\log_2$ ratio of single-copy gains	τ_j^2	Prior variance of μ_j where $j = 1, \ldots, 4$
μ_4	Expected $\log_2$ ratio of multiple-copy gains	θ_{ij}	Constants determining the prior distribution of row i of the matrix A for $i, j = 1, \ldots, 4$
σ_j^2	Measurement error variance associated with the la-		

Chapter 7

Bayesian Hidden Markov Modeling of Array CGH Data

Subharup Guha

Assistant Professor, Department of Statistics, University of Missouri-Columbia, Columbia, MO 65211 (E-mail: GuhaSu@missouri.edu)

7.1 Introduction

The normal genomic code of human females consists of 23 matched pairs of chromosomes. The genomic code of human males consists of 22 matched pairs of autosomal chromosomes and an unmatched pair of sex chromosomes. The *copy number* of normal DNA is therefore 2 for the autosomal chromosomes. The end of the chromosome, or *telomere*, corresponding to the short chromosomal arm is called the *p* telomere. The chromosomal end corresponding to its long arm is called the *q* telomere. Human cells can be classified into *body* and *germ* cells. The life cycle of most somatic cells includes a growth phase followed by cell division through mitosis. In order to proceed to a subsequent stage of the cell cycle, all cells must satisfy a regulation procedure that ensures that the cells develop normally, that DNA is correctly copied during cell division, and that cell defects, if any, are repaired.

Two kinds of genes, proto-oncogenes and tumor suppressors, play key roles in the cell regulation procedure. Proto-oncogenes tend to push the cells through the quality control check points onto a subsequent stage of the cell cycle. Tumor-suppressors tend to hold the cells back, inhibiting cell division when there are defects and signaling the cells to die when their lifespans have ended or when they have defects that cannot be repaired. Pasternak (1999) provides an excellent overview of the relevant biology for this problem.

Occasionally, proto-oncogenes mutate into oncogenes. New cells inherit the mutations through mitosis and typically acquire multiple copies of oncogenes after several stages of mitosis. Oncogenes cause cells to divide at a rapid rate resulting in the development of tumors. Tumors may also develop due to mutations in tumor suppressors that cause them to become nonfunctional, allowing the proto-oncogenes to play a dominant role in the cell function and, ultimately, resulting in the loss of one copy or both copies (i.e. *deletion*) of the tumor suppressors. A person may eventually acquire a cancer phenotype after a number of complex biological events. An example of such an event, although not a necessary condition, is the ability of tumor cells to metastasize, making the tumor malignant. In addition to individual variability, not all the cells in a tumor specimen necessarily exhibit the same kind of genomic alteration. As the disease progresses, there are larger scale changes in tumor DNA because of the breakdown of the cell regulation procedure.

Copy number changes, or alterations in the number of copies in tumor DNA, are therefore closely associated with the development and progression of cancer. Comparative genomic hybridization (CGH) has emerged as a powerful technique for detecting copy number change because it has fairly high resolution of a few million bases (with the entire genome consisting of approximately 3 billion bases) and, additionally, has the ability to span the entire genome in a single experiment (Kallioniemi et al. 1992). The method is briefly described as follows. Fragmented DNA from a test sample is labeled with fluorochrome (typically Cy3) and mixed with normal DNA that is identically fragmented but differentially labeled (typically using fluorochrome Cy5). The normal and tumor DNA fragments are simultaneously hybridized to a normal metaphase spread. Subsequent image analysis yields data consisting of fluorescence intensity ratios along the test and reference DNA sample genomes. Array CGH techniques (Solinas-Toldo et al., 1997; Pinkel et al., 1998; Snijders et al., 2001; Pinkel and Albertson, 2005) hybridize the DNA fragments or "clones" to mapped array fragments. CGH arrays have a resolution of the order of 1 Mb (1 million base pairs). Oligonucleotide and cDNA arrays (Pollack et al., 1999; Brennan et al., 2004) have a higher resolution of 50–100 kb (1 kb = 1,000 base pairs). The fluorescence intensity ratios are normalized as part of a pre-processing step to correct for non-biological sources of error such as intensity fluctuations, background noise and fabrication artifacts (Brown, Goodwin and Sorgeret, 2001; McLachlan, Do, and Ambroise, 2004). For a review and comparison of different normalization methods for array CGH, refer to Khojasteh, Lam, Ward, and MacAulay (2005).

The intensity ratios of array CGH data (equivalently, their transformation on the $\log_2$ scale) are informative about genomic changes in copy number. In an imaginary scenario where all tumor cells have identical genomic changes in the absence of normalization errors, measurement errors and contamination from surrounding normal tissue, the normal (or *copy-neutral*) clones correspond to a $\log_2$ ratio of 0 because the normal and tumor DNA fragments

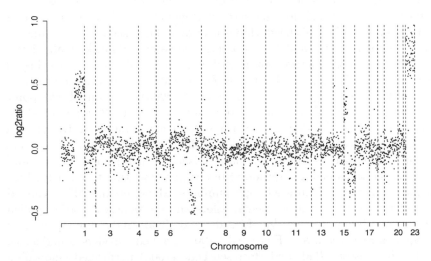

FIGURE 7.1: Normalized copy number ratios of a comparison of DNA from cell strain S0034 (Snijders et al., 2001) with normal DNA. The BACs are ordered by position in the genome beginning at 1p and ending at Xq. The vertical bars indicate borders between chromosomes.

have two copies. In such a situation, the $\log_2$ ratios of single-copy losses would be $\log_2 1/2 = -1$ and the $\log_2$ ratios of single-copy gains would be $\log_2 3/2 = 0.58$. Multiple-copy gains or *amplifications* associated with onco-gene mutations would correspond to $\log_2$ ratios belonging to the sequence $\log_2 4/2, \log_2 5/2, \ldots$, whereas loss of both copies or *deletions*, often associated with tumor-suppressor mutations, would correspond to a value of $-\infty$. In this hypothetical situation, the genomic alterations could be easily deduced from array CGH data without the help of statistical techniques.

Fortunately for statisticians, this is not the case! Figure 7.1 plots the normalized $\log_2$ ratios of breast cancer specimen S0034 that were informally analyzed by Snijders et al. (2001) and are available from table J at *http:// www.nature.com /ng/journal/ v29/n3/suppinfo/ ng754_S1.html*. Although it represents very clean data, the graph highlights some of the key statistical issues and departures from the idealized scenario. For example, even after accounting for measurement error, the $\log_2$ ratios of gains and losses differ considerably from the theoretical values. In particular, the mean values are shrunk toward 0. There are several reasons for this phenomenon, including the contamination of the tumor sample with normal cells. There is also a more subtle effect of the 0 changing slightly from chromosome to chromosome. Additionally, there is dependence among the $\log_2$ ratios of neighboring clones.

As increasing amounts of array CGH data become available, there is a need for automated algorithms for characterizing genomic profiles. A number of techniques aim to fulfill this need. Hodgson et al. (2001) used a normal mixture of three components to model the observed emissions. Pollack et al. (2002) proposed a threshold method for identifying clones with extreme value of emissions. Cheng, Kimmel, Neiman, and Zhao (2003) introduced a regression-based test for altered copy numbers. Jong et al. (2003) relied on a break point model. Fridlyand, Snijders, Pinkel, Albertson, and Jain (2004) applied an unsupervised hidden Markov model. Hupe, Stransky, Thiery, Radvanyi, and Barillot (2004) used a likelihood function with adaptively determined weights based on a smoothed version of the data. Olshen, Venkatraman, Lucito, and Wigler (2004) developed a variation of binary segmentation to identify chromosomal segments with altered copy numbers. Myers et al. (2004) applied an edge filter to detect the segments. Eilers and de Menezes (2005) applied quantile smoothing, Huang, Wu, Lizardi, and Zhao (2005) used penalized least squares regression, and Hsu et al. (2005) applied wavelets. Lingjaerde, Baumbusch, Liestol, Glad, and Borresen-Dale (2005) performed smoothing using the signs of neighboring data values, inspecting the width and magnitude of the segments to detect regions of copy number change. Picard, Robin, Lavielle, Vaisse, and Daudin (2005) discussed a penalized likelihood function. Wang, Kim, Pollack, Balasubramanian, and Tibshirani (2005) built hierarchical clustering-style trees along each chromosome and selected interesting clusters by controlling the false discovery rate.

While comparing some of the above algorithms, Lai, Johnson, Kucherlapati, and Park (2005) commented that "a particularly helpful feature for future implementations of some algorithms would be to estimate the statistical significance of the detected copy number changes and then rank them accordingly." Lai and co-authors pointed out that only two of the algorithms (those of Wang et al., 2005 and Lingjaerde et al., 2005) detected copy number changes based on statistical significance and both relied on false discovery rates.

In this article, we present the statistical framework for detecting copy number gains and losses developed in Guha, Li and Neuberg (2008). Section 7.3 discusses a framework for simulation-based posterior inference. Section 7.4 demonstrates the performance of the technique using publicly available data. Section 7.4.2 compares the proposed Bayesian HMM with some of the existing algorithms using the framework of Lai et al. (2005). Section 7.5 uses simulation studies to compare our Bayesian HMM with alternative techniques for analyzing array CGH data.

Unlike the other statistical methods including the HMM approach of Fridlyand et al. (2004), the proposed model allows the use of objective decision rules based on posterior probabilities to detect copy number alterations. The biologist does not have to make a subjective decision about plausible thresholds for identifying copy number change. The classification scheme of Section 7.3.1 is motivated by biological considerations that facilitate the interpretation of the algorithm's output.

7.2 A Bayesian Model

We model each chromosome separately because the propensity for genomic alterations varies across chromosomes. Let $L_1, \ldots, L_n$ represent the clones on the chromosome arranged from the p telomere to the q telomere. Let Y_k denote the normalized $\log_2$ ratio observed at L_k where $k = 1, \ldots, n$. Associated with each clone is a latent variable called the *copy number state*. The copy number state s_k takes values in the set $\{1, 2, 3, 4\}$. The value $s_k = 1$ corresponds to a copy number loss at L_k that could either be a loss of one copy or a deletion; the value $s_k = 2$ represents the copy-neutral state; the value $s_k = 3$ represents a single-copy gain; $s_k = 4$ represents an amplification (i.e. multiple-copy gain). An *altered state* refers to a copy number state that is different from 2. The parameters of interest are then $s_1, \ldots, s_n$.

The statistical dependence among adjacent clones is modeled using a hidden Markov model (Rabiner, 1989; MacDonald and Zucchini, 1997; Durbin, Eddy, Krogh, and Michison, 1998). For any m indices for which $1 \leq k_1 \leq \ldots \leq k_m \leq n$, a Markov model assumes that $\Pr\left[s_{k_m} \mid s_1, \ldots, s_{k_{m-1}}\right] = \Pr\left[s_{k_m} \mid s_{k_{m-1}}\right]$. The hidden Markov model (HMM) assumes that the conditional probabilities of neighboring clones is $\Pr\left[s_{k+1} \mid s_k\right] = a_{s_k s_{k+1}}$ where $\boldsymbol{A} = ((a_{ij}))$ is the matrix of stationary transition probabilities. We assume that the elements of $\boldsymbol{A}$ are strictly positive, which implies that the hidden Markov process is aperiodic and irreducible, and its four states are positive recurrent. The transition matrix $\boldsymbol{A}$ then has a unique stationary distribution denoted by $\pi_{\boldsymbol{A}}$ (Karlin and Taylor, 1975). We assume that s_1, the copy number state of the first clone, is distributed as $\pi_{\boldsymbol{A}}$. This defines a stochastic process on the index set of positive integers that we denote by $\mathcal{H}(\boldsymbol{A})$.

Let $X \sim F \cdot I(c < X < d)$ imply that X has the distribution F restricted to the interval (c, d) with the density suitably rescaled to make it a random variable. Given a positive real number ϵ and positive constants θ_{ij}, with $\boldsymbol{a}_i$ representing row i of the matrix $\boldsymbol{A}$, we assume the following model:

$$Y_k \overset{indep}{\sim} N(\mu_{s_k}, \sigma_{s_k}^2) \cdot I\left(\left|\frac{Y_k - \mu_{s_k}}{\sigma_{s_k}}\right| \leq 3\right) \qquad k = 1, \ldots, n \qquad (7.1)$$

$$\boldsymbol{s} \sim \mathcal{H}(\boldsymbol{A})$$

$$\mu_1 \sim N\left(-1, \tau_1^2\right) \cdot I\left(\mu_1 < -\epsilon\right) \qquad\qquad\qquad (7.2)$$

$$\mu_2 \sim N\left(0, \tau_2^2\right) \cdot I\left(-\epsilon < \mu_2 < \epsilon\right)$$

$$\mu_3 \sim N\left(0.58, \tau_3^2\right) \cdot I\left(\epsilon < \mu_3 < 0.58\right)$$

$$\mu_4 \sim N\left(1, \tau_4^2\right) \cdot I\left(\mu_4 > \mu_3 + 3\sigma_3\right) \qquad\qquad (7.3)$$

$$\sigma_j^{-2} \sim \text{gamma}\,(1, 1) \cdot I(\sigma_j^{-2} > 6) \qquad j = 1, 2, 3 \qquad (7.4)$$

$$\sigma_4^{-2} \sim \text{gamma}\,(1, 1)$$

$$\boldsymbol{a}_i \overset{indep}{\sim} \mathcal{D}_4\left(\theta_{i1}, \theta_{i2}, \theta_{i3}, \theta_{i4}\right) \qquad i = 1, \ldots, 4. \qquad (7.5)$$

The normal likelihood in (7.1) is restricted to the interval $[\mu_{s_k} - 3\sigma_{s_k}, \mu_{s_k} + 3\sigma_{s_k}]$ because a distribution with thinner tails often provides a better fit than the unrestricted normal. Parameter μ_j is the expected $\log_2$ ratio of all clones for which $s_k = j$. As mentioned in Section 7.1, the μ_j's typically differ from their theoretical values and are therefore regarded as random. The biological interpretation associated with the copy number states allows us to assume the ordering, $\mu_1 < \mu_2 < \mu_3 < \mu_4$, and our knowledge of array CGH allows the specification of informative priors. For example, we know that some $\log_2$ ratios corresponding to copy number losses may be positive in noisy data sets, although the mean μ_1 is never positive. Similar considerations for the other μ_j's motivate the truncated supports in (7.2) through (7.3), which also enforce the relative ordering among the μ_j's. The lower endpoint of the support of μ_4 is chosen to be a distance of $3\sigma_3$ units from μ_3 so that only a small fraction of single-copy gains are erroneously classified as multiple-copy gains. For μ_1 (μ_4), the untruncated part of the prior is centered at the theoretical value for a loss (gain) of one copy. For μ_2 and μ_3, the untruncated parts of the priors are set equal to the theoretical values for pure samples.

The constant ϵ determines the supports of the μ_j's rather than boundaries for the $\log_2$ ratios, unlike threshold-based approaches for detecting copy number change. In fact, to match actual array CGH data, our model allows positive log-intensity ratios for copy number losses especially with large measurement error, although the mean μ_1 does not exceed $-\epsilon$. As demonstrated by simulation studies in Section 7.5.2, the results are robust to choices of ϵ in the range $[0.05, 0.15]$. We set $\epsilon = 0.1$ for all our analyses. The results are also not sensitive to choices of τ_1, τ_2 and τ_3 belonging to the interval $[0.5, 2]$. Setting $\tau_4 \leq 2$ guarantees sufficiently high prior probability to large mean values for high-level amplifications. In Sections 7.4 and 7.5, we set $\tau_1 = \tau_2 = \tau_3 = 1$ and $\tau_4 = 2$.

The assumption $\sigma_j^{-2} > 6$ in (7.4) is equivalent to $\sigma_j < 0.41$ for $j = 1, 2, 3$. The restriction is mild because typical data sets exhibit much smaller within-group variability for these copy number states. The prior support of σ_4^{-2} is not bounded below because state 4 is an aggregation of multiple-copy gains resulting in large within-group variability. Independent Dirichlet priors on $\mathcal{R}^4$ are assumed for the rows of the stochastic matrix $\boldsymbol{A}$ in (7.5). As we see in Section 7.5.2, the results are robust to choices of θ_{ij} that are small in comparison to n. We set the θ_{ij}'s equal to 1 in Sections 7.4 and 7.5.

The informative priors are found to work consistently well for all kinds of data. At the same time, they are flexible enough to allow Bayesian learning and information sharing across the clones. We demonstrate in Sections 7.4 and 7.5 that posterior inference is reasonable and sensitive to the characteristics of the data. The independent priors for the chromosomes result in independent marginal posteriors of interest, $[s_1, \ldots, s_n \mid Y_1, \ldots, Y_n]$.

For latent class models like HMMs, a key issue is label switching (refer to Scott, 2002 for a discussion), which is an identifiability problem where the likelihood is invariant under permutations of the state space labels. This results in inefficient exploration of the posterior by simulation. Our model avoids the problem because the condition $\mu_1 < \mu_2 < \mu_3 < \mu_4$ is violated on permuting the labels.

7.3   Characterizing Array CGH Profiles

We rely on simulation-based methods for inference because the posterior distribution cannot be investigated by mathematical analysis or numerical integration. Section 7.7 describes a Metropolis-within-Gibbs algorithm for generating posterior samples. The copy number states are simulated via a stochastic version of the forward-backward algorithm (Chib, 1996; Robert, Ryden and Titterington, 1999) since it mixes faster than a Gibbs sampler. (Refer to Scott 2002 for a comparison of the two methods.) The transition matrix A is generated using an independent-proposal Metropolis-Hastings algorithm. The likelihood means and variances in (7.1) are generated via Gibbs sampling.

7.3.1   Classification Scheme

The post–burn-in copy number states generated by the MCMC procedure represent draws from the marginal posterior of interest, $[s_1, \ldots, s_n \mid Y_1, \ldots, Y_n]$. For each MCMC draw, the copy number states $s_1, \ldots, s_n$ are labeled as *focal aberrations, transition points, amplifications, deletions* or *whole chromosomal changes* as described below. The classification scheme is neither mutually exclusive nor exhaustive because a clone may belong to none, exactly one, or more than one of the following categories.

1. **Focal aberrations**   Focal aberrations (Fridlyand et al., 2004) represent localized regions of altered copy number. Specifically, a focal aberration is one of the following:

    (i) A single clone that does not belong to a telomere and has an altered state different from its neighbors.

    (ii) Two end-clones of a telomere that share a common altered state different from that of the third clone from the telomere. That is, the clones L_1 and L_2 belonging to the p-telomere are focal aberrations if $s_1 = s_2$, $s_2 \neq 2$ and $s_2 \neq s_3$. Similarly, the clones L_{n-1} and L_n belonging to the q-telomere are focal aberrations if $s_{n-1} = s_n$, $s_{n-1} \neq 2$ and $s_{n-1} \neq s_{n-2}$.

    (iii) Two or more adjacent clones mapped within a small region of the genome (e.g. 5 Mb) and having a common altered copy number state different from their neighbors.

2. **Transition points**   Transition points are associated with the $(n-1)$ inter-clonal spaces. An inter-clonal space is a transition point if it borders on two large regions associated with different copy number states. In contrast, focal aberrations represent small regions of altered copy number. In particular, a transition point must satisfy two conditions:

   (i) It should not be adjacent to a telomere.

   (ii) After excluding all focal aberrations on the chromosome, the neighboring clones on both sides of the inter-clonal space have different copy number states.

   Transition points differ from the "segments" defined by the CBS algorithm of Olshen et al. (2004). Whereas a transition point is associated with large-scale regions of gains and losses, the CBS algorithm classifies clones into segments regardless of their chromosomal distances. A transition point is declared only when the width of the altered region exceeds 5 Mb. For example, five contiguous clones that are highly amplified would generally be identified as a segment by the CBS algorithm (although there are examples in Section 7.4 where the procedure ignores obvious amplifications and deletions to control the false positive rate). In contrast, if these five clones are all located within 5 Mb, we label them as focal aberrations rather than identify them as a region bordered by two transition points.

3. **High-level amplifications**   A clone for which $s_k = 4$.

4. **Deletions**   Deletions are focal aberrations for which $s_k = 1$ and $(Y_k - \mu_1)/\sigma_1 < -2.5$.

7.3.2   Posterior Inference

For $k = 1, \ldots, n$, we declare the clone L_k to be a focal aberration if the posterior probability that it is a focal aberration exceeds 0.5. This probability can be estimated using a post–burn-in MCMC sample as follows. The classification scheme of Section 7.3.1 labels the clone L_k as "1" (i.e. focal aberration) for some MCMC draws and as "0" for the remaining draws. The average of these binary outcomes is a simulation-consistent estimate of the posterior probability that L_k is a focal aberration. A similar method is used to identify deletions and high-level amplifications.

We detect the transition points based on the configuration having the highest joint posterior probability. Formally, let $\boldsymbol{\nu}(\boldsymbol{s}) = (g_1, \ldots, g_{n-1})$ represent the configuration of transition points where g_j equals 1 if the jth inter-clonal gap is a change point, and equals 0 otherwise. The mapping from $\boldsymbol{s}$ to $\boldsymbol{\nu}(\boldsymbol{s})$ is many-one. Let $\boldsymbol{\nu}^*$ denote the global maximum of the posterior of $\boldsymbol{\nu}(\boldsymbol{s})$. A simulation-consistent estimate of ν^* is computed on the basis of the MCMC sample and used to detect the transition points.

We can now construct summary tables and plots that are of direct interest to the end-user. As the following examples illustrate, localized as well as large-scale genomic regions of copy number change identified by the Bayesian HMM can be important tools for identifying candidate genes associated with various kinds of cancer.

7.4 Illustrations

7.4.1 Pancreatic Adenocarcinoma Data

Pancreatic adenocarcinoma is characterized by high levels of genomic instability from the earliest stages of the disease (Gisselsson et al., 2000 and 2001; van Heek et al., 2002), including early-stage mutations in the oncogene $KRAS$ and later-stage losses of the tumor supressors $p16^{INK4A}$, $p53$ and $SMAD4$ (Bardeesy and DePinho, 2002). Frequent gains and losses have been mapped to regions on chromosomes 3–13, 17, 18, 21 and 22 (Johansson et al., 1992; Solinas-Toldo et al., 1997; Mahlamaki et al., 1997 and 2002; Seymour et al., 1994, among many others).

Using the CBS algorithm of Olshen et al. (2004), Aguirre et al. (2004) individually analyzed 24 pancreatic adenocarcinoma cell lines and 13 primary tumor specimens. The CBS algorithm segments the data and computes the within-segment means. However, it does not detect copy number gains or losses. The CBS algorithm was first run by Aguirre and co-authors on the unnormalized $\log_2$ ratios to plot the histogram of the within-segment means. The tallest mode of the histogram was subtracted from the data to compute the normalized data which are available at *http://genomic.dfci.harvard.edu/ array_cgh.htm*. Then, setting thresholds in an ad-hoc manner, Aguirre et al. declared normalized $\log_2$ ratios greater than 0.13 in absolute value as copy number changes (gains or losses), greater than 0.52 as high-level amplifications, and less than −0.58 as deletions. Objective criteria were defined for comparing the copy number alterations of the 37 array CGH profiles and for identifying 54 frequently altered *minimal common regions* (MCRs) associated with the disease. Candidate genes located within the MCRs were subsequently confirmed by expression profile analysis.

The Bayesian HMM was applied to analyze these data. The algorithm was found to compare very favorably with the CBS procedure. Some examples are plotted in Figure 7.2. We discuss these in detail:

Upper left panel (Chromosome 8 of specimen 30): The vertical lines correspond to the transition points identified by the Bayesian HMM. The

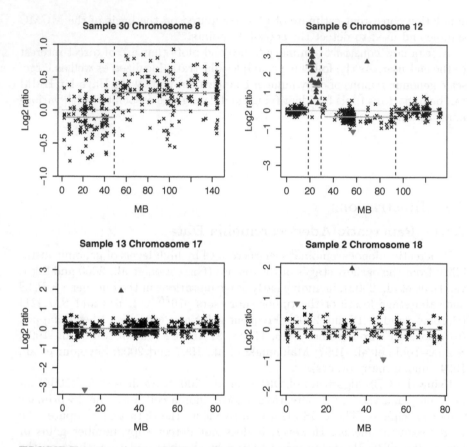

FIGURE 7.2:   Array CGH profiles of some pancreatic cancer specimens. In each panel, the clonal distance in Mb from the p telomere has been plotted on the x-axis. High-level amplifications and outliers are respectively indicated by ▲ and ▼. The broken vertical lines represent transition points. For comparison, the bold horizontal lines display the segment means computed by the CBS algorithm. See Section 7.4.1 for further discussion.

bold horizontal lines represent the within-segment means computed by the CBS algorithm. We find that both algorithms picked up the overall trend in the data. However, while the end-user decides whether or not the CBS algorithm's within-segment means correspond to copy number changes, the Bayesian HMM automatically identifies the first region as primarily copy-neutral and the second region as primarily consisting of single-copy gains.

Upper right panel (Chromosome 12 of specimen 6): The CBS procedure declared the first group of high intensity ratios as two separate segments. In contrast, the Bayesian HMM is motivated from the perspective of detecting copy number change. It declared these clones as a single region of high-level amplifications. The next set of clones with lower $\log_2$ ratios were identified as focal aberrations because they are localized changes less than 2 Mb in width. The two amplified regions detected by the Bayesian HMM correspond to the two minimal common regions (MCRs) on chromosome 12 associated with copy number gains (see Table 1 of Aguirre et al.). The first MCR contains the *KRAS2* gene. Point mutations on this gene occur in more than 75% of pancreatic cancer cases (Almoguera et al., 1988). The CBS algorithm failed to detect the second MCR. The MCR has been verified by Aguirre et al. using quantitative PCR techniques.

Bottom left panel (Chromosome 17 of specimen 13): The region from 17p13.3 to 17q11.1 (10.36 Mb to 12.8 Mb) contains the tumor supressors *p*53 and *MKK*4. Mutations on the gene *p*53 occur in at least 50% of pancreatic adenocarcinoma cases (Caldas et al., 1994). The single probe corresponding to this region was easily detected by the Bayesian HMM. In contrast, the CBS algorithm effectively declared the *entire* chromosome as copy-neutral.

Bottom right panel (Chromosome 18 of specimen 2): At around 48 Mb, the Bayesian HMM algorithm detected an outlier associated with a copy number loss. This outlier corresponds to the *SMAD4* tumor suppressor gene located at 18q21. A mutation on this gene has a known association with pancreatic cancer (Bardeesy and DePinho, 2002). Aguirre et al. mention that the CBS procedure missed the well-established association even though single-probe losses at this location were present in several specimens of the data set.

The above examples show that the CBS procedure often ignores obvious single-probe aberrations to control the False Discovery Rate. Single-probe aberrations frequently observed across tumor specimens provide one of the most cost-effective avenues for cancer research because subsequent gene validation studies are more expensive than CGH. There are many other instances of this difference between the CBS and Bayesian HMM algorithms. For example, the MCR from 68.27 to 68.85 Mb on chromosome 12 maps to highly amplified clones in 34 out of 37 tumor specimens. In every case, the Bayesian HMM declared the clones as high-level amplifications, but the CBS procedure detected only the amplification in specimen 8.

The results demonstrate the effectiveness of the Bayesian HMM in detecting global trends as well as localized changes in copy number. This feature is important in identifying genes for which point mutations do not progressively become large-scale genomic changes as the disease advances (e.g. *SMAD4* in the foregoing example). The Bayesian HMM has considerable potential as a

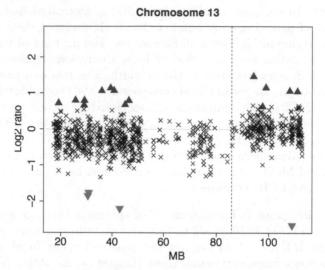

FIGURE 7.3: Array CGH profile of chromosome 13 of GBM31. The clonal distance in Mb from the *p* telomere is plotted on the x-axis. High-level amplifications and outliers are respectively indicated using ▲ and ▼. The broken vertical line represents a transition point.

diagnostic tool during the early stage of cancer when genomic alterations are localized to small parts of the genome.

7.4.2 Comparisons with Some Existing Methods

Using the Glioblastoma Multiforme data of Bredel el al. (2005), Lai et al. (2005) evaluated 11 algorithms for array CGH data. The data were normalized using the Limma package (Smyth, 2004) and are available at http://www.chip.org/~ppark/*Supplements*/Bioinformatics05b.html. Graphical summaries were presented in Lai et al. as Figures 3 and 4. Sample GBM31 (Figure 3 of Lai et al.) exhibits low signal-to-noise ratio. There is a large region of losses on chromosome 13. Lai and co-authors found that the algorithms CBS of Olshen et al. (2004), CGHseg of Picard et al. (2005), GA of Jong et al. (2003) and GLAD of Hupe et al. (2004) segmented chromosome 13 into two regions and detected the region of copy number loss. Smoothing-based methods like lowess, the quantreg algorithm of Eilers and de Menezes (2005) and the wavelet method of Hsu et al. (2005) were sensitive to local trends but less effective in detecting global trends. The HMM algorithm of Fridlyand et al. (2004) did not detect any segment.

We used an identical evaluation procedure to evaluate the Bayesian HMM. Figure 7.3 displays the result for sample GBM31. The partitioned regions are the same as those identified by the CBS, CGHseg, GA and GLAD algorithms.

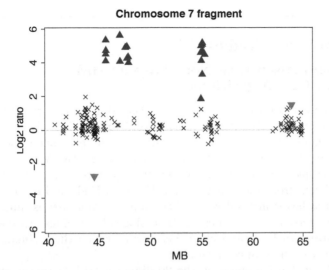

FIGURE 7.4: Partial array CGH profile of chromosome 7 of GBM29. The clonal distance in Mb from the p telomere is plotted on the x-axis. High-level amplifications and outliers are respectively indicated using ▲ and ▼.

Local changes in the copy number, identical to those collectively detected by the CGHseg and GLAD algorithms, are marked as high-level amplifications (▲) and deletions (▼).

The second example considered by Lai et al. (2005) is a fragment of chromosome 7 from sample GBM29 (refer to Figure 4 of that paper). The data reveal high $\log_2$ intensity ratios around the EGFR locus. The algorithms CGHseg, GA, GLAD, quantreg and wavelet separated the data into three distinct amplification regions. The algorithms ACE (Lingjaerde et al., 2005), CBS and CLAC (Wang et al. 2005) detected two distinct regions instead of three. ChARM (Myers et al., 2004) grouped all the high $\log_2$ intensity ratios into a single region. The HMM algorithm of Fridlyand et al. (2004) did not detect the amplifications.

Figure 7.4 displays the results for the Bayesian HMM algorithm. The high $\log_2$ ratios are identified as high-level amplifications (▲). Unlike the algorithms investigated in Lai et al. (2005), the single clone having a highly negative value is detected by our algorithm and marked as a deletion. The amplifications are identified as focal aberrations rather than separate regions because both clusters are less than 5 Mb in width.

We find that the Bayesian HMM algorithm combines the strength of the smoothing-based algorithms in detecting local features with the strength of the segmentation-based methods in detecting global trends. The success of the technique is especially impressive with noisy data.

7.5   Simulation Studies

7.5.1   Comparison with Non-Bayesian HMM
and CBS Algorithms

The frequentist equivalent of the above Bayesian procedure estimates the hyperparameters using the Baum-Welch EM algorithm, iteratively incrementing the likelihood until relative changes in the hyperparameters become sufficiently small. Conditional on the hyperparameters, the Viterbi algorithm computes the *aposteriori* most likely sequence of states $s_1, \ldots, s_n$. This technique is different from the non-Bayesian HMM of Fridlyand et al. (2004). In particular, the latter method regards the number of states as unknown and the states as exchangeable. There is no biological meanings associated with the latent states in Fridlyand et al. because of which the technique cannot directly detect changes in copy number.

To find the global maximum in the 20-dimensional hyperparameter space, the EM algorithm has to be run from several starting points. For typical array CGH data, each run often requires hundreds of iterations to converge. Because of this, the computational costs associated with the frequentist and Bayesian analyses are comparable. When R is used as the computing platform, the CBS algorithm is considerably faster than either method. However, all three approaches are computationally feasible and have negligible costs compared to the experimental resources necessary to process the tumor specimens and produce the data.

A comparison of the Bayesian and non-Bayesian profiles reveals that the two procedures often gave similar results. However, many profiles are noticeably different, including the chromosome–specimen pairs $(5, 2)$, $(5, 7)$, $(12, 10)$, $(7, 13)$, $(15, 13)$, $(5, 19)$, $(18, 31)$ and $(19, 34)$. Two profiles are displayed in Figure 7.5. In all these examples, the non-Bayesian hyperparameter estimates correspond to a larger likelihood than the Bayes estimates. However, the Bayesian profiles look more reasonable when we compare the smallest $\log_2$ ratios labeled as amplifications by the two methods.

We performed a simulation study of the differences between the methods. For each of the afore-mentioned chromosome–specimen pairs, we obtained signal-to-noise ratios that were typical of array CGH data by setting the hyperparameters equal to their estimated values. Using the model described in Section 7.2, we then generated the underlying copy number states and log-ratios for $n = 200$ clones. The Bayesian and non-Bayesian HMMs were applied to infer the latent copy number states. The procedure was independently replicated 100 times. Table 7.1 displays the percentage of correctly labeled copy number states for the two methods. The Bayesian HMM outperforms the non-Bayesian HMM in all the cases.

Using eight *randomly* selected chromosome–specimen pairs, but an otherwise identical simulation strategy, Table 7.2 compares the CBS algorithm

TABLE 7.1: Estimated Percentages of Correctly Discovered Copy Number States for the Bayesian and Non-Bayesian Methods, along with the Estimated Standard Errors

Source		Bayesian HMM		Non-Bayesian HMM	
Chromosome	Specimen	% accuracy	SE	% accuracy	SE
5	2	94.81	0.789	86.89	1.685
5	7	91.99	1.188	81.44	1.942
12	10	95.22	0.390	89.08	1.378
7	13	92.41	1.019	80.09	2.333
15	13	92.42	1.322	82.55	1.649
5	19	88.02	2.189	73.09	2.873
18	31	84.95	2.512	71.17	2.448
19	34	88.13	2.000	72.10	2.124

Note: The estimates were based on 100 independently generated data sets. The first two columns specify the chromosome and specimen numbers of the Section 7.4.1 data set whose the estimated hyperparameters were used to generate the data. See the text for an explanation.

TABLE 7.2: Estimated Percentages of Correctly Discovered Copy Number States for the Bayesian and Non-Bayesian Methods

Source		Bayesian HMM	Non-Bayesian HMM	CBS
Chromosome	Specimen	% accuracy	% accuracy	% accuracy
13	33	94.38 (1.203)	72.01 (2.634)	67.72 (3.512)
19	4	88.20 (1.129)	87.94 (0.534)	75.36 (1.726)
14	1	87.35 (1.893)	76.47 (1.834)	86.70 (0.426)
12	17	80.84 (1.736)	76.11 (1.453)	44.12 (1.791)
1	24	40.64 (2.512)	54.31 (1.460)	35.37 (2.470)
3	35	96.03 (0.239)	72.06 (2.509)	92.43 (0.488)
23	12	74.31 (3.417)	65.2 (2.420)	58.08 (3.311)
15	34	90.79 (2.164)	68.3 (2.798)	55.22 (4.175)

Note: The estimated standard errors are in parentheses. The estimates were based on 100 independently generated data sets. The first two columns specify the chromosome and specimen numbers of the Section 7.4.1 data set whose the estimated hyperparameters were used to generate the data. See the text for an explanation.

with the Bayesian and non-Bayesian HMMs. The method used by Aguirre et al. (2004) was applied to declare copy number gains and losses for the CBS algorithm. The Bayesian HMM outperforms the CBS algorithm, often substantially, in seven cases. The difference is inconclusive in one case. In six out of eight cases, the Bayesian HMM outperforms the non-Bayesian HMM, with the difference being inconclusive in one case. These results provide significant evidence in favor of the Bayesian HMM. We would like to emphasize here that the simulated data were generated from the Section 7.2 model, a strategy that is likely to favor the HMM-based methods. There may be alternative simulation procedures where the reliability of the CBS algorithm is greater than that of the proposed Bayesian HMM.

The Bayesian HMM benefits from the informative priors of Section 7.2. Prior knowledge about array CGH helps the procedure distinguish between competing sets of hyperparameter values that are equally plausible under the likelihood but not under the posterior. For example, consider the frequently encountered situation where very few $\log_2$ ratios are assigned to one or more copy number state. In such a situation, the likelihood alone may be unable to distinguish between the matching non-Bayesian HMM and a model having fewer than four states. This results in likelihood-based estimates where one or more of the μ_j's are approximately equal. Because of the well-defined meanings assigned to the four states of the HMM, the copy number states detected by the non-Bayesian model often seem incorrect in such cases. The Bayesian approach is more robust in such situations because the informative priors prevent even states having very few probes and $\log_2$ ratios having considerable amount of overlap from being classified as a common state. While it may be true that for some data a model with fewer than four states is better-fitting from a statistical viewpoint, the states may not have a simple biological interpretation. The detection of copy number gains and losses, which is one of the main goals of the analysis, is also less straightforward.

Several examples in Section 7.4.1 suggest that our Bayesian HMM is better than the CBS algorithm in detecting amplifications localized to a few probes. This is a useful feature of our approach because single-probe amplifications common to different specimens are often the focus of future, more expensive gene validation studies. To investigate the difference between the two algorithms by a controlled simulation, we independently generated 25 data sets using the following strategy: *(i)* Fifty out of $n = 200$ clones were randomly chosen and assigned as amplifications with a mean signal of 2 on the $\log_2$ scale. *(ii)* The remaining clones were assumed to be copy-neutral with a mean signal of zero. *(iii)* The data were generated by adding Gaussian noise with a standard deviation of 0.1 to these mean values.

Figure 7.6 displays one of the 25 simulated data sets. The simulated data are quite atypical because the signal-to-noise ratio of 20 and percentage of amplified probes (25%) are uncharacteristically high. In spite of these features that simplified the discovery of copy number change, the CBS algorithm failed to detect amplifications in *any* of the 25 data sets. The Bayesian HMM on

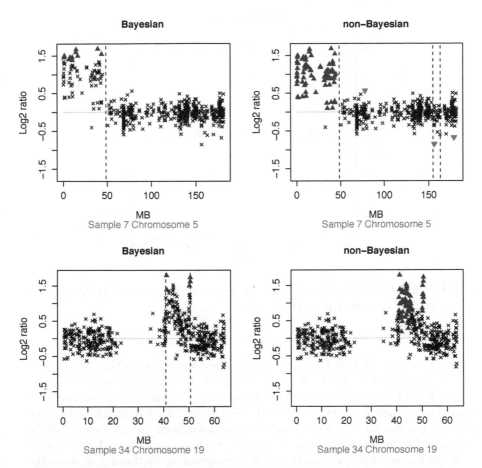

FIGURE 7.5: Examples from Section 7.4.1 where the Bayesian and non-Bayesian array CGH profiles are different. The upper panels correspond to chromosome 5 of sample 7 and the lower panels correspond to chromosome 19 of sample 34. The clonal distance in Mb from the p telomere has been plotted on the x-axis. High-level amplifications and outliers are indicated using ▲ and ▼ respectively. The broken vertical lines represent transition points.

the other hand, correctly identified all the amplifications. The false discovery rate of our Bayesian HMM was zero for all the data sets and the average true discovery rate exceeded 99%.

7.5.2 Prior Sensitivity

The preceding analyses assumed $\epsilon = 0.1$ for the prior supports of the μ_j's (refer to Section 7.2) and $\theta_{ij} = 1$ for the priors of the transition matrix rows,

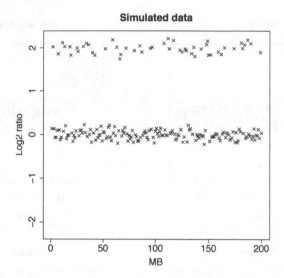

FIGURE 7.6: Simulated data set. The clonal distance in Mb from the p telomere has been plotted on the x-axis.

where $i = 1, \ldots, 4$ and $j = 1, \ldots, 4$. To alleviate concerns that the results are sensitive to the choice of ϵ, we generated 100 data sets with $n = 500$ clones each. For each data set, the true means $\mu_1, \ldots, \mu_4$ were uniformly generated from narrow intervals centered respectively at -0.5, 0, 0.5 and 1. The standard deviations σ_j were uniformly generated from the interval $[0.2, 0.25]$ which is typical of noisy data. The true transition matrices $\boldsymbol{A}$ were simulated as follows. For the matrix row 2 corresponding to transitions from the copy-neutral state, the off-diagonal elements were uniformly generated from $[0.01, 0.02]$; for the remaining rows, the off-diagonal elements were uniformly generated from $[0.02, 0.05]$. These nine elements uniquely determined the row-stochastic matrix. For $k = 1, \ldots, 500$, the copy number states s_k were then generated and the artificial data computed by adding Gaussian noise to the means μ_{s_k}.

For ϵ belonging to a grid of points in the interval $[0.05, 0.15]$, the Bayesian HMM was used to analyze each simulated data set. The posterior expectations of the means μ_j, the true discovery rates and false discovery rates were all robust to the choice of ϵ. Figure 7.7 plots the estimates of $\mu_1, \ldots, \mu_4$ for three randomly chosen data sets versus values of ϵ. The flat lines are indicative of the lack of sensitivity to ϵ. Using a similar technique, the results were found to be insensitive to choices of $\{\theta_{ij}\}_{i,j}$ small in comparison to n.

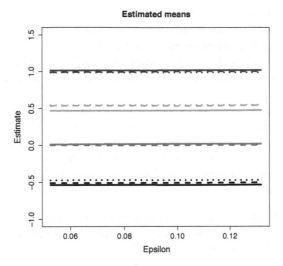

FIGURE 7.7: Estimated means $\hat{E}[\mu_j|\boldsymbol{Y}]$ for three independently generated data sets (shown by solid, dashed, and dotted lines) plotted against ϵ.

7.6 Conclusion

We propose a Bayesian hierarchical approach for analyzing array CGH data based on a hidden Markov model. The informative priors allow Bayesian learning from the data. One of the strengths of the fully automated approach is the ability to detect copy number changes like gains, losses, amplifications, outliers and transition points based on the posterior distribution. Summaries of the array CGH profiles are then generated. The profiles can be compared across individuals to identify the genomic alterations involved in the disease pathogenesis.

Several examples presented in Section 7.4 demonstrate the reliability of our Bayesian HMM. The sensitivity of the algorithm to individual probes allows it to locate candidate genes that may be missed by other algorithms. The performance of the algorithm is impressive for the Glioblastoma data set of Bredel el al. (2005) having high measurement error. Combined with the results presented in Lai et al. (2005), the latter analysis reveals a very favorable comparison with such outstanding algorithms as those of Picard et al. (2005) and Olshen et al. (2004). Section 7.5 compares our Bayesian HMM with alternative algorithms using simulations. The results confirm the accuracy of the approach.

The simulation study in Section 7.5.2 and our own experience with the algorithm suggest that the output is robust to variations in the constant ϵ used in the prior specification of the means μ_j and in the constants θ_{ij} used in the transition matrix priors of Section 7.2. Although the informative priors for the means μ_j substantially influence the results, the order constraints on the μ_j's and the biological meanings associated with the copy number states allow us to specify priors that work consistently well across different data sets. In all our analyses, we have used the default parameterizations specified in Section 2.2.

An advantage of our method is that it has virtually no tuning parameters. Unlike other algorithms (see Lai et al., 2005), the user is only required to input the normalized $\log_2$ ratios to produce reliable array CGH profiles. This is a convenient feature for an end-user with little or no statistical training. The technique has been implemented by Bioinformatics Toolbox and is available at *http://www.mathworks.com*.

Acknowledgment. Reprinted with permission from *The Journal of the American Statistical Association*. Copyright 2008 by the American Statistical Association. All rights reserved.

7.7 Appendix: An MCMC Algorithm

The following algorithm is independently run for each chromosome to generate an MCMC sample for the chromosomal parameters. We group the model parameters into four blocks, namely, $B_1 = A$, $B_2 = (s_1, \ldots, s_n)$, $B_3 = (\mu_1, \mu_2, \mu_3, \mu_4)$, and $B_4 = (\sigma_1^2, \sigma_2^2, \sigma_3^2, \sigma_4^2)$. The starting values of the parameters are generated from the priors. The algorithm iteratively generates each of the four blocks conditional on the remaining blocks and the data. Let $B_1^{(v-1)}, \ldots, B_4^{(v-1)}$ denote the values of the blocks at the $(v-1)^{st}$ iteration. In the next iteration, the blocks are generated as follows:

Updating block B_1. The transition matrix is generated using a Metropolis-Hastings step because the normalizing constant of the full conditional cannot be computed in closed form. This step makes independent proposals from a distribution that closely approximates the full conditional of the transition matrix. The proposal is accepted or rejected with a probability that compensates for the approximation. Typically, most of the Metropolis-Hastings proposals are accepted. Using the copy number states generated at iteration $v-1$, we compute the number of transitions from state i to state j, denoted by $u_{ij}^{(v)} = \sum_{k=1}^{n-1} I\left(s_k^{(v-1)} = i, s_{k+1}^{(v-1)} = j\right)$, where $i, j = 1, \ldots, 4$. We generate a proposal C for the transition matrix from the distributions $[c_i \mid Y, B_{-1}] \sim D_3\left(1 + u_{i1}^{(v)}, 1 + u_{i2}^{(v)}, 1 + u_{i3}^{(v)}, 1 + u_{i4}^{(v)}\right)$, where row $i = 1, \ldots, 4$, and B_{-1} denotes the blocks, $\{B_2, B_3, B_4\}$. The proposal ignores the marginal distribution of state s_1 and so it differs from the full conditional of the transition matrix. To compensate for this, we accept the proposal (in other words, set $A^{(v)}=C$) with probability β, where $\beta = \min\left\{1, \pi_C(s_1^{(v-1)})/\pi_{A^{(v-1)}}(s_1^{(v-1)})\right\}$, and otherwise reject the proposal (in other words, set $A^{(v)}=A^{(v-1)}$). As defined earlier, $\pi_D(s)$ denotes the probability of state s under the stationary distribution of a given transition matrix D.

Updating block B_2. The copy number states are generated by a stochastic version of the forward-backward algorithm. We compute the distribution $[s_n \mid B_{-2}, Y_1, \ldots, Y_n]$ at the beginning of the backward step. We generate s_n from this distribution. The backward step is continued to compute and generate a draw from the distribution $[s_{n-1} \mid s_n, B_{-2}, Y_1, \ldots, Y_n]$. The sequence of computing and generating a draw from $[s_k \mid s_{k+1}, B_{-2}, Y_1, \ldots, Y_n]$ is iterated for $k = n-2$ down to $k = 1$. This produces a sample from the joint distribution $[s_1, \ldots, s_n \mid B_{-2}, Y_1, \ldots, Y_n]$.

Updating block B_3. For $s = 1, \ldots, 4$, let δ_{0s} be the center of the untruncated normal distribution in the prior specification of μ_s. Compute the sums $n_s = \sum_{k=1}^n I\left(s_k^{(v)} = s\right)$, averages $\bar{Y}_s = \frac{1}{n_s}\sum_{k=1}^n Y_k \cdot I(s_k^{(v)} = s)$, precisions $\theta_s^2 = \tau_s^{-2} + \left(\sigma_s^{(v-1)}/\sqrt{n_s}\right)^{-2}$ and weighted means

$$\gamma_s = \frac{1}{\theta_s^2}\left[\delta_{0s} \cdot \tau_s^{-2} + \bar{Y}_s \cdot \left(\frac{\sigma_s^{(v-1)}}{\sqrt{n_s}}\right)^{-2}\right].$$

For $s = 1, \ldots, 4$, generate $\left[\mu_s^{(v)} \mid Y, B_{-3}\right] \sim N\left(\gamma_s, \theta_s^2\right) \cdot I_s$, where the intervals I_s denotes the support of the μ_s (see prior specification).

Updating block B_4. For $j = 1, \ldots, 4$, compute $n_j = \sum_{k=1}^n I\left(s_k^{(v)} = j\right)$ and $V_j = \sum_{k=1}^n \left(Y_k - \mu_{s_k}^{(v)}\right)^2 \cdot I\left(s_k^{(v)} = j\right)$. Generate

$$\left[\sigma_j^{(v)} \mid Y, B_{-4}\right] \sim \left[\text{gamma}\left(1 + \frac{n_j}{2}, \epsilon + \frac{V_j}{2}\right)\right]^{-0.5}.$$

References

Aguirre, A. J., Brennan, C., Bailey, G., Sinha, R., Feng, B., Leo, C., Zhang, Y., Zhang, J., Gans, J. D., Bardeesy, N., Cauwels, C., Cordon-Cardo, C., Redston, M. S., DePinho, R. A. and Chin, L. (2004). High-resolution characterization of the pancreatic adenocarcinoma genome. *Proceedings of the National Academy of Sciences USA* **101**, 9067–9072.

Almoguera, C., Shibata, D., Forrester, K., Martin, J., Arnheim, N., Perucho, M. (1988). Most human carcinomas of the exocrine pancreas contain mutant c-K-ras genes. *Cell* **53**, 549–554.

Bardeesy, N. and DePinho, R. A. (2002). Pancreatic cancer biology and genetics. *Nature Reviews Cancer* **2**, 897–909.

Bredel, M., Bredel, C., Juric, D., Harsh, G. R., Vogel, H., Recht, L. D., and Sikic, B. I. (2005). High-resoluton genome-wide mapping of genetic alternations in human glial brain tumors. *Cancer Research* **65**, 4088–4096.

Brennan, C., Brennan, C., Zhang, Y., Leo, C., Feng, B., Cauwels, C., Aguirre, A. J., Kim, M., Protopopov, A., and Chin, L. (2004). High-resolution global profiling of genomic alterations with long oligonucleotide microarray. *Cancer Research* **64**, 4744–4748.

Brown, C. S., Goodwin, P. C. and Sorger P. K. (2001). Image metrics in the statistical analysis of DNA microarray data. *Proceedings of the National Academy of Sciences USA* **98**, 8944–8949.

Caldas, C., Hahn, S. A., da Costa, L. T., Redston, M. S., Schutte, M., Seymour, A. B., Weinstein, C. L., Hruban, R. H., Yeo, C. J., Kern, S. E. (1994). Frequent somatic mutations and homozygous deletions of the p16 (MTS1) gene in pancreatic adenocarcinoma. *Nature Genetics* **8**, 27–32.

Cheng, C., Kimmel, R., Neiman, P. and Zhao, L. P. (2003). Array rank order regression analysis for the detection of gene copy-number changes in human cancer. *Genomics* **82**, 122–129.

Chib, S. (1996). Calculating posterior distributions and modal estimates in Markov mixture models. *Journal of Econometrics* **75**, 79–97.

Durbin, R., Eddy, S., Krogh, A., and Michison, G. (1998). *Biological Sequence Analysis*. Cambridge University Press.

Eilers, P. H. C. and de Menezes, R. X. (2005). Quantile smoothing of array CGH data. *Bioinformatics* **21**, 1146–1153.

Freeman, J. L., Perry, G. H., Feuk, L., Redon, R., McCarroll, S. A., Altshuler, D. M., Hiroyuki, A., Jones, K. W., Tyler-Smith, C., Hurles, M. E., Carter, N. P., Scherer, S. W. and Lee, C. (2006). Copy number variation: New insights in genome diversity. *Genome Research* **16**, 949–961.

Fridlyand, J., Snijders, A. M., Pinkel, D., Albertson, D. G., Jain, A. N. (2004). Application of Hidden Markov Models to the analysis of the array CGH data. *Journal of Multivariate Analysis* **90**, 132–153.

Gisselsson, D., Pettersson, L., Hoglund, M., Heidenblad, M., Gorunova, L., Wiegant, J., Mertens, F., Dal Cin, P., Mitelman, F. and Mandahl, N. (2000). Chromosomal breakage-fusion-bridge events cause genetic intratumor heterogeneity. *Proceedings of the National Academy of Sciences, USA* **97**, 5357–5362.

Gisselsson, D., Jonson, T., Petersen, A., Strombeck, B., Dal Cin, P., Hoglund, M.,Mitelman, F., Mertens, F. and Mandahl, N. (2001). Telomere dysfunction triggers extensive DNA fragmentation and evolution of complex chromosome abnormalities in human malignant tumors. *Proceedings of the National Academy of Sciences, USA* **98**, 12683–12688.

Guha, S., Li, Y. and Neuberg, D. (2008). Bayesian Hidden Markov Modeling of array CGH data. *Journal of the American Statistical Association* **108**, 485–497.

van Heek, N. T., Meeker, A. K., Kern, S. E., Yeo, C. J., Lillemoe, K. D., Cameron, J. L., Offerhaus, G. J., Hicks, J. L., Wilentz, R. E., Goggins, M. G., et al. (2002). Telomere shortening is nearly universal in pancreatic intraepithelial neoplasia. *American Journal of Pathology* **161**, 1541–1547.

Hodgson, G., Hager, J., Volik, S., Hariono, S., Wernick, M., Moore, D., Nowak, N., Albertson, D., Pinkel, D., Collins, C. et al. (2001). Genome scanning with array CGH delineates regional alterations in mouse islet carcinomas. *Nature Genetics* **29**, 459–464.

Huang, T., Wu, B., Lizardi, P., and Zhao, H. (2005). Detection of DNA copy number alterations using penalized least squares regression. *Bioinformatics* **21**, 3811–3817.

Hupe, P., Stransky, N., Thiery, J.-P., Radvanyi, F. and Barillot, E. (2004). Analysis of array CGH data: from signal ratio to gain and loss of DNA regions. *Bioinformatics* **20**, 3413–3422.

Hsu, L., Self, S. G., Grove, D., Randolph, T., Wang, K., Delrow, J. J., Loo, L. and Porter, P. (2005). Denoising array-based comparative genomic hybridization data using wavelets. *Biostatistics* **6**, 211–226.

Johansson, B., Bardi, G., Heim, S., Mandahl, N., Mertens, F., Bak-Jensen, E., Andren- Sandberg, A. and Mitelman, F. (1992). Nonrandom chromosomal rearrangements in pancreatic carcinomas. *Cancer* **69**, 1674–1681.

Jong, K., Marchiori, E., Vaart, A., Ylstra, B., Weiss, M. and Meijer, G. (2003). Chromosomal breakpoint detection in human cancer. In *Applications of evolutionary computing: Evolutionary computation and bioinformatics*, Vol. 2611, Springer, pp. 54–65.

Kallioniemi, A., Kallioniemi, O. P., Sudar, D., Rutovitz, D., Gray, J. W., Waldman, F. and Pinkel, D. (1992). Comparative genomic hybridization for molecular cytogenetic analysis of solid tumors. *Science* **258**, 818–821.

Karlin, S. and Taylor, H. M. (1975). *A First Course in Stochastic Processes*, 2nd edition. Academic Press, New York.

Khojasteh, M., Lam, W. L., Ward, R. K., MacAulay, C. (2005). A stepwise framework for the normalization of array CGH data. *BMC Bioinformatics* **6**, 274.

Lai, W., Johnson, M. J., Kucherlapati, R., Park, P. J. (2005). Comparative analysis of algorithms for identifying amplifications and deletions in array CGH data. *Bioinformatics* **21**, 3763–3770.

Lengauer, C., Kinzler, K. and Vogelstein, B. (1998). Genetic instabilities in human cancers. *Nature* **396** 643–649.

Lingjaerde, O. C., Baumbusch, L. O., Liestol, K., Glad, I. K. and Borresen-Dale, A.-L. (2005). CGH-Explorer: a program for analysis of array-CGH data. *Bioinformatics* **21**, 821–822.

Myers, C. L., Dunham, M. J., Kung, S. Y. and Troyanskaya, O. G. (2004). Accurate detection of aneuploidies in array CGH and gene expression microarray data. *Bioinformatics* **20**, 3533–3543.

MacDonald, I.L. and Zucchini, W. (1997). *Hidden Markov and Other Models for Discrete-value Time Series*. Boca Raton: Chapman & Hall, Inc.

McLachlan, G. J., Do, K.-A. and Ambroise C. (2004). *Analyzing Microarray Gene Expression Data*. Hoboken, New Jersey: John Wiley & Sons, Inc.

Mahlamaki, E. H., Barlund, M., Tanner, M., Gorunova, L., Hoglund, M., Karhu, R. and Kallioniemi, A. (2002). Frequent amplification of 8q24, 11q, 17q, and 20q-specific genes in pancreatic cancer. *Genes Chromosomes Cancer* **35**, 353–358.

Mahlamaki, E. H., Hoglund, M., Gorunova, L., Karhu, R., Dawiskiba, S., Andren-Sandberg, A., Kallioniemi, O. P. and Johansson, B. (1997). Comparative genomic hybridization reveals frequent gains of 20q, 8q, 11q, 12p, and 17q, and losses of 18q, 9p, and 15q in pancreatic cancer. *Genes Chromosomes Cancer* **20**, 383–391.

Murphy, K. M., Brune, K. A., Griffin, C., Sollenberger, J. E., Petersen, G. M., Bansal, R., Hruban, R. H. and Kern, S. E. (2002). Evaluation of candidate genes MAP2K4, MADH4, ACVR1B, and BRCA2 in familial pancreatic cancer. *Cancer Research* **62**, 3789–3793.

Olshen, A. B., Venkatraman, E. S., Lucito, R., Wigler, M. (2004). Circular binary segmentation for the analysis of array-based DNA copy number data. *Biostatistics* **4**, 557–572.

Pasternak, J. J. (1999). *An Introduction to Human Molecular Genetics: Mechanism of Inherited Diseases*. Fitzgerald Science Press, Bethesda, MD.

Picard, F., Robin, S., Lavielle, M., Vaisse, C. and Daudin, J.-J. (2005). A statistical approach for array CGH data analysis. *BMC Bioinformatics* **6**, 27.

Pinkel, D. and Albertson, D. G. (2005). Array comparative genomic hybridization and its applications in cancer. *Nature Genetics* **37**, Suppl. 11–17.

Pinkel, D., Segraves, R., Sudar, D., Clark, S., Poole, I., Kowbel, D., Collins, C., Kuo, W., Chen, C., Zhai, Y. et al., (1998). High resolution analysis of DNA copy number variation using comparative genomic hybridization to microarrays. *Nature Genetics* **20**, 207–211.

Pollack, J.R., Perou, C. M., Alizadeh, A. A., Eisen, M. B., Pergamenschikov, A., Williams, C. F., Jeffrey, S. S., Botstein, D., Brown, P. O. (1999) Genome-wide analysis of DNA copy-number changes using cDNA microarrays. *Nature Genetics* **23**, 41–46.

Pollack, J., Sorlie, T., Perou, C., Rees, C., Jeffrey, S., Lonning, P., Tibshirani, R., Botstein, D., Borresen-Dale, A. and Brown, P. (2002). Microarray analysis reveals a major direct role of DNA copy number alteration in the transcriptional program of human breast tumors. *Proceedings of the National Academy of Sciences, USA* **99**, 12963–12968.

Rabiner, L.R. (1989). A tutorial on Hidden Markov Models and selected applications in speech recognition. *Proceedings of the IEEE* **77**, 257–286.

Robert, C. P., Ryden, T., and Titterington, D. M. (1999). Convergence controls for MCMC algorithms, with applications to hidden Markov chains. *Journal of Statistical Computing and Simulation* **64**, 327–355.

Scott, S. (2002). Bayesian methods for hidden Markov models: Recursive computing in the 21st century. *Journal of the American Statistical Association* **97**, 337–351.

Seymour, A. B., Hruban, R. H., Redston, M., Caldas, C., Powell, S. M., Kinzler, K. W., Yeo, C. J. and Kern, S. E. (1994). Allelotype of pancreatic adenocarcinoma. *Cancer Research* **54**, 2761–2764.

Smyth, G. K. (2004). Linear models and empirical Bayes methods for assessing differential expression in microarray experiments. *Statistical Applications in Genetics and Molecular Biology* **3**, article 3.

Snijders, A. M., Nowak, N., Segraves, R., Blackwood, S., Brown, N., Conroy, J., Hamilton, G., Hindle, A. K., Huey, B., Kimura, K., Law, S., Myambo, K., Palmer, J., Ylstra, B., Yue, J. P., Gray, J. W., Jain, A. N., Pinkel, D., Albertson, D. G. (2001). Assembly of microarrays for genome-wide measurement of DNA copy number. *Nature Genetics* **29** 4370–4379.

Solinas-Toldo, S. et al. (1997). Matrix-based comparative genomic hybridization: biochips to screen for genomic imbalances. *Genes Chromosomes Cancer* **20**, 399–407.

Wang, P., Kim, Y., Pollack, J., Balasubramanian, N. and Tibshirani, R. (2005). A method for calling gains and losses in array CGH data. *Biostatistics* **6**, 45–58.

Symbol Description

T	A phylogenetic tree.	$\mathbf{X}$	Data matrix representation of a sample of data.
V	Vertices or nodes of a phylogenetic tree.	Y	Augmented data matrix labeling of internal nodes and leaves with character states.
E	Edges or branches of a phylogenetic tree.		
ρ	Root of a phylogenetic tree.	$\mathcal{Y}$	Set of all possible labelings of internal nodes and leaves character states.
ν_i	Expected number of sequence changes on a branch of a phylogenetic tree.		
t_i	Duration of branch i of a phylogenetic tree, in terms of time.	ψ	Character evolution model parameters.
		$\mathbf{Q}(\psi)$	Instantaneous rate of change matrix with parameters ψ.
r_i	Rate of sequence change on branch i of a phylogenetic tree.	$\mathbf{P}$	Matrix of transition probabilities for character states across an edge in the tree.
c_i	Number of character state changes on a branch i.	$a(b)$	Ancestral node of branch b.
$\mathbf{h}$	Set of ages or divergence times of nodes.	$d(b)$	Descendent node of branch b.
N	Number of leaves in the tree (leaves can represent species or individuals).	ω	Rate multiplier associated with nonsynonymous substitutions.
M	Number of characters (or sites) in a sample of data.	Θ	Vector of population sizes.
		$\mathbf{G}$	Set of genealogies.

Chapter 8

Bayesian Approaches to Phylogenetic Analysis

Mark T. Holder, Jeet Sukumaran, and Rafe M. Brown
Department of Ecology and Evolutionary Biology, University of Kansas

8.1 Background

A phylogeny is the set of genealogical relationships between species. There are many definitions of what a species is and how we should recognize species (Mayden, 1999). Nevertheless, there is general agreement among the differing species concepts that a species will be made up of a lineage of populations that are connected by ancestor-descendant relationships between the individuals. Significantly, different species do not represent independent samples from some pool of possible species. Rather, species typically arise by a process of diversification in which one ancestral species splits to form two or more descendant species. Hybridization between existing species is another speciation mechanism (see Mallet, 2007, for a recent review). In either case, different species can be classified based on how closely they are related – how far back in time one must go to find their common ancestor.

Virtually every analysis of biological data must confront the fact that the individuals that we see today are the result of historical process of evolution (e.g. see the analytical techniques reviewed by Harvey & Pagel, 1991). For example, when comparing mouse, chimp, and human we must take into account the fact human and chimp share a much more recent common ancestor with each other than they do with rodents. Thus, the estimation of phylogenetic relationships has become a crucial first step in the analysis of all types of comparative data.

8.1.1 Calculating the Likelihood for a Phylogenetic Model

In principle, any heritable property of an organism could be used as data for phylogenetic inference. In the past two decades, however, molecular sequence data have become the most widely-used source of data for phylogenetic analyses. We will focus on the analysis of such data here (though much of what is said is applicable to the analysis of any character data). As a result of mutation, differences accumulate between alleles within a population, and, over time, some mutations will be fixed within a population. Because the barrier between different species prevents a new mutation from spreading from one species into another, genetic differences accumulate between different species. The sharing of novel sequence variants between species allows us to infer the history of the species from genetic data.

On the basis of the simple comparison of DNA sequences (and often on the basis of how the data are collected), we can usually be confident that we have identified a comparable gene[1] among different species – thus we assume that the copies of the gene from different species have all descended from a single gene in a common ancestor of the current species. A more fine-grained statement of homology is provided by the process of multiple sequence alignment (MSA). MSA procedures insert gaps in sequences such that the total length of the sequences is identical for all species in the matrix. Typically MSA is performed prior to phylogenetic inference, and for the purpose of tree inference the alignment is considered to be a fixed statements of homology of particular positions within a DNA sequence. We assume not only that the genes in the analysis share a common ancestor, but that the residues in a particular column of our data matrix are derived from the same nucleotide within this ancestral gene. As will be discussed later, joint estimation of the alignment and phylogeny should be a preferable to sequential alignment then tree inference (Sankoff, 1975).

After alignment, our data can be expressed in a $N \times M$ table of sequences, **X**, where N is the number of individuals in our sample, and M is the sum of the number of nucleotides and the number of gaps for any sequence. This table is frequently referred to as the data matrix. The columns of the matrix are referred to as "characters" (in the general context) or "sites" (in the case of sequence data). The phylogeny itself can be represented as a tree, $T = (V, E)$, where V represents the set of nodes (also called vertices) in the graph, and E is the set of branches (or "edges" in the mathematical literature). The leaves of the graph correspond to the observed sequences. The branches of the tree can be thought of as representing genealogical lineages; branch lengths are often chosen to reflect opportunity for character change across that branch. It is common to use a parameterization in which ν_i is the expected number of

---

[1] We use typically use the term "gene" in a very loose way, to indicate a contiguous stretch of DNA sequence. Phylogenetic analyses do not require that the input data come from sequences that code for a protein.

changes along branch i under a particular model of character evolution. The expected number of changes along branch i is given by $\nu_i = r_i t_i$, where r_i is the rate of change and t_i is the duration of the branch (in terms of time). In the most general case we are unwilling make any constraints on the rate of evolution, and thus r_i and t_i are not identifiable.

To employ a Bayesian perspective to learning the evolutionary tree for a set of sequences, we must specify a model which describes the probabilities of different pattern of sequence substitution. For all but the simplest models (e.g. the model of Jukes & Cantor, 1969), this calculation will be a function that includes nuisance parameters, ψ, that characterize the process of molecular evolution. If we assume that the the process of evolution acts independently on each site in our gene, then the probability of any particular set of sequences can be calculated as a product over every column in the data matrix (denoted $\mathbf{x}$):

$$\Pr(\mathbf{X}|T, \nu, \psi) = \prod_{j=1}^{M} \Pr(\mathbf{x_j}|T, \nu, \psi) \tag{8.1}$$

The specification can be dramatically simplified by assuming that events on one branch of the tree are independent of the process of character evolution on other branches. In many cases the lineages of a gene tree will be evolving in separate species for most of the duration of the phylogenetic history, thus the assumption of independence across branches seems reasonable.

If we knew the ancestral sequence at every internal node in the tree then we would have a full labeling of the nodes the tree with sequences. We can think of working with such a node-labelled tree as having an augmented data matrix, Y, that has size $|V| \times M$. In this matrix, $y_{i,j} = x_{i,j}$ for all j whenever i represents a leaf of the tree. When i indexes an internal node then $y_{i,j}$ can take the value of any character state; for DNA data $y_{i,j} \in \{A, C, G, T\}$. We can use $a(b)$ to denote the row index in our augmented matrix that corresponds to the node that is the ancestor of branch b, and we can use $d(b)$ for the index of the descendant node. Thus, in the context of a particular site, j, and a specific labeling, Y, the sequence at site j changes from $y_{a(b),j}$ to $y_{d(b),j}$ across the branch b. This allows us to express the likelihood of a labeling of the tree for a site as a product over all branches, and the likelihood for an entire of a labeling is a product over all sites:

$$\Pr(Y|T, \nu, \psi) = \prod_{j=1}^{M} \left[\Pr(y_{\rho,j}|\psi) \prod_{b \in E} \Pr(y_{a(b),j} \to y_{d(b),j}|\nu_b, \psi) \right] \tag{8.2}$$

Here $\Pr(y_{\rho,j}|\psi)$ represents the prior probability of a particular nucleotide at the root of the tree, a node that will be denoted ρ.

In real data analyses, we do *not* know the ancestral sequences, so we must sum over all possible internal node labelings. The set of all possible labellings, $\mathcal{Y}(\mathbf{X})$, is large ($4^{|V|-N}$ for DNA data), but a dynamic programming routine known as the pruning algorithm of Felsenstein (1981) allows us to do the summation in an amount of computation that is linear with respect to the number of internal nodes. Thus, we can efficiently calculate the likelihood of the observed data $\mathbf{X}$ by summing over all labelings that are consistent with $\mathbf{X}$ (all labelings that agree with $\mathbf{X}$ on the sequences at the leaves of the tree):

$$\Pr(\mathbf{X}|T, \nu, \psi) = \prod_{j=1}^{M} \sum_{Y \in \mathcal{Y}(\mathbf{X})} \left[\Pr(y_{\rho,j}|\psi) \prod_{b \in E} \Pr(y_{a(b),j} \to y_{d(b),j}|\nu_b, \psi) \right] \quad (8.3)$$

To finish the specification of the likelihood function, we have to describe how to calculate the probability of a particular state transition across a branch. We usually express models of sequence evolution as a matrix consisting of the instantaneous rates of change between the four states, $\mathbf{Q}(\psi)$. For a particular branch length, we can calculate a transition probability matrix, $\mathbf{P}(\nu_i, \psi) = e^{\mathbf{Q}(\psi)\nu_i}$. The elements of this matrix give us the probabilities of the form $\Pr(y_{a(b),j} \to y_{d(b),j}|\nu_b, \psi)$ that we need to calculate equation (8.3). Often we will use the equilibrium state frequencies for the process of character evolution to give us the prior probabilities for the ancestral sequence at the root of the tree.

8.1.2 Maximum Likelihood Approaches

The conceptual framework for calculating likelihoods on trees dates back to the early 1970s and 1980s (Neyman, 1971; Felsenstein, 1981), but the usage of likelihood-base phylogenetic techniques was not widely used by systematists until the 1990s. For even a moderate number of species, the number of possible phylogenetic trees is enormous, so maximum likelihood (ML) estimation under these models usually relies on heuristic searches that are not guaranteed to find the tree with the highest likelihood. Typically, the maximum likelihood estimate (MLE) of the tree topology is found by iteratively examining trees within a small neighborhood of a current tree. For each tree considered, numerical optimization routines can be used to approximate the MLE's of $\hat{\nu}$ and $\hat{\psi}$. Nonparametric bootstrapping Felsenstein (1985) is the most commonly-used method of assessing the strength of support for different parts of the inferred tree. Other means of assessing the support for phylogenetic hypothesis tests have been developed (Goldman *et al.*, 2000, provides a nice review of these), and this area continues to be an area of active research (Shimodaira, 2002; Anisimova & Gascuel, 2006). Recently new algorithms and implementations (e.g. Stamatakis, 2006; Zwickl, 2006; Whelan, 2007) have made it feasible to approximate the ML tree estimate for hundreds of taxa in a moderate amount of time.

The flexibility and power of ML approaches to phylogenetic inference offer significant advantages over non-statistical approaches to tree estimation and inference procedures that rely on summary statistics rather than the full data matrix. Nevertheless, there are some notable weaknesses of the inference approach outlined above. For example: 1. accommodating uncertainty in the phylogenetic estimates into subsequent comparative analyses can be awkward if we cannot assign a probability of a tree being correct; 2. the simple, widely-used models assume that different sites are independent, and identically-distributed; 3. while systematists have long known that evolutionary tree for a particular gene may differ from the phylogeny of the species sampled, the processes governing this source of error are not modeled in the system depicted in equations above; 4. the model described above does not distinguish between the rate of evolution and the duration of branches on the tree – this make it impossible to assign times to nodes on the tree; and 5. multiple sequence alignment should be performed jointly with tree inference to more accurately account for the uncertainty in our tree estimates.

We will discuss developments toward addressing these weaknesses. In some cases the advances have been aided by adopting a Bayesian approach to phylogenetics, but several of the improvements to models of sequence evolution could be used in either Bayesian or ML estimation.

8.2 Bayesian Version of "Standard" Model-Based Phylogenetics

Bayesian methods provide a helpful and intuitive method for summarizing the uncertainty in phylogenetic estimates. A Bayesian phylogenetic analysis can provide statements about the probability that a particular tree (or a portion of the trees) are a correct representation of the true phylogeny. This supplies evolutionary biologists with a well-justified weighting of trees for further inference procedures. The review of Huelsenbeck *et al.* (2001) emphasized this point and greatly increased the awareness of the benefits of Bayesian inference to the community of evolutionary biologists, while the availability of free software (see Table 8.1) was also crucial in the adoption of Bayesian tree estimation.

Significant early work in developing a fully Bayesian approach to phylogenetics (Mau, 1996; Rannala & Yang, 1996; Mau & Newton, 1997; Li *et al.*, 2000; Mau *et al.*, 1999; Newton *et al.*, 1999; Larget & Simon, 1999) introduced Markov chain Monte Carlo (MCMC) approaches for exploring a parameter space that includes tree topologies as well as branch lengths and the parameters of the model of sequence evolution. For example, Larget & Simon (1999) introduced an MCMC proposal that changes the length of three branches in the tree, can result in a local rearrangement in the tree topology referred to as an nearest-neighbor interchange (see also Holder *et al.*, 2005). Repeated application of Metropolis-Hastings updates allow the posterior probability surface

TABLE 8.1: Software for Bayesian Phylogenetic Analyses

MrBayes	Ronquist & Huelsenbeck (2003)	Phylogeny estimation under a variety of standard models for sequence and morphological data
BEST	Liu & Pearl (2007)	Phylogeny estimation under a hiearchical structured coalescent model using multi-locus gene data
BEAST	Drummond & Rambaut (2007a)	Co-estimation of phylogeny, divergence time and evolutionary rates using sequence data
SimulFold	Meyer & Miklós (2007)	Co-estimation of alignment, RNA structure, and phylogeny using RNA data
PhyloBayes	Lartillot & Phillipe (2004)	Phylogeny estimation with site-specific mixture models using sequence data
BayesPhylogenies	Pagel & Meade (2004)	Phylogeny estimation using sequence and morphological data, with standard or mixture models
PHASE	Jow *et al.* (2002)	Phylogeny estimation using RNA data with conserved secondary structure
P4	Foster (2004)	Phylogeny estimation with variety of models, including support for non-homogenous mixture models
BAli-Phy	Suchard & Redelings (2006)	Simultaneous alignment and phylogeny estimation

to be explored by the MCMC simulation. Given the complexity of tree space, it is not surprising that there have been many suggestions of Metropolis-Hastings proposal mechanisms to explore the posterior probability surface. Lakner *et al.* (2008) provide a helpful analysis of several of these MCMC "moves" and conclude that the efficiency of an MCMC sample is improved by using a mixture of moves that slightly perturb the tree and moves that make more drastic changes to the topology.

Improving the efficiency of MCMC samplers for phylogenetic inference will almost certainly continue to be an active area of research for several years. Some recent promising work exploits the general framework of SAMC introduced by Liang *et al.* (2007). Cheon & Liang (2008) introduced a sequential stochastic approximation Monte Carlo (sequential SAMC) that avoids poor mixing caused by local optima in the posterior distribution. The method samples incomplete trees (those that lack some of the leaves of the tree), as well as the tree with the full leaf set. Extrapolation (addition of a leaf) and projection (removal of a leaf) operators connect samplers with differing leaf sets. Experimental results indicate that the technique can outperform traditional MCMC simulations. Rodrigue *et al.* (2008) have recently developed a sampler based on data augmentation. Rather than use transition probabilities that consider multiple substitutions across a branch of the tree, their MCMC sampler explores the state space of substitutional mappings (the complete specification of every substitution on the tree). By constructing a uniformized rate matrix which includes "self-substitutions" in addition to substitutions that change the state, they are able to efficiently explore this state space. The technique is particularly useful when analysing data that has a large number of states (Rodrigue *et al.*, 2008, applied it to codon models). Thus far the technique has been used in contexts in which the tree is fixed, but the technique could be used in conjunction with tree inference.

Of course, the adoption of a Bayesian perspective on the phylogenetic estimation requires the specification of prior probability distributions over all parameters of the model. In most cases in phylogenetics, vague priors have been used for the tree topology, branch lengths, and the parameters of the models of character evolution. For example, many practitioners assume that all tree topologies are equally likely and that the branch length can be assumed to be independent of each other with each branch length assigned a exponential distribution as the prior probability distribution (e.g. Li *et al.*, 2000). Models of diversification of species provide another possible source of prior probability distributions over trees and branch lengths. For example, Rannala & Yang (1996) used a prior on trees derived from considering the phylogeny to be the result of a process of birth (speciation) and death (extinction); while Larget & Simon (1999) used a pure-birth process as the source of prior probability densities for different trees. As will be discussed later, when inferring trees of genes within a population, then priors derived from the coalescent process of population genetics have been used.

8.3   Current Directions in Model-Based Phylogenetics

Equation (8.3) was based an a very simple model of sequence evolution in which each site evolves according to the same process, and all events are treated as if they were independent of each other. Since the early 1990's there has been a dramatic expansion in the types of evolutionary processes that have been modeled in phylogenetics. Much of this work has been enabled by the relative ease of implementing models in the context of Bayesian MCMC. In the next sections, we will briefly mention some important advances in models used to infer trees from sequences.

8.3.1   Modeling Variability in the Rate of Sequence Substitution

The rate of sequence evolution is a crucial quantity in phylogenetics because it is a strong determinant of how much "noise" is predicted to occur in the data. If rates of character evolution are very low, then models of sequence evolution will predict very little homoplasy (convergence, parallelism, or reversion that leads to the same character state being acquired more than one time over the evolutionary history of a group). If a character is associated with a high rate of change, then evolutionary scenarios which imply repeated acquisitions of the same nucleotide in different parts of the tree may only have slightly lower likelihoods than more "parsimonious" scenarios (in phylogenetics, the term parsimony refers to scoring trees based on the minimum number of character state transitions required to fit the data). Thus, it is crucial that inference accommodates the fact that different sites have differing rates of substitution. Recognition of the nearly ubiquitous nature of among-site rate variation was a crucial step in improving the fit of models of sequence evolution(reviewed by Felsenstein, 2001). An important early contribution was the development of a simple method for approximating gamma-distributed among-site rate variation by using a discrete number of rate categories (Yang, 1994). However Yang's gamma-distributed models are not rich enough to accommodate the complexity of real data (Kosakovsky Pond & Frost, 2005, see).

Recently, Huelsenbeck & Suchard (2007) addressed among-site rate variation using a Dirichlet process prior model that views sequence data as having arisen as a mixture of sites evolving at different rates. Their framework creates a posterior probability surface over all possible mixtures, encompassing a range of model complexity from a simple model that assumes the all sites are identically distributed up to a model in which each site has a different rate of substitution. Their MCMC sampler considers mixtures with differing numbers of components and also considers different assignments of sites to models.

In addition to extending the number of discrete rate categories in order to analyze complex data sets, there has been some work towards exploiting more sources of information about the rates of different sites. Yang (1995) extended the discrete approximation to gamma-distributed rates of Yang (1994) by allowing for correlations in rate at adjacent sites in a sequence. This model was implemented using a first-order hidden Markov model to describe the rate category assignments for sites in the sequence. This approach was recently extended by Stern & Pupko (2006) to allow for more complex correlations of rates among sites that are close to each other in the sequence. Stern & Pupko (2006) were motivated to produce a model of rate variation that comes closer to mimicking the patterns of rate variation that arise from constraints on protein structure, and they found that their model fits most protein data sets better.

8.3.2 Modeling Variability in Rates of Different Types of Substitution

In addition to variation in the overall rate of change, the types of substitutions will often differ across sites as well. In recognition of this, one can allow for multiple processes by inferring different sequence substitution models for different subsets of sites. An extreme example is the model presented by Halpern & Bruno (1998); in their model each position in a sequence of amino acids has a distinct set of equilibrium state frequencies.

If one is willing to assign different sites into different model categories based on *a priori* considerations, then the mixture model becomes what is referred to as a "partitioned model" in the phylogenetics literature (see MrBayes, Ronquist & Huelsenbeck, 2003, for a software implementation). Partitioned analyses have become a common method for phenomena such as differences in base frequencies and transition/transversion rates when comparing the different codon positions of a coding sequence.

The phrase "mixture models" or "mixed model" is used in the phylogenetics literature to refer to a model consisting of collection of models for a site in the sequence. Each site is assumed to evolve according to a single model (one that is constant over the tree), but not all sites are generated from a model with the same parameter values and the assignment of a site to its "submodel" is not known *a priori*. Instead, each site has a probability p_m of being assigned to model m (where $1 = \sum_m p_m$, and the vector $\mathbf{p}$ can be treated as a free parameter). Thus, the entire sequence is viewed as a convoluted mixture of sites, each of which is drawn from one of several processes. Pagel & Meade (2004) introduce a method (and software) for allowing for multiple instantaneous rates of change to explain a set of sites; they propose using Bayes Factors to choose the number of mixture components to use.

In many cases, we expect their to be some form of spatial patterning to the process of molecular evolution. For example, a transmembrane region of a protein will span several sites each of which will be characterized by hydrophobic amino acids. The evolutionary pattern of sites within this region will reflect selective constraints which prevent the fixation of mutations that lead to hydrophilic amino acids. So, a pattern of only hydrophobic amino acids at one site in a protein sequence should increase the probability that a neighboring site will evolve according to a similar, hydrophobic amino-acid model. Liò *et al.* (1998) present a hidden Markov model that uses several model of amino acid replacement to explain the data, but also takes into account the spatial structuring of sites in a protein sequence into different categories (see also Thorne *et al.*, 1996; Goldman *et al.*, 1996, 1998). While the method of Liò *et al.* (1998) is implemented in the context of maximum likelihood, it is clearly transferable to Bayesian context. One of the appealing aspects of their approach is that it allows for complexity of the evolutionary process (they use eight different matrices of replacement rates rather than a single process) while exploiting prior knowledge – the parameter values for the eight matrices are drawn from a large database of sequences that are independent of the sequences being studied. In a Bayesian implementation, the rates of replacement that are applicable for different secondary structures could be treated as random variables, but the parameters from analyses of other sequences could serve as the basis for informative prior.

Recently Gowri-Shankar & Rattray (2006) introduced another model to accommodate the spatial patterning of molecular forces within sequences, but their approach is more descriptive than to the model of Liò *et al.* (1998). In the model by Gowri-Shankar & Rattray (2006), sites fall into one of several categories that vary with respect to rate of evolution and base composition. They analyzed RNA sequences which fold into secondary structures. The requirement to fold into a particular conformation constrains the evolution of some sites (those that fall in the base-paired stems regions of the structure); as a result, these sequences have strong spatial patterning with respect to the rate of evolution – high rate sites are clustered in contiguous regions that correspond to unpaired loops in the structure. The "clumping" of high-rate sites leads to spatial pattern in the base composition parameters of the Gowri-Shankar and Rattay's (2006) model. To avoid overparameterization artifacts associated with inferring multiple sets of base frequencies, they introduce a Gaussian process prior to smooth the base frequency parameters in similar rate categories.

8.3.3   Allowing Process of Molecular Evolution to Change over the Tree

The previous sections have discussed some attempts to recognize the among-site variability in the process of molecular evolution. It is widely recognized that a particular site can change its rate of evolution over the course of the phylogeny; Lopez *et al.* (2002) documented this process as a common phenomenon and named it "heterotachy." In the section on divergence time estimation, we will consider models that allow for the overall rate of substitution to change over the course of a phylogeny. If the phylogeny is the target of inference (rather than the ages of the nodes in the tree), then it can be helpful to consider models that attempt to prevent heterotachy from distorting phylogenetic inference. The traditional approach (e.g. equations 8.2 and 8.3), parameterizes the branch lengths in terms of expected number of substitutions, and treats each of these branch lengths as an independent parameters. This makes tree inference robust to changes in the rate of evolution, *if* the changes in rate affect all sites in the sequence equally. If different subsets of the sequence (or even individual sites) change rate but do not do so in a concerted fashion, then methods that use a single branch length for all sites can be misled by heterotachy.

Covarion models allow for interactions among sites leading to each site changing rate; they can be formulated in a way that tries to determine the interacting sites (Fitch & Markowitz, 1970), or with reference to a single site that changes its rate in response to an unknown factor (Tuffley & Steel, 1997; Huelsenbeck, 2002). In the simplest version, a site toggles between "on" and "off" states, but the covarion-style models has also been generalized (Galtier, 2001) to allow for a site that changes among several rates of evolution, or to allow a site to change between models that differ by parameters other than just the total rate of evolution (Whelan, 2008).

A more parameter-rich approach to inferring trees from data that may display heterotachy is to treat the data as a mixture generated from models that share the same tree topology, but have different sets of branch lengths. vStefankovivc & Vigoda (2007) demonstrated that, under some relatively simple models of sequence evolution, assuming a mixture of just two sets of branch lengths can lead to non-identifiability of the tree topology. This result calls into question the prospects for inferring trees using general models and mixture-of-branch-lengths approach. Interestingly, Zhou *et al.* (2007) have compared the performance of the covarion and mixed branch length approach to dealing with heterotachy, and found that the simpler covarion-style model tends to outperform the mixed-branch length approach. In that study, performance was assessed by the Bayesian Information Criterion (Schwarz, 1978) and cross-validation to choose among alternative models.

Another interesting result in this area is the work of Wu *et al.* (2008). Their model extends the discrete approximation to gamma-distributed among-site rate variation (Yang, 1994), in which one uses a discrete number of rates of evolution to explain the sites in a sequence. Simply by allowing sites to change their rate categories on each branch of the tree, Wu *et al.* (2008) produce a class of models for among-site rate variation that have significantly better fit to sequences (as long as the model is not too simple – under a Jukes-Cantor model with gamma-distributed rates, the model is unchanged by allowing sites to switch categories). The result is intriguing because it suggests that relatively simple approaches may adequately deal with heterotachy or at least increase the robustness of phylogenetic analysis to some forms of heterotachy.

It can also be important to consider models which allow for the changes over the tree of values of parameters other than just the overall rate of substitutions. For example, the base composition among taxa is a property of sequences that differs among species. Failing to account for variability in the base composition across species could mislead phylogenetic inference procedures because two species that happen to evolve similar base compositions may then acquire similar nucleotides at a large number of sites which are tolerant to new mutations. Thus, one change in the forces that drive molecular evolution can lead to a large number of convergent changes among sequences. This pitfall has long been recognized (see Foster & Hickey, 1999; Conant & Lewis, 2001; Rosenberg & Kumar, 2003, for assessments of the severity of the problem), leading to corrections to distance-based approaches (Lockhart *et al.*, 1994; Galtier & Gouy, 1995; Yang & Roberts, 1995). Methods for calculating likelihoods under models of sequence evolution with nonhomogenous nucleotide compositions have also been proposed (Galtier, 1998), but are not widely used. Overparameterization can be a concern in non-homogenous models. We do not know how many times the base compositions have changed over the tree, nor do we observe the substitutions themselves leading to considerable uncertainty about the complexity of the process underlying the observed sequences. Foster (2004) introduced Bayesian (and ML) approaches to modeling changes in nucleotide composition over the tree, and showed that model selection procedures could be used to select a model that is less parameter-rich than a full model in which every branch has a separate vector of state frequencies. Recent Bayesian approaches to this problem (Gowri-Shankar & Rattray, 2007) use a model space which allows the number of distinct sets of base composition parameters to be treated as a random variable. (Gowri-Shankar & Rattray, 2007) use reversible-jump Markov chain Monte Carlo (Green, 1995) to sample over this space of varying dimension. This approach is a promising method for tuning the complexity of the model to the information in the data at hand.

In another interesting application, Guindon *et al.* (2004) implemented a mixture of models that differ in the value of the parameter ω, the rate multiplier associated with nonsynonymous substitutions. Because nonsynonymous changes are more likely to cause measurable phenotypic effects, this parameter has an interpretation in the context of natural selection; $\omega > 1.0$ is interpreted as evidence that selection favors amino acid substitutions at a site, while values less than 1.0 indicate purifying selection. In the approach of Guindon *et al.* (2004) a codon is permitted to change its assignment to submodel within the mixture. This models changes in the amount of selective constraint over the course of the tree. They conclude that allowing ω vary over the tree not only improves the fit of the model (to the HIV sequences that they examined), but is also important to correctly estimating the strength of selection.

8.3.4   Context-Dependent Models

One of the most exciting developments in the models of sequence evolution for phylogenetic inference has been the creation of models that do not assume that the evolution at a site is independent of neighboring sites. Assuming that the evolutionary process is the same regardless of context greatly simplifies calculation of the likelihood – it allows us to assume independence across sites and write equation (8.1) as a product over all sites in a sequence. Unfortunately, neither the forces of natural selection nor mutation are thought to occur in a manner that acts on one site alone; thus, the context of a site matters in evolution and should be addressed in our models.

By treating the type and timing of each substitution as a latent variable that can be varied during the course of an MCMC simulation, one can implement a model that allows for context-dependence (Jensen & Pedersen, 2000). Thus far, these techniques have been used to model the effects of secondary structure on RNA evolution (Yu & Thorne, 2006); mutational context-dependence in mutational processes (Hwang & Green, 2004; Siepel & Haussler, 2004; Hobolth

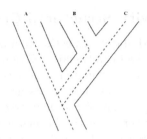

FIGURE 8.1: Deep coalescence, with the contained gene tree (dashed line) failing to coalesce in the ancestral population of species B and C, leading to the allele sampled in B being more closely related to allele in A, rather than the true sister species, C.

et al., 2006; Hobolth, 2008); and structural constraints on protein evolution (Robinson *et al.*, 2003; Rodrigue *et al.*, 2005, 2006; Choi *et al.*, 2007). Thorne (2007) provides a more thorough review of these topics than we will attempt.

Fairly simple, context-dependent processes of substitution can result in complex evolutionary patterns. The rate of evolution at a particular site and the types of mutations that fix at that site may change over the course of the tree (if there are changes in the neighborhood of the site). In models that evaluate mutants according to secondary and 3D structure (Yu & Thorne, 2006; Robinson *et al.*, 2003; Rodrigue *et al.*, 2005, 2006; Choi *et al.*, 2007), the amount of heterogeneity across sites can be very large. Thus, using context-dependent models may simultaneously provide us with a tool that deals with spatial and temporal variability in the evolutionary process. Significantly, this complexity can be accomplished while adding relatively few parameters (e. g. the model of Robinson *et al.*, 2003, adds only two parameters to the independent-sites codon model). Exploiting information from research on the determinants of protein structure not only helps inform these phylogenetic methods, but also provides the potential to allow evolutionary analyses to be used to enlighten research into the constraints on protein folding. It is clear that there is still a great deal of research to be done in this area. For example, Rodrigue *et al.* (2006) found that while context-dependent models did improve the fit of sequences on a tree, the models still needed to be used in association with information from empirically-derived independent sites models. Similarly, when Thorne *et al.* (2007) examined context-dependent models as a bridge toward uniting phylogenetic and population models, their analyses revealed noticeable failures to predict enough sites with large deleterious effects. Clearly the use of context-dependent models in evolutionary analyses is still at an early stage.

8.4   Integration with Structured Coalescent

The methods described above can estimate a gene tree rather than a phylogeny (a tree depicting the evolutionary relationships between species). It is common practice to use the estimated gene tree as a proxy for the species tree. If the times between speciation events are long enough relative to the population sizes, the topologies of the gene tree and species tree should be identical and the relative error in the branch lengths (i.e., the difference in ages between gene divergence and actual population divergence) are insignificant. If, however, the duration between speciation events are short (or the population sizes are large), then it is possible for polymorphism to be retained through multiple episodes of speciation. This situation can result in a discordance between the gene tree topology and species tree topology, and is called *deep coalescence*[2] (see Figure 8.1). If we only have data from a single locus, then we will not see evidence for deep coalescence even if it does occur. In such cases we must be cautious about treating a gene tree as if it were identical to the species tree. A further complication arises from the fact that we do not observe the gene trees for our data, but must infer them. So, conflicting gene trees can result from estimation error as well as biological processes. Recombination within a locus can further complicate the situation because it can lead to a different genealogical histories for opposite ends of the same gene (see Chan *et al.*, 2006; Fang *et al.*, 2007; Paraskevis *et al.*, 2005; Sinsheimer *et al.*, 2003; Suchard *et al.*, 2002, for recent work on detecting recombination using phylogenetic approaches). When we have data from multiple loci then recombination can actually be helpful. Having multiple unlinked loci gives us multiple samples of gene trees and this allows us to assess the impact of deep coalescence on our species tree estimation.

The potential discordance between a gene tree and the phylogeny has long been appreciated within phylogenetics (see discussion in Page & Holmes, 1998, p31-33, for example). This discordance is typically assessed through sequential analysis: gene trees are produced, and then the researcher looks for well-supported groupings that conflict with each other. If no such groupings are found, then the gene trees are judged to be reliable estimates of the species tree. Numerous approaches have been proposed to obtain a species tree when there is conflict among gene trees. Unfortunately, these procedures do not capture all of the relevant information about how deep coalescence occurs. For example, gene tree parsimony (Slowinski & Page, 1999) selects a species tree by finding the tree that minimizes the number of deep coalescences. The assumption that deep coalescence is a rare event is not always justified, and, even if deep coalescences are rare, they may occur more frequently in some

---

[2]The phrase "incomplete lineage sorting" is often used for the phenomenon that we will call "deep coalescence."

parts of the phlyogeny (those with ancestral species that have large populations or short durations). This information is not adequately incorporated into a gene tree parsimony analysis.

The concatenating of the sequence data then treating it as if it were single linked locus, or, alternatively, basing the species tree on a majority-rule consensus tree of a set of gene trees, are other ways to infer a phylogeny when estimates of different gene trees conflict with each other(see Gadagkar *et al.*, 2005, for a comparison of the performance of the latter two methods). If the most common gene tree topology always agreed with the species tree, then concatenation would converge to the species tree (given enough sequence data). However, Degnan & Rosenberg (2006) have recently demonstrated that in the case of species trees of some shapes, the most common gene tree will *not* be identical in topology to the species tree shapes. This phenomenon is not just the result of random errors associated with sampling too few gene trees – it persists even when a large (even unbounded) number of independent loci are sampled.

The results of Degnan & Rosenberg (2006) argue for "coalescent-aware" approaches to inferring the species tree from a set of gene sequences. Coalescent theory is a branch of population genetics that is used to calculate a probability distribution over the space of all possible genealogies for a given sample of alleles within a population (see Ewens, 2004; Hein *et al.*, 2005, for much more thorough treatments than the brief overview that follows). As the result of either genetic drift or selection, lineages of genes sweep through populations until they become fixed. Depicting the ancestor-descendant relationships of genes within a population over time results in a tree of gene sequences. Kingman (1982a,b) made the crucial realization that one could calculate a probability density for any gene tree shape by looking backwards in time and viewing the problem as one of ancestors of the sampled genes coalescing until they form one lineage. From the traditional perspective (looking forward in time) the coalescent events are actually divergence points in the gene tree caused by the same allele in a parent being passed on to multiple offspring. Kingman demonstrated that the probability density for a tree can be built up by considering the waiting time between coalescence events. The size of the population is a crucial parameter – larger populations lead to larger times to coalescence.

Coalescent theory becomes important in phylogenetic analysis when there is a non-negligible probability of a polymorphism being maintained for the duration of a branch of the phylogeny. In such a case, coalescence has not completely occurred, and alleles in one species may be more closely related to an allele in a different species than they are to another allele from the same species. This can be modeled using a hierarchically-structured coalescent model in which the species phylogeny defines the boundaries for the coalescent processes and specifies the timing of the splitting of gene pools of a parent species into those of the daughter species (see Rannala & Yang, 2003). Edwards *et al.* (2007) and Liu & Pearl (2007) describe a method to estimate the joint posterior probability of species trees and contained gene trees, given data from multiple unlinked gene loci. They calculate posterior probability of a species phylogeny with branch lengths and vector populations sizes (T, ν, Θ), via the posterior probability of the set of genealogies for all of the loci, $\mathbf{G}$:

$$\Pr(T, \nu, \Theta | \mathbf{X}) \propto \int \Pr(T, \nu, \Theta | \mathbf{G}) \Pr(\mathbf{G} | \mathbf{X}) d\mathbf{G} \tag{8.4}$$

$$\propto \int \int \Pr(T, \nu, \Theta | \mathbf{G}) \Pr(\mathbf{G} | \psi, \mathbf{X}) \Pr(\psi) d\mathbf{G} d\psi \tag{8.5}$$

where:

$$\Pr(T, \nu, \Theta | \mathbf{G}) = \frac{\Pr(\mathbf{G} | T, \nu, \Theta) \Pr(T, \nu, \Theta)}{\Pr(\mathbf{G})} \tag{8.6}$$

Conditional on the set of genealogies, the posterior probability of the species tree is independent of sequence data. This hierarchical nature of the formulation allowed Edwards *et al.* (2007) and Liu & Pearl (2007) to implement an MCMC sampler for the posterior probability of the species tree by using multiple layers of MCMC sampling. They make a slight change to MrBayes (Ronquist & Huelsenbeck, 2003) to collect a sample of genealogies from the posterior probability distribution, $\Pr(\mathbf{G}^\dagger | \psi, \mathbf{X})$, where the dagger indicates that the prior distribution for used to collect this sample is not the correct prior over genealogies. This is done because they can construct an artificial prior, $\Pr(\mathbf{G}^\dagger)$, that is easier to implement than the correct prior, $\Pr(\mathbf{G})$. Given this set of sampled genealogies, they can use an MCMC sampler similar to the one proposed by Rannala & Yang (2003), to sample a set of species trees $\Pr(T, \nu, \Theta | \mathbf{G}^\dagger)$. In the final step of their algorithm they use importance sampling to correct for the previous use of an incorrect prior; thus they obtain $\Pr(T, \nu, \Theta | \mathbf{G})$ from $\Pr(T, \nu, \Theta | \mathbf{G}^\dagger)$ using importance weighting to subsample their collection of species trees. The method of Liu & Pearl (2007) and Edwards *et al.* (2007) has been implemented in a program called "'BEST'", and the approach represents a significant advance in estimating a species tree from sequence data from multiple unlinked gene loci.

8.5 Divergence Time Estimation

For many evolutionary or systematic questions, it is of interest to estimate a tree with branch lengths proportional to units of time. This allows one to assign dates to ancestral nodes in the tree; because a node on a phylogeny represents the point at which two species began to diverge, this procedure is often referred to as divergence time estimation.

The simplest model for assigning dates to a tree assumes a constant rate of change across all branches of the tree. When dealing with molecular sequences, this is commonly referred to as the "molecular clock" assumption (Zuckerland & Pauling, 1965). To estimate the global rate of change of under a molecular clock, calibration dates are used to specify the ages of at least one internal nodes. Calibration points are based on external information, such as fossils or geological events. Unfortunately, the molecular clock assumption is not a biologically realistic. As mentioned in earlier sections, the rate of evolution can vary over the tree.

If we allow the mean rate of evolution to assume *any* value on a branch, then the rate and time parameters are not identifiable (recall that the rate and time for a branch always occur together in the likelihood equation as the product, $\nu = rt$). Thus, to estimate times on a tree, the rates of evolution on each branch must be constrained in some way.

Several distinct approaches have been taken to relaxing the assumption of a molecular clock but still estimating the divergence times of nodes on the tree. "Local clock" models (see Yoder & Yang, 2000; Yang & Yoder, 2003) allow for a small number of rates to be in effect on different parts of the tree. These models shall not be discussed in any detail here. Instead, we will focus on relaxed-clock models which use a prior on the rates of evolution to force rates to be similar across the tree.

In autocorrelated rate models, rates of evolution on a branch depend on the rates of neighboring branches in the tree; this structure embodies the idea that the factors that control the rate of sequence evolution (for example the error rates of DNA polymerases, or the effective population size) are themselves heritable factors that can evolve on the tree. One of the earliest examples of this approach was the nonparameteric rate-smoothing method of Sanderson (1997). As input, this method takes a tree with estimates of number of changes on each branch, **c**. Then one tries to find a set of times, **h**, for the nodes of the tree. The duration (in time) of a branch can be obtained as the difference in heights of the parent and child node for the branch. Thus, from **h** we can obtain the set of branch durations, **t**. For the number of changes across a branch and the duration, one can set the rate for a branch to be $\hat{r}_i = c_i/(Mt_i)$. In nonparametric rate smoothing, we seek values of **h** that minimizes the variation between rates in ancestral and descendent branches. If we interpret the rate assigned to a node in the tree as the rate of preceding branch, then we can express nonparametric rate smoothing as seeking rates that minimize:

$$\sum_{b \in E} \left(r_{a(b)} - r_{d(b)} \right)^2 \qquad (8.7)$$

(note that we do not have a number of changes or a rate for the branch leading to the root of the tree; choosing a rate for this branch to minimize the changes in rate is equivalent to adding variance of the rates of the branches that diverge from the root of the tree). Sanderson (2002) later introduced a more flexible approach based on penalized likelihood. Changes in rates across the tree are penalized, but the strength of the penalty can be tuned. In Sanderson's penalized likelihood approach the rates on the branches and the timing of the nodes are parameters. The likelihood comes from assuming that $c_i \sim \text{Poisson}(r_i t_i)$. As discussed by Sanderson (2002), the method could be extended to use a negative binomial to calculate the likelihood (this may be more appropriate when there is among-site rate variation). The penalty term is the squared difference in rates for an ancestor-descendant pair of branches plus the variance of the rates for branches that are adjacent to the root. This penalty term is scaled by a coefficient, λ, and subtracted from the log likelihood for a set of rates. Thus, the method attempts to maximize:

$$\sum_{b \in E} \left[\ln \text{Pr}(c_b | r_b t_b) - \lambda \left(r_{a(b)} - r_{d(b)} \right)^2 \right] \tag{8.8}$$

As λ approaches ∞, the model approaches a molecular clock model; the penalty effectively forces the rates to be equal across all branches. As λ approaches 0, the likelihood increases as rates are allowed to vary dramatically. Cross-validation is used to find the optimal value of λ.

The approaches of Sanderson (2002, 1997) are elegant and very popular, but they do have drawbacks. The calibration points are taken as point estimates, and there is no way to accommodate uncertainty or error in these estimates. Sanderson (2002) allows for some calibration points to be specified as boundary conditions (either a maximum or minimum age for a particular node), but there is no mechanism for placing a probability distribution over all plausible calibration points. As currently implemented, these methods do not accommodate uncertainty in the estimation of topology or the error in branch length estimates. The amount of rate variation across long branches may be expected to be larger than across short branches; this suggests that the penalty used in Sanderson (2002) may be improved by weighting the squared difference in rate according to the branch duration.

Fully Bayesian approaches to divergence time estimation with autocorrelated rate variation were first introduced by Thorne *et al.* (1998), who present a method that models the rate of evolution of the rate of evolution. The rate of evolution is allowed to vary among branches in the tree, but the rate of evolution along a descendant branch is assumed to be log-normally distributed with mean equal to the rate of the ancestral branch and variance proportional to the time difference between the midpoints of the ancestral and descendent branches. Kishino *et al.* (2001) reparameterized this model to assign rates to nodes rather than branches. This updated model better reflects the expected correlations of rates across a tree if the factors that control the rate of evolution were gradually changing within each species (see Thorne & Kishino, 2002, for extensions to multigene datasets). Kishino *et al.* (2001) also changed the prior distribution on the rates of evolution at the end of a branch such that the expectation of the rate for a descendant node is equal to the rate at the ancestral node.

Huelsenbeck *et al.* (2000) used a compound-Poisson process model in which the rate of evolution "jumps" to a new rate at distinct points on the tree. The number of changes in the rates of evolution are assumed to follow a Poisson process. When a change in the rate of evolution occurs, the new rate is obatained by multiplying the previous rate by a factor drawn from a gamma distribution that has been tuned to keep the expectation of the rate from tending toward 0 or ∞.

Aris-Brosou & Yang (2002) try a variety of a parameteric-distributions for the prior of rates on a branch, and incorporate an Ornstein-Uhlenbeck process to model changes in the means and variances of the distributions. Welch *et al.* (2005) criticized the priors used for rates and divergence times as biologically unreasonable and causing a systematic bias toward higher rates toward the tail of the branch and younger dates being estimated for nodes. Lepage *et al.* (2006) (also see Lepage *et al.*, 2007) addressed these concerns by using a Cox–Ingersoll–Ross process instead of the Ornstein–Uhlenbeck to model the evolution of means and variances of the prior distribution of rates.

Many software implementations of divergence time estimation methods do not allow for uncertainty in the timing of the calibration points on the tree. This situation is unsatisfactory from a biological perspective, as there inevitably is some (often substantial) uncertainty in any given calibration. Yang & Rannala (2006) address this by allowing the date of a calibrated node to be described by a probability distribution.

Drummond *et al.* (2006) present a model in which the rates are drawn from from a prior distribution, but rates are not assumed to be autocorrelated along the tree. While many of the methods mentioned treat the tree topology as fixed, the implementation of their method also allows for simultaneous co-estimation of phylogeny and divergence times, as well as specification of calibration times in terms of distributions rather than points, thus explicitly accommodating uncertainty in both phylogeny and calibration times. Lepage *et al.* (2007) introduce another uncorrelated relaxed clock model which they term the "white-noise" model; as in the model of Drummond *et al.* (2006), the rate for each branch is drawn from a prior distribution that is independent of other rates on the tree. The variance of this prior distribution varies between the models; in the white-noise model of Lepage *et al.* (2006) the variance is smaller for long branches (which is consistent with the action of an uncorrelated process over a long period of time).

A simulation study by Ho *et al.* (2005) indicates that the divergence time estimates are often quite sensitive to the priors chosen. They argue that, in light of their results, the uncorrelated relaxed clock methods (Drummond *et al.*, 2006, e.g.) may offer a conservative method of estimating dates. Lepage *et al.* (2007) use Bayes factors and real datasets to compare the behavior of methods; they also examined the effects of using different prior probabilities for the times associated with internal nodes of the tree. They also find that divergence time estimates are quite sensitive to the model assumed, but their results indicate that autocorrelated approaches outperform the uncorrelated relaxed clocks. The sensitivity of divergence time estimation to the details of the models used and the priors is unsurprising in light of work by Britton (2005) showing that when the molecular clock assumption does not hold, estimation of divergence times of nodes is not consistent. Even as the length of sequence data collected goes to ∞, there some uncertainty about the ages of the nodes will remain. This is true even if the process controlling the rate of evolution of the rate of evolution is modeled accurately. Adding more loci can help if each locus can be viewed as another sample from a process that controls the evolution of the rates of molecular change; but if the changes in rates across loci are strongly correlated, then even sequencing of the entire genome may not provide a precise estimate of divergence times.

An intriguing recent study (Ho *et al.*, 2007), found evidence for time-dependency of the rates of evolution; rates of sequence evolution appear to be faster when recently diverged sequences are compared. A proposed mechanism for this phenomenon is the presence with slightly deleterious alleles; such polymorphisms would contribute a great deal of the sequences differences between closely related sequences, but would be negligible for comparisons of deeper divergences because they would rarely fix in the population. The implications of this study are that researchers should be very cautious about inferring dates on nodes of a tree when the dates of the calibration points are very different from the node of interest (such inferences are also susceptible to artifacts arising from lack of fit of the model of sequence substitution).

8.6   Simultaneous Alignment and Inference

A full discussion of statistical approaches to multiple sequence alignment are beyond the scope of this chapter (we refer readers to the excellent review by Lunter *et al.*, 2005b, for more information). Nevertheless, it is important to note that Bayesian approaches have made an enormous impact on the state-of-the-art methods for simultaneous alignment and tree inference. Most phylogenetic analysis is conducted after an alignment has been constructed, and the alignment is treated as fixed for the purpose of tree estimation. This is unfortunate because the data that we observe are unaligned, and the alignment is merely an estimate of the true alignment (see Sankoff, 1975, for an early approach to this problem). Performing alignment then tree inference sequentially makes it difficult to recognize the full amount of uncertainty in our outcome. Even worse, many of the commonly used alignment procedures use progressive alignment strategies that rely on a phylogenetic tree to some extent. Thus there is the potential for overconfidence in the final tree estimate arising from the fact that the input data for phylogenetic inference was actually inferred assuming a particular guide tree.

Alignment is necessary because during the course of sequence evolution nucleotides are inserted and deleted in addition to undergoing replacement by another nucleotide. It may appear that we could treat these insertion and deletion events (often referred to as "indels") as simply another type of evolutionary event in the likelihood calculations that we outlined in equation (8.3). Unfortunately, this is not the case (but see Rivas *et al.*, 2008, for recent progress in techniques that treat gaps as an extra state). When we cannot confidently establish the homology of the nucleotides in our data then we are unsure of which nucleotides should be placed in a column of in our data matrix; so, we cannot break the likelihood calculation down into a product over columns. The likelihood of the sequences should be a sum of probabilities over all possible sequences for ancestral nodes in the tree. When insertions and deletions are allowed, we do not even know the length of the sequences for ancestral species. The number of possible ancestral sequences becomes infinite. Even if we crop this calculation by forcing ancestral sequences to be close the length of the observed sequences the number of possible evolutionary scenarios that could generate the leaf sequences is enormous. Much of the progress that has been made in statistical alignment on phylogenies has been made possible by reformulating models that include indels as hidden Markov Models that emit possible alignments for an ancestral node and its descendants (see Durbin *et al.*, 1998; Lunter *et al.*, 2005b, for more and discussion) We will mention some of the advances that have been made in improving the richness and realism of models of the indel process that can be analyzed in the context of a tree.

Thorne *et al.* (1991) presented the fundamental model for incorporating indels events in a phylogenetic context; their "TKF91" model was restricted to insertions and deletions of single bases, but it introduced methods for calculating the probability of a pair of sequences (ancestor and descendant) across a single branch. Thorne *et al.* (1992) extended the model to allow for insertions and deletions with the length of these events corresponding to a geometrical distribution (rather than being fixed at a length of 1 base). Unfortunately, for computational reasons, the model enforces the unrealistic restriction that multiple indel events cannot overlap with each other. For pairwise alignment approximations to a model that allows for overlapping indel events have been developed (e.g. Knudsen & Miyamoto, 2003; Miklós *et al.*, 2004).

While models of the indel process have been improved, there has also been work on developing MCMC samplers that can jointly infer alignment and the evolutionary tree. Lunter *et al.* (2005a) developed a method to infer trees under the TKF91 model. Treating ancestral sequences as latent variables and allowing an MCMC simulation to sample over all possible states can ease the construction of a sampler over alignments and trees. Unfortunately, increasing the state space in this way comes at the cost of increasing the number of MCMC iterations that must be performed for accurate approximations. Lunter *et al.* (2005a) introduce recursions from dynamic programming to avoid augmenting the state space of the MCMC sampler; they also implement a method of updating blocks of the alignment at a single time to improve mixing in their MCMC simulation.

Redelings & Suchard (2005) present method for inferring trees from unaligned data using a novel indel model that allows for multiple base indels but does not consider the length of a branch when calculating the probabilities of indel events. In later work (Redelings & Suchard, 2007), extend their model to one in which the branch length does affect the probability of indel events, but the distribution of lengths of a sequence is identical for the ancestor and descendant of a branch. This allows them to continue to implement their approach on unrooted trees. Redelings & Suchard (2007) also introduce more drastic proposals for changing the tree topology in an alignment-aware way; these improvements shorten the number of simulation iterations that must be performed to accurately approximate the posterior probability distribution.

The recent work of Meyer & Miklós (2007) exemplifies many of the new tools that are being used for phylogenetic modeling. They present a modeling approach that exploits knowledge of the thermodynamic properties of RNA folding to evaluate possible secondary structures for a set of RNA sequences. Their approach treats the evolutionary tree relating the sequences and the alignment as random variables to be estimated during the analysis.

8.7 Example Analysis

As an example application of Bayesian methods in phylogenetic inference, we focus on frogs of the subfamily Raninae (Ranidae, Anura). While recent studies of Anuran phylogeny have relied on nuclear gene fragments (San Mauro *et al.*, 2005; Roelants *et al.*, 2007; Wiens, 2007; Wiens *et al.*, In press), the majority of the publically avaliable molecular data for frogs are small fragments of mitochondrial DNA (mtDNA). Thus, we will use ~ 2500 base pairs of mitochondrially-encoded DNA to estimate phylogenetic relationships within this group.

Previous hypothesized relationships for the group have been based of morphological analyses. Of particular interest are the phylogenetic placement of *Rana sanguinea*, *Rana macrops*, and *Rana tipanan*. The Palawan wood frog, *Rana sanguinea* is endemic to Palawan Island, Philippines, and has previously been considered to be a "Papuan element" in Philippine herpetofauna (Inger, 1954; Brown, 2007). Thus, one would predict that this species would be related to the *Rana papua* group (i.e. *Papurana* of Dubois (1992), *Sylvirana* of Frost *et al.* (2006)). The rare central Sulawesi Masarang rock frog *Rana macrops* exhibits a distinctive combination of habitat preference and morphology: it a rocky stream species that is not arboreal and yet has widely expanded finger and toe disks. These characteristics have led to the hypothesis that it is closely related to Asian stream frogs *Rana chalconota*, i.e., *Chalcorana* of

Dubois (1992), *Hydrophylax* of Frost *et al.* (2006), or, by implication, *Hylarana* of Che *et al.* (2007). Alternately, its conspicuous bright green coloration has led others to suggest that it may be most closely related to the green rock frogs of Indochina, the *Rana livida* complex (i.e., *Odorrana* of Stuart, 2008). The bright and contrasting coloration of *Rana tipanan* (the Philippine Sierra Madres frog), and microhabitat preference for high-gradient streams resemble many of the torrent frogs of mainland Asia (genus *Amolops*). If *Rana tipanan* is a member of this group, then it may represent an independent invasion of the Philippine archipelago. Another possibility is that *Rana tipanan* is a member of the Philippine stream frogs clade – the *Rana everetti* group, *Hydrophylax* of Frost *et al.* (2006) or, by implication *Hylarana* of Che *et al.* (2007).

DNA sequences of the 12S and 16S mitochondrial genes were were assembled for 178 ranine taxa. Sequences for 157 of these taxa were downloaded from GenBank, where they were initially deposited as part of several large-scale phylogenetic studies of frogs (Hillis & Wilcox, 2005; Bossuyt *et al.*, 2006; Frost *et al.*, 2006; Roelants *et al.*, 2007; Che *et al.*, 2007; Wiens *et al.*, In press). Sequences for the remaining 21 taxa were original sequences from the collections of RMB at the University of Kansas Natural History Museum. Sequences were aligned using MUSCLE (Edgar, 2004), and manually adjusted by RMB, resulting in a final alignment of 2529 characters. We conducted phylogenetic analyses under the general time reversible model of sequences evolution with an unknown proportion of invariant sites and gamma-distributed rate heterogeneity among variable sites. We used MrBayes (Ronquist & Huelsenbeck, 2003) to conduct analyses in the absence of a molecular clock, and BEAST (Drummond & Rambaut, 2007b) to examine effects of using a relaxed molecular clock that uses a log-normal prior distribution for the rates of evolution with no expectation of phylogenetic correlations in the rate of evolution Drummond *et al.* (2006). Under this approach, branch lengths are proportional to time, but we did not attempt to calibrate the clock. For each of the models, we conducted four independent MCMC simulations of 200 million iterations each. We verified that the effective sample size of the MCMC samples for each parameter in the model was greater than 150, and that the the average standard deviation of the split frequencies betweens runs was less than 0.01 (Lakner, 2008). Samples from the initial 10% of each MCMC simulation discarded.

Figure 8.2 shows the majority-rule consensus tree summary for the analysis with no molecular clock. Clades labelled A and B are composed of a diverse group of mainly Asian and Eurasian frog taxa currently assigned to several genera. Clade C is composed primarily of species now referred to the genus *Odorrana*. Clade D contains frogs from the Americas and Clade E consists of European and Asian species. Branch lengths are drawn in proportion to the posterior mean of the length parameter for each branch. The tree shows large apparent differences in the rate of evolution (the tip to root distances are quite variable), but the phylogenetic relationships are quite robust to differing assumptions about changes in the rate of evolution. The posterior probabilities of splits based on the analysis without a molecular clock enforced

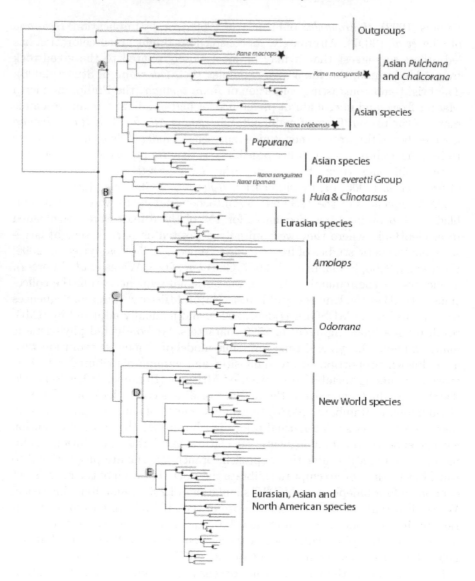

FIGURE 8.2: Majority-rule consensus tree phylogeny of 178 species of true frogs of the subfamily Raninae. Branches with posterior probability > 0.9 are denoted by black dots at the descendant node. Empty circles denote nodes on branches with posterior probability between 0.7 and 0.9. Stars indicate three species of frogs endemic to the Indonesian island of Sulawesi. An arrow indicates the only clade that was in the no-clock consensus tree but not in the relaxed-clock consensus tree.

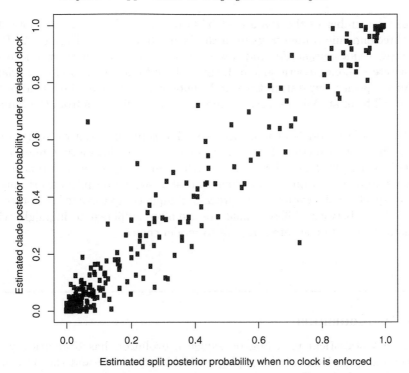

Estimated split posterior probability when no clock is enforced

FIGURE 8.3:   Comparison of posterior probabilities of clades under the no-clock and relaxed clock approaches, as estimated under 4 independent runs of 200 million generations each under each approach.

were tightly correlated with the estimates of the posterior probability for the same split under the relaxed clock model (Figure 8.3). Only one split in the no-clock majority-rule consensus tree conflicted with a split in relaxed-clock majority-rule consensus tree. This split is indicated with an arrow in Figure 8.2, and these conflicting splits are responsible for the two points furthest from the diagonal in Figure 8.3

Despite low support for some aspects of the phylogenetic relationships, we are able to conclusively address the hypotheses about the three species mentioned above. First, the Palawan wood frog, *Rana sanguinea*, is not closely related to the genus *Papurana* (the *Rana papua* Group of frogs from the Papuan realm), but instead is part of the *Rana everetti* Group, a small radiation endemic to the Philippines. Second, *Rana macrops* is clearly not related to the morphologically similar Indochinese genus *Odorrana*, but may actually be a lineage sister to the remaining ranines, consisting of southeast Asian taxa referred to the genera *Pulchrana* (i.e., *Rana signata* Group) and *Chalcorana* (i.e., *Rana chalconota*). Third, the Philippine Sierra Madre frog *Rana tipanan* is clearly not related to members of the Indochinese genus *Amolops* but is instead a member of the *Rana everetti* Group as well. Finally, we note that the three endemic ranine frogs from the Island of Sulawesi (Figure 8.2: *Rana macrops*, *Rana mocquardii*, and *Rana celebensis*, denoted with stars) do not constitute a monophyletic group. Instead these taxa are each closely related to Asian species, suggesting three independent invasions of the island of Sulawesi, all from an Asian source, to the exclusion of lineages from the Papuan realm.

Clearly other models of sequence evolution could be used to analyze this data. The rapid growth of Bayesian approaches to phylogenetic inference has made it relatively straightforward to conduct an analysis, such as this one, in which specific evolutionary questions can be answered despite uncertainty in many aspects of the evolutionary process. Simple comparison of the posterior probabilities between different analyses are often sufficient to highlight what aspects of the inference are sensitive to model choice.

8.8   Conclusions

Explicit statistical modeling of sequence evolution has dramatically improved our knowledge of the phylogenetic history of life and the process of molecular evolution. Despite the progress, there are many open questions and areas of active research in the field of statistical phylogenetics. We have attempted to summarize of some of these active areas. Bayesian approaches promise to be at the center of much of this work. In many cases, progress is being made by incorporating relatively weak information from a variety of sources into a phylogenetic perspective. For example, insights from protein structure prediction provide some clues about the patterns of amino acid replacement that we might expect to see, but these predictions are not so specific that the information completely identifies the evolutionary history of a sequence.

Biologists are interested in very complex evolutionary analyses. For example, Maddison *et al.* (2007) discuss a new method for inferring whether or not a characteristic of an organisms (such as carnivory vs omnivory) affects the speciation or extinction rate of species that display that trait. A few years ago, this sort of analysis would have been done by inferring a species tree, mapping the carnivorous/omnivorous condition onto the tree to find the most parsimonious explanation of how that trait evolved, and finally looking for evidence that groups that are inferred to have been carnivorous have a lower rate of diversification. This type of sequential analysis uses point estimates in each step. Such a procedure may correctly identify the direction of an effect and give a very coarse estimate of the magnitude, but it fails to yield the appropriate statements about the degree of support for the hypotheses. This was a particularly unsatisfying situation because in many cases the biologist was already aware that there was some signal for a particular hypothesis – the entire point of conducting an explicit statistical analysis was to assess the magnitude of the effect and the strength of support for a hypothesis. A full analysis of such a question would consider many sources of variability and account for uncertainty in: how the character evolves on the tree; the topology of the phylogeny (given that the gene trees that we sample to estimate the species tree may not coincide with the species phylogeny); the gene tree (given that complex patterns of sequence indels and substitutions may make the gene tree difficult to reconstruct); and the depth (in time) of the internal nodes of the species tree. The ability to use a Bayesian analysis to investigate the joint posterior probability distribution of all unknowns means that such an analysis may soon be feasible. The flexibility of Bayesian inference in terms of model specification and use of external information promises to make phylogenetic analyses an increasingly rich and collaborative discipline at the core of bioinformatic analyses.

References

Anisimova, Maria, & Gascuel, Olivier. 2006. Approximate Likelihood-Ratio Test for Branches: A Fast, Accurate, and Powerful Alternative. *Systematic Biology*, **55**(4), 539–552.

Aris-Brosou, Stephane, & Yang, Ziheng. 2002. Effects of models of rate evolution on estimation of divergence dates with special reference to the metazoan 18S ribosomal RNA phylogeny. *Systematic Biology*, **51**(5), 703–714.

Bossuyt, F., Brown, R. M., Hillis, D. M., Cannatella, D. C., & Milinkovitch, M. C. 2006. Late Cretaceous diversification resulted in continent-scale regionalism in the cosmopolitan frog family Ranidae. *Systematic Biology*, **55**, 579–594.

Britton, Tom. 2005. Estimating Divergence Times in Phylogenetic Trees Without a Molecular Clock. *Systematic Biology*, **54**(3), 500–507.

Brown, R. M. 2007. Introduction to Robert F. Inger's Systematics and Zoogeography of Philippine Amphibia. *Pages 1–17 of: Systematics and Zoogeography of Philippine Amphibia*. Natural History Publications.

Chan, Cheong Xin, Beiko, Robert G, & Ragan, Mark A. 2006. Detecting recombination in evolving nucleotide sequences. *BMC Bioinformatics*, **7**(Jan), 412.

Che, J., Pang, J., Zhao, H., Wu, G. F., Zhao, E. M., & Zhang, Y. P. 2007. Phylogeny of Raninae (Anura: Ranidae) inferred from mitochondria and nuclear sequences. *Molecular Phylogenetics and Evolution*, **43**, 1–13.

Cheon, Sooyoung, & Liang, Faming. 2008. Phylogenetic tree construction using sequential stochastic approximation Monte Carlo. *Biosystems*, **91**(Aug), 94–107.

Choi, S, Hobolth, A, Robinson, D, Kishino, H, & Thorne, Jeffery L. 2007. Quantifying the Impact of Protein Tertiary Structure on Molecular Evolution. *Molecular Biology and Evolution*, Jan.

Conant, Gavin C, & Lewis, Paul O. 2001. Effects of Nucleotide Composition Bias on the Success of the Parsimony Criterion in Phylogenetic Inference. *Molecular Biology and Evolution*, **18**(6), 1024–1033.

Degnan, J, & Rosenberg, N. 2006. Discordance of species trees with their most likely gene trees. *PLoS Genet*, **2**(5).

Drummond, A., & Rambaut, A. 2007a. BEAST: Bayesian evolutionary analysis by sampling tree. *BMC Evolutionary Biology*, **7**, 214.

Drummond, A. J., & Rambaut, A. 2007b. BEAST v1.4. *http://beast.bio.ed.ac.uk/*.

Drummond, Alexei J., Ho, Simon Y. W., Phillips, Matthew J., & Rambaut, Andrew. 2006. Relaxed Phylogenetics and Dating with Confidence. *PLOS Biology*, **4**(5), e88.

Dubois, A. 1992. Notes sur la classification des Ranidae (Amphibiens, Anoures). *Bulletin de la Société Linnéenne de Lyon*, **51**, 305–352.

Durbin, R., Eddy, S., Krogh, A., & Mitchison, G. 1998. *Biological Sequence Analysis*. Cambridge University Press.

Edgar, Robert C. 2004. MUSCLE: multiple sequence alignment with high accuracy and high throughput. *Nucleic Acids Research*, **32**(5), 1792–1797.

Edwards, Scott V., Liu, Liang, & Pearl, Dennis K. 2007. High-resolution species trees without concatenation. *Proceedings of the National Academy of Sciences*, **104**(14), 5936–5941.

Ewens, Warren J. 2004. *Mathematical Population Genetics*. 2nd edn. Springer.

Fang, Fang, Ding, Jing, Minin, Vladimir N, Suchard, Marc A, & Dorman, Karin S. 2007. cBrother: relaxing parental tree assumptions for Bayesian recombination detection. *Bioinformatics*, **23**(4), 507–8.

Felsenstein, J. 2001. Taking Variation of Evolutionary Rates Between Sites into Account in Inferring Phylogenies. *J Mol Evol*, **53**(Jan), 447–455.

Felsenstein, Joseph. 1981. Evolutionary trees from DNA sequences: a maximum likelihood approach. *Journal of Molecular Evolution*, **17**, 368–376.

Felsenstein, Joseph. 1985. Confidence Limits on Phylogenies: An Approach Using the Bootstrap. *Evolution*, **39**(4), 783–791.

Fitch, Walter M., & Markowitz, E. 1970. An improved method for determining codon variability in a gene and its application to the rate of fixation of mutations in evolution. *Biochemical Genetics*, **4**, 579–593.

Foster, P. 2004. Modeling Compositional Heterogeneity. *Systematic Biology*, **53**(3), 485–495.

Foster, Peter G, & Hickey, Donal A. 1999. Compositional Bias May Affect Both DNA-Based and Protein-Based Phylogenetic Reconstructions. *Journal of Molecular Evolution*, **48**(Jan), 284–290.

Frost, D. R., Grant, T., Faivovich, J., Bain, R. H., Haas, A., Haddad, C. F. B., De Sá, R. O., Channing, A., Wilkinson, M., Donnellan, S. C., Raxworthy, C. J., Campbell, J. A., Blotto, B. L., Moler, P., Drewes, R. C., Nussbaum,

R. A., Lynch, J. D., Green, D. M., & Wheeler, W. C. 2006. The amphibian tree of life. *Bulletin of the American Museum of Natural History*, **297**, 1–370.

Gadagkar, Sudhindra R, Rosenberg, Michael S, & Kumar, Sudhir. 2005. Inferring species phylogenies from multiple genes: concatenated sequence tree versus consensus gene tree. *J Exp Zool B Mol Dev Evol*, **304**(1), 64–74.

Galtier, N. 1998. Inferring pattern and process: maximum-likelihood implementation of a nonhomogeneous model of DNA Sequence Evolution for Phylogenetic Analysis. *Molecular Biology and Evolution*, Jan.

Galtier, N. 2001. Maximum-Likelihood Phylogenetic Analysis Under a Covarion-like Model. *Molecular Biology and Evolution*, **18**(5), 866–873.

Galtier, N, & Gouy, M. 1995. Inferring Phylogenies From DNA Sequences of Unequal Base Compositions. *Proceedings of the National Academy of Sciences of the USA*, **92**(24), 11317–11321.

Goldman, N, Thorne, Jeffery L, & Jones, D T. 1996. Using evolutionary trees in protein secondary structure prediction and other comparative sequence analyses. *J Mol Biol*, **263**(2), 196–208.

Goldman, N, Thorne, Jeffery L, & Jones, D T. 1998. Assessing the impact of secondary structure and solvent accessibility on protein evolution. *Genetics*, **149**(1), 445–58.

Goldman, Nick, Anderson, J. P., & Rodrigo, A. G. 2000. Likelihood-based tests of topologies in phylogenetics. *Systematic Biology*, **49**, 652–670.

Gowri-Shankar, V, & Rattray, M. 2006. On the Correlation Between Composition and Site-Specific Evolutionary Rate: Implications for Phylogenetic Inference. *Molecular Biology and Evolution*, **23**(2), 352–364.

Gowri-Shankar, V, & Rattray, M. 2007. A Reversible Jump Method for Bayesian Phylogenetic Inference with a Nonhomogeneous Substitution Model. *Molecular Biology and Evolution*, **24**(6), 1286–1299.

Green, Peter J. 1995. Reversible jump Markov chain Monte Carlo computation and Bayesian model determination. *Biometrika*, **82**(4), 711–732.

Guindon, S, Rodrigo, A, Dyer, K, & Huelsenbeck, J. 2004. Modeling the site-specific variation of selection patterns along lineages. *Proc Natl Acad Sci US A*, **101**(35), 12957–12962.

Halpern, Aaron L., & Bruno, W. J. 1998. Evolutionary distances for protein-coding sequences: modeling site-specific residue frequencies. *Molecular Biology and Evolution*, **15**(7), 910–917.

Harvey, Paul H., & Pagel, Mark D. 1991. *The Comparative Method in Evolutionary Biology*. Oxford University Press.

Hein, Jotun, Schierup, Mikkel H., & Wiuf, Carsten. 2005. *Gene Genealogies, Variation and Evolution: A Primer in Coalescent Theory.* Oxford University Press.

Hillis, D. M., & Wilcox, T. P. 2005. Phylogeny of the New World true frogs. *Molecular Phylogenetics and Evolution*, **34**, 299–314.

Ho, Simon Y W, Phillips, Matthew J, Drummond, Alexei J, & Cooper, Alan. 2005. Accuracy of rate estimation using relaxed-clock models with a critical focus on the early metazoan radiation. *Molecular Biology and Evolution*, **22**(5), 1355–63.

Ho, Simon Y W, Shapiro, Beth, Phillips, Matthew J, Cooper, Alan, & Drummond, Alexei J. 2007. Evidence for time dependency of molecular rate estimates. *Systematic Biology*, **56**(3), 515–22.

Hobolth, A, Nielsen, R, Wang, Y, Wu, F, & Tanksley, S. 2006. CpG+ CpNpG Analysis of Protein-Coding Sequences from Tomato. *Molecular Biology and Evolution*, **23**(6), 1318–1323.

Hobolth, Asger. 2008. A Markov chain Monte Carlo expectation maximization algorithm for statistical analysis of DNA sequence evolution with neighbor-dependent substitution rates. *Journal of Computational & Graphical Statistics*, **17**(1), 138–162.

Holder, Mark T., Lewis, Paul O., Swofford, David L., & Larget, Bret. 2005. Hastings Ratio of the LOCAL Proposal Used in Bayesian Phylogenetics. *Systematic Biology*, **54**, 961–965.

Huelsenbeck, J, & Suchard, Mark A. 2007. A Nonparametric Method for Accommodating and Testing Across-Site Rate Variation. *Systematic Biology*, **56**(6), 975–987.

Huelsenbeck, John P. 2002. Testing a covariotide model of DNA substitution. *MBE*, **19**(5), 698–707.

Huelsenbeck, John P., Larget, Bret, & Swofford, David. 2000. A compound Poisson process for relaxing the molecular clock. *Genetics*, **154**, 1879–1892.

Huelsenbeck, John P, Ronquist, Frederik, Nielsen, R, & Bollback, J. 2001. Bayesian Inference of Phylogeny and Its Impact on Evolutionary Biology. *Science*, **294**(Jan), 2310–2314.

Hwang, D, & Green, P. 2004. Bayesian Markov Chain Monte Carlo Sequence Analysis Reveals Varying Neutral Substitution Patterns in Mammalian Evolution. *Proceedings of the National Academy of Science USA*, **101**(Jan), 13994–14001.

Inger, R. F. 1954. Systematics and zoogeography of Philippine Amphibia. *Fieldiana*, **33**, 181–531.

Jensen, Jens Ledet, & Pedersen, Anne-Mette Krabbe. 2000. Probabilistic models of DNA sequence evolution with context dependent rates of substitution. *Advances in Applied Probability*, **32**, 499–517.

Jow, H, Hudelot, C, Rattray, M, & Higgs, P. 2002. Bayesian Phylogenetics Using an RNA Substitution Model Applied to Early Mammalian Evolution. *Molecular Biology and Evolution*, **19**(9), 1591–1601.

Jukes, T. H., & Cantor, C. R. 1969. *Evolution of protein molecules*. Academic Press. Pages 21–132.

Kingman, J.F.C. 1982a. The coalescent. *Stochastic Processes and their Applications*, **13**, 235–248.

Kingman, J.F.C. 1982b. On the genealogy of large populations. *Journal of Applied Probability*, **19**, 27–43.

Kishino, Hirohisa, Thorne, Jeffrey L., & Bruno, W. J. 2001. Performance of a divergence time estimation method under a probabilistic model of rate evolution. *Molecular Biology and Evolution*, **18**, 352–361.

Knudsen, Bjarne, & Miyamoto, Michael M. 2003. Sequence alignments and pair hidden Markov models using evolutionary history. *Journal of Molecular Biology*, **333**(2), 453–60.

Kosakovsky Pond, Sergei L., & Frost, Simon. 2005. A Simple Hierarchical Approach to Modeling Distributions of Substitution Rates. *Molecular Biology and Evolution*, **22**(2), 223.

Lakner, Clemens. 2008. splitsmb v0.1.1. *http://sourceforge.net/projects/splitsmb*.

Lakner, Clemens, van der Mark, Paul, Huelsenbeck, John P, Larget, Bret, & Ronquist, Fredrik. 2008. Efficiency of Markov Chain Monte Carlo Tree Proposals in Bayesian Phylogenetics. *Systematic Biology*, **57**(1), 86–103.

Larget, Bret, & Simon, Donald L. 1999. Markov chain Monte Carlo algorithms for the Bayesian analysis of phylogenetic trees. *Molecular Biology and Evolution*, **16**(6), 750–759.

Lartillot, N., & Phillipe, H. 2004. A Bayesian mixture model for across-site heterogeneities in the amino-acid replacement process. *Molecular Biology and Evolution*, **21**(6), 1095–1109.

Lepage, Thomas, Lawi, S., Tupper, P. F., & Bryant, D. 2006. Continuous and tractable models for the evolutionary rate. *Mathematical Biosciences*, **199**, 216–233.

Lepage, Thomas, Bryant, David, Philippe, Herve, & Lartillot, Nicolas. 2007. A general comparison of relaxed molecular clock models. *Molecular Biology and Evolution*, **24**(12), 2669–2680.

Li, S. Y., Pearl, D. K., & Doss, H. 2000. Phylogenetic tree construction using Markov chain Monte Carlo. *Journal of the American Statistical Association*, **95**(450), 493–508.

Liang, Faming, Liu, Chuanhai, & Carroll, Raymond. 2007. Stochastic Approximation in Monte Carlo Computation. *Journal of the American Statistical Association*, **102**(Mar), 305–320.

Liò, P, Goldman, N, Thorne, Jeffery L., & Jones, D T. 1998. PASSML: combining evolutionary inference and protein secondary structure prediction. *Bioinformatics*, **14**(8), 726–33.

Liu, Liang, & Pearl, D. K. 2007. Species trees from gene trees: reconstruction Bayesian posterior distributions of a species phylogeny using estimated gene tree distributions. *Systematic Biology*, **56**(3), 504–514.

Lockhart, P, Steel, M, Hendy, M, & Penny, D. 1994. Recovering evolutionary trees under a more realistic model of sequence evolution. *Molecular Biology and Evolution*, **11**(4), 605–612.

Lopez, P, Casane, D, & Philippe, H. 2002. Heterotachy, an Important Process of Protein Evolution. *Molecular Biology and Evolution*, **19**(1), 1–7.

Lunter, G., Miklós, I., Drummond, A., Jensen, J. L., & Hein, J. 2005a. Bayesian coestimation of phylogeny and sequence alignment. *BMC Bioinformatics*, **6**, 83.

Lunter, G., Drummond, A.J., Miklós, I., & Hein, J. 2005b. Statistical Alignment: Recent Progress, New Applications, and Challenges. *Pages 375–406 of:* Nielsen, Rasmus (ed), *Statistical Methods in Molecular Evolution (Statistics for Biology and Health)*. Berlin: Springer.

Maddison, Wayne, Midford, Peter E., & Otto, S. E. 2007. Estimating a binary character's effect on speciation and extinction. Systematic Biology. *Systematic Biology*, **56**(5), 701–710.

Mallet, J. 2007. Hybrid Speciation. *Nature*, **466**(Jan), 279–282.

Mau, B, & Newton, M. 1997. Phylogenetic Inference for Binary Data on Dendograms Using Markov Chain Monte Carlo. *Journal of Computational and Graphical Statistics*, **6**(1), 122–131.

Mau, Bob. 1996. *Bayesian Phylogenetic Inference via Markov Chain Monte Carlo Methods*. Ph.D. thesis, University of Wisconsin-Madison, Dept. of Statistics.

Mau, Bob, Newton, Michael A, & Larget, Bret. 1999. Bayesian Phylogenetic Inference via Markov Chain Monte Carlo Methods. *Biometrics*, **55**(1), 1–12.

Mayden, Richard L. 1999. Consilience And A Hierarchy Of Species Concepts: Advances Toward Closure On The Species Puzzle. *Journal of nematology*, **31**(2), 95–116.

Meyer, Irmtraud M, & Miklós, I. 2007. SimulFold: Simultaneously Inferring RNA Structures Including Pseudoknots, Alignments, and Trees Using a Bayesian MCMC Framework. *PLoS Comput Biol*, **3**(8), e149.

Miklós, I., Lunter, G. A., & Holmes, I. 2004. A "long indel model" for evolutionary sequence alignment. *Mol. Biol. Evol.*, **21**(3), 529–540.

Newton, M, Mau, B, & Larget, B. 1999. Markov Chain Monte Carlo for the Bayesian Analysis of Evolutionary Seqeuences from Aligned Molecular Sequences. *Statistics in Molecular Biology IMS Lecture Notes*, **33**(Jan), 143–162.

Neyman, J. 1971. Molecular studies of evolution: a source of novel statistical problems. *Pages 1–27 of:* Gupta, S. S., & Yackel, J. (eds), *Statistical decision theory and related topics*. New York: Academic Press.

Page, Roderic D. M., & Holmes, Edward C. 1998. *Molecular Evolution: A Phylogenetic Approach*. Malden, MA, USA: Blackwell Science Ltd.

Pagel, Mark, & Meade, Andrew. 2004. A Phylogenetic Mixture Model for Detecting Pattern-Heterogeneity in Gene Sequence or Character-State Data. *Systematic Biology*, **53**(4), 571–581.

Paraskevis, D, Deforche, K, Lemey, P, Magiorkinis, G, Hatzakis, A, & Vandamme, A-M. 2005. SlidingBayes: exploring recombination using a sliding window approach based on Bayesian phylogenetic inference. *Bioinformatics*, **21**(7), 1274–5.

Rannala, Bruce, & Yang, Ziheng. 1996. Probability Distribution of Molecular Evolutionary Trees: A New Method of Phylogenetic Inference. *Journal of Molecular Evolution*, **43**, 304–311.

Rannala, Bruce, & Yang, Ziheng. 2003. Bayes estimation of species divergence times and ancestral population sizes using DNA sequences from mutiple loci. *Genetics*, **164**, 1645–1656.

Redelings, Benjamin, & Suchard, Marc. 2005. Joint Bayesian Estimation of Alignment and Phylogeny. *Systematic Biology*, **54**(3), 401–418.

Redelings, Benjamin D, & Suchard, Marc A. 2007. Incorporating indel information into phylogeny estimation for rapidly emerging pathogens. *BMC Evolutionary Biology*, **7**(Jan), 40.

Rivas, Elena, Eddy, Sean R, & Haussler, David. 2008. Probabilistic Phylogenetic Inference with Insertions and Deletions. *PLoS Computational Biology*, **4**(9), e1000172.

Robinson, Douglas M., Jones, David T., Kishino, Hirohisa, Goldman, Nick, & Thorne, Jeffrey L. 2003. Protein Evolution with Dependence Among Codons Due to Tertiary Structure. *Molecular Biology and Evolution*, **20**, 1692–1704.

228   *Bayesian Modeling in Bioinformatics*

Rodrigue, N, Philippe, H, & Lartillot, N. 2008. Uniformization for sampling realizations of Markov processes: applications to Bayesian implementations of codon substitution models. *Bioinformatics*, **24**(1), 56–62.

Rodrigue, Nicolas, Lartillot, Nicolas, Bryant, David, & Philippe, Hervé. 2005. Site interdependence attributed to tertiary structure in amino acid sequence evolution. *Gene*, **347**, 207–217.

Rodrigue, Nicolas, Philippe, Hervé, & Lartillot, Nicolas. 2006. Assessing site-interdependent phylogenetic models of sequence evolution. *MBE*, **23**(9), 1762–75.

Roelants, K., Gower, D. J., Wilkinson, M., Loader, S. P., Biju, S. D., Guillaume, K., Moriau, L., & Bossuyt, F. 2007. Global patterns of diversification in the history of modern amphibians. *Proceedings of the National Academy of the Sciences*, **104**, 887–892.

Ronquist, Fredrik R., & Huelsenbeck, John P. 2003. MRBAYES 3: Bayesian phylogenetic inference under mixed models. *Bioinformatics*, **19**(12), 1574–1575.

Rosenberg, M, & Kumar, S. 2003. Heterogeneity of Nucleotide Frequencies Among Evolutionary Lineages and Phylogenetic Inference. *Molecular Biology and Evolution*, Jan.

San Mauro, D., Vences, M., Alcobendas, M., Zardoya, R., & Meyer, A. 2005. Initial diversification of living amphibians predated the breakup of Pangaea. *American Naturalist*, **165**, 590–599.

Sanderson, M. J. 1997. A nonparametric approach to estimating divergence times in the absence of rate constancy. *Molecular Biology and Evolution*, **14**, 1218–1231.

Sanderson, M. J. 2002. Estimating absolute rates of molecular evolution and divergence times: a penalized likelihood approach. *Molecular Biology and Evolution*, **19**, 101–109.

Sankoff, David. 1975. Minimal Mutation Trees of Sequences. *SIAM Journal on Applied Mathematics*, **28**(1), 35–42.

Schwarz, Gideon. 1978. Estimating the Dimension of a Model. *Annals of Statistics*, **6**(2), 461–464.

Shimodaira, Hidetoshi. 2002. An approximately unbiased test of phylogenetic tree selection. *Systematic Biology*, **51**(3), 492–508.

Siepel, Adam, & Haussler, David. 2004. Phylogenetic estimation of context-dependent substitution rates by maximum likelihood. *MBE*, **21**(3), 468–88.

Sinsheimer, Janet S, Suchard, Marc A, Dorman, Karin S, Fang, Fang, & Weiss, Robert E. 2003. Are you my mother? Bayesian phylogenetic inference of recombination among putative parental strains. *Applied Bioinformatics*, **2**(3), 131–44.

Slowinski, J. B., & Page, Roderic D. M. 1999. How Should Species Phylogenies Be Inferred from Sequence Data? *Systematic Biology*, **48**(4), 814–825.

Stamatakis, Alexandros. 2006. RAxML-VI-HPC: Maximum Likelihood-based Phylogenetic Analyses with Thousands of Taxa and Mixed Models. *Bioinformatics*, **22**(21), 2688–2690.

Stefankovivc, Daniel, & Vigoda, Eric. 2007. Pitfalls of Heterogeneous Processes for Phylogenetic Reconstruction. *Systematic Biology*, **56**(1), 113–124.

Stern, Adi, & Pupko, Tal. 2006. An evolutionary space-time model with varying among-site dependencies. *MBE*, **23**(2), 392–400.

Stuart, B. L. 2008. The phylogenetic problem of *Huia* (Amphibia: Ranidae). *Molecular Phylogenetics and Evolution*, **46**, 49–60.

Suchard, M. A., & Redelings, B. D. 2006. BAli-Phy: simultaneous Bayesian inference of alignment and phylogeny. *Bioinformatics*, **22**, 2047–2048.

Suchard, Marc A, Weiss, Robert E, Dorman, Karin S, & Sinsheimer, Janet S. 2002. Oh brother, where art thou? A Bayes factor test for recombination with uncertain heritage. *Systematic Biology*, **51**(5), 715–28.

Thorne, J. 2007. Protein evolution constraints and model-based techniques to study them. *Curr Opin Struct Biol*, **17**(Jan), 337–341.

Thorne, J. L., Kishino, H., & Felsenstein, J. 1991. An evolutionary model for maximum likelihood alignment of DNA sequences. *Journal of Molecular Evolution*, **33**, 114–124.

Thorne, Jeffery L., & Kishino, H. 2002. Divergence time and evolutionary rate estimation with multilocus data. *Systematic Biology*, **51**(5), 689–702.

Thorne, Jeffery L, Kishino, H, & Felsenstein, Joseph. 1992. Inching toward reality: an improved likelihood model of sequence evolution. *J Mol Evol*, **34**(1), 3–16.

Thorne, Jeffery L, Goldman, N, & Jones, D T. 1996. Combining protein evolution and secondary structure. *MBE*, **13**(5), 666–73.

Thorne, Jeffery L., Choi, Sang Chul, Yu, J, Higgs, PG, & Kishino, H. 2007. Population genetics without intraspecific data. *Molecular Biology and Evolution*, **24**(8), 1667–1677.

Thorne, Jeffrey L., Kishino, Hirohisa, & Painter, Ian S. 1998. Estimating the rate of evolution of the rate of molecular evolution. *Molecular Biology and Evolution*, **15**, 1647–1657.

Tuffley, Chris, & Steel, Mike. 1997. Modelling the covarion hypothesis of nucleotide substitution. *Mathematical Biosciences*, **147**, 63–91.

Welch, J. J., Fontanillas, E., & Bromham, L. 2005. Molecular dates for the "cambrian explosion": the influence of prior assumptions. *Systematic Biology*, **54**, 672–678.

Whelan, Simon. 2007. New Approaches to Phylogenetic Tree Search and Their Application to Large Numbers of Protein Alignments. *Systematic Biology*, **56**(5), 727–740.

Whelan, Simon. 2008. The genetic code can cause systematic bias in simple phylogenetic models. *Philos Trans R Soc Lond, B, Biol Sci*, **363**, 4003–4011.

Wiens, J. J. 2007. Global patterns of species richness and diversification in amphibians. *American Naturalist*, **170**, S86–S106.

Wiens, J. J., A., Sukumaran R., Pyron, A., & Brown, R. M. In press. Parallel causes drive the latitudinal diversity gradient in old world and new world frog clades (Ranidae and Hylidae). *Evolution*.

Wu, Jihua, Susko, Edward, & Roger, Andrew J. 2008. An independent heterotachy model and its implications for phylogeny and divergence time estimation. *Molecular Phylogenetics and Evolution*, **46**(2), 801–6.

Yang, Z. 1995. A space-time process model for the evolution of DNA sequences. *Genetics*, **139**(2), 993–1005.

Yang, Z., & Yoder, A. D. 2003. Comparison of likelihood and bayesian methods for estimating divergence times using multiple gene loci and calibration points, with application to a radiation of cute-looking mouse lemu species. *Systematic Biology*, **52**(5), 705–716.

Yang, Ziheng. 1994. Maximum likelihood phylogenetic estimation from DNA sequences with variable rates over sites: approximate methods. *Journal of Molecular Evolution*, **39**.

Yang, Ziheng, & Rannala, Bruce. 2006. Bayesian estimation of species divergence times under a molecular lock using multiple fossil calibrations with soft bounds. *Molecular Biology and Evolution*, **23**(1), 212–226.

Yang, Ziheng, & Roberts, Dave. 1995. On the use of nucleic acid sequences to infer early branchings in the tree of life. *Molecular Biology and Evolution*, **12**(3), 451–458.

Yoder, A. D., & Yang, Z. 2000. Estimation of primate speciation dates using local molecular clocks. *Molecular Biology and Evolution*, **17**(7), 1081–1090.

Yu, J, & Thorne, J. 2006. Dependence among Sites in RNA Evolution. *Molecular Biology and Evolution*, **23**(8), 1525–1536.

Zhou, Y, Rodrigue, N, Lartillot, N, & Philippe, H. 2007. Evaluation of the models handling heterotachy in phylogenetic inference. *BMC Evolutionary Biology*, **2007**(7), 206.

Zuckerland, E., & Pauling, L. 1965. Evolutionary divergence and convergence. *Pages 97–166 of:* Bryson, V., & J., Vogel H. (eds), *Evolving genes and Protein*. Academic Press.

Zwickl, Derrick J. 2006. *Genetic algorithm approaches for the phylogenetic analysis of large biological sequence datasets under the maximum likelihood criterion*. Ph.D. thesis, The University of Texas at Austin.

Chapter 9

Gene Selection for the Identification of Biomarkers in High-Throughput Data

Jaesik Jeong, Marina Vannucci, Kim-Anh Do, Bradley Broom, Sinae Kim, Naijun Sha, Mahlet Tadesse, Kai Yan, and Lajos Pusztai
Indiana University, Indianapolis, IN; Rice University, Houston, TX; UT Texas MD Anderson Cancer Center, Houston, TX; University of Michigan, Ann Arbor, MI; University of Texas at El Paso, TX and Georgetown University, Washington, DC

9.1　Introduction

In this chapter we review our contribution to the development of Bayesian methods for variable selection. In particular, we review linear regression settings where the response variable is either continuous, categorical or a survival time. We also briefly describe how some of the key ideas of the variable selection method can be used in a different modeling context, i.e. model-based sample clustering. In the linear settings, we use a latent variable selection indicator to induce mixture priors on the regression coefficients. In the clustering setting, the group structure in the data is uncovered by specifying mixture models. In both the linear and the clustering settings, we specify conjugate priors and integrate out some of the parameters to accelerate the model fitting. We use Markov chain Monte Carlo (MCMC) techniques to identify the high probability models. The methods we describe are particularly relevant for the analysis of genomic studies, where high-throughput technologies allow thousands of variables to be measured on individual samples. The amount of measured variables in such data is in fact often substantially larger than the sample size. A typical example with this characteristic, and one that we use to illustrate our methodologies, is DNA microarray data.

The practical utility of variable selection is well recognized and this topic has been the focus of much research. Variable selection can help assess the importance of explanatory variables, improve prediction accuracy, provide a better understanding of the underlying mechanisms generating data, and reduce the cost of measurement and storage for future data. A comprehensive account of widely used classical methods, such as stepwise regression with forward and backward selection can be found in Miller.[22] In recent years, procedures that specifically deal with very large number of variables have been proposed. One such approach is the least absolute shrinkage and selection operator (lasso) method of Tibshirani,[36] which uses a penalized likelihood approach to shrink to zero coefficient estimates associated with unimportant covariates. For Bayesian variable selection methods, pioneering work in the univariate linear model setting was done by Mitchell & Beauchamp,[23] and George & McCulloch,[15] and in the multivariate setting by Brown *et al.*.[6] The key idea of the Bayesian approach is to introduce a latent binary vector to index possible subsets of variables. This indicator is used to induce a mixture prior on the regression coefficients, and the variable selection is performed based on the posterior model probabilities.

In high-throughput genomic and proteomic studies, there is often interest in identifying markers that discriminate between different groups of tissues. The distinct classes may correspond to different subtypes of a disease or to groups of patients who respond differently to treatment. The problem of locating relevant variables could arise in the context of a supervised or an unsupervised analysis. In the supervised setting, a training data is available in which the group membership of all samples is known. The goal of variable selection in this case is to locate sets of variables that relate to the pre-specified groups, so that the class membership of future samples can be predicted accurately. Standard methods for classification include linear and quadratic discriminant analyses. When dealing with high-dimensional data, where the sample size n is smaller than the number p of variables, a dimension reduction step, such as partial least squares[26] or singular value decompositions,[41] is often used. Methods that directly perform variable selection in classification, such as support vector machines[40] and the shrunken centroid approach,[37] have also been developed in recent years. The Bayesian methods reviewed here address the selection and prediction problems in a unified manner.

Variable selection is a major goal in the unsupervised setting as well, where the samples' outcomes are not observed and the goal of the analysis is to identify variables with distinctive expression patterns while uncovering the latent classes. In statistics this unsupervised framework is referred to as clustering. The problem of variable selection is this setting is fundamentally different from the supervised setting and requires different modeling strategies. Here we adopt a model-based approach that views the data as coming from a mixture of probability distributions.[21] Diebolt & Robert[9] present MCMC strategies when the number of mixture components is known. Richardson & Green[28] &

Stephens[33] propose methods for handling finite mixture models with an unknown number of components. We also look into infinite mixture models that use Dirichlet process priors.[4,12] A few procedures that combine the variable selection and clustering tasks have been proposed. For instance, Fowlkes *et al.*[13] use a forward selection approach in the context of hierarchical clustering. Recently, Friedman, & Meulman[14] have proposed a hierarchical clustering procedure that uncovers cluster structure on separate subsets of variables.

This chapter is organized as follows. In Section 9.2, we briefly review the Bayesian variable selection methods for linear model settings. In Section 9.3 we discuss variable selection in the context of clustering, using both finite mixture models with an unknown number of components and infinite mixture models and Dirichlet process priors. We conclude the chapter in Section 9.4 with an application of the methods to a DNA microarray dataset.

9.2   Variable Selection in Linear Settings

Let us start from the classical linear regression setting with a continuous response. Categorical responses and survival times will be addressed next.

9.2.1   Continuous Response

In the multivariate linear model setting an $n \times q$ continuous response $\boldsymbol{Y}$ is related to the $n \times p$ covariate matrix $\boldsymbol{X}$ via a model of the form

$$\boldsymbol{Y}_i = \mathbf{1}\alpha' + \boldsymbol{X}_i'\boldsymbol{B} + \boldsymbol{\varepsilon}, \quad \boldsymbol{\varepsilon} \sim \mathcal{N}(\mathbf{0}, \Sigma), \quad i = 1, \dots, n, \tag{9.1}$$

where $\boldsymbol{B}$ is the $p \times q$ matrix of regression coefficients and $\boldsymbol{X}_i$, $\boldsymbol{Y}_i$ are $p \times 1$, $1 \times q$ vectors, respectively. Basically, there are two things we have to consider for Bayesian analysis: Prior elicitation and posterior inference.

Often, not all the covariates in $\boldsymbol{X}$ play an important role of explaining changes in $\boldsymbol{Y}$ and the goal is to identify the promising subset of predictors. For instance, in DNA microarray studies where thousands of variables (genes) are measured, a large number of them provide very little, if any, information about the outcome. This is a problem of variable selection. In the Bayesian framework, variable selection is accomplished by introducing a latent binary vector, $\boldsymbol{\gamma}$, which is used to induce mixture priors on the regression coefficients[15,6]

$$\boldsymbol{B}_j \sim \gamma_j \mathcal{N}(0, \tau_j^2 \Sigma) + (1 - \gamma_j)\mathcal{I}_0, \tag{9.2}$$

where $\boldsymbol{B}_j$ indicates the j-th row of $\boldsymbol{B}$ and $\mathcal{I}_0$ is a vector of point masses at 0. If $\gamma_j = 1$, variable X_j is included in the model and is considered meaningful in explaining the outcome; if $\gamma_j = 0$, the corresponding vector of regression coefficients has a prior with point mass at 0 and variable X_j is therefore excluded from the model and deemed unimportant.

Suitable priors can be specified for γ, the simplest choice being independent Bernoulli priors

$$p(\gamma) = \prod_{j=1}^{p} \theta^{\gamma_j}(1-\theta)^{1-\gamma_j}, \tag{9.3}$$

where $\theta = |\gamma|/p$ and $|\gamma| = \sum_{j=1}^{p} \gamma_j$ is the number of variables expected *a priori* to be included in the model. This prior can be relaxed by using hyperprior. For example, a beta distribution, $Beta(a,b)$ can be used as a hyper prior for θ where a, b are some constants. Conjugate normal and inverse-Wishart priors can be specified for the parameters B and Σ, respectively.

9.2.2 Categorical Response

In classification, the observed outcome, Z_i is a categorical variable that takes one of K values identifying the group from which each sample arises. A multinomial probit model can then be used to link the categorical outcome, Z, to the linear predictors, X, by using a data augmentation approach, as in Albert and Chib.[1] This approach introduces a latent matrix $Y_{n \times (K-1)}$ where the row vector $Y_i = (y_{i,1}, \cdots, y_{i,K-1})$ indicates the propensities of sample i to belong to one of the K classes. According to the type of response (nominal or ordinal), there are two different mappings between the categorical outcome and the latent continuous outcome. A correspondence between the nominal categorical outcome, Z_i, and the latent continuous outcome, Y_i, is defined by

$$z_i = \begin{cases} 0 \text{ if } \max_{1 \leq k \leq K-1}\{y_{i,k}\} \leq 0 \\ j \text{ if } \max_{1 \leq k \leq K-1}\{y_{i,k}\} > 0 \text{ and } j = \operatorname*{argmax}_{1 \leq k \leq K-1}\{y_{i,k}\}. \end{cases} \tag{9.4}$$

Here we view the first category as a "baseline". The data augmentation allows us to express the classification problem as a multivariate regression model of the form (9.1), in terms of the latent continuous outcome Y.

If Z is an ordered categorical outcome, such as the stage of a tumor, we account for the ordering by introducing a latent continuous vector Y and modifying the correspondence between y_i and z_i to

$$z_i = j \quad \text{if} \quad \delta_j < y_i \leq \delta_{j+1}, \quad j = 0, \cdots, K-1, \tag{9.5}$$

where the boundaries δ_j are unknown and $-\infty = \delta_0 < \delta_1 < \cdots < \delta_{K-1} < \delta_K = \infty$. In this case the classification problem reduces to a linear regression model that is univariate in the response.

Bayesian variable selection can then be implemented as described in Section 9.2.1 by introducing a latent binary vector γ that induces mixture priors on the regression coefficients, see Sha *et al.*[31,32]

9.2.3 Survival Time

In survival analysis, censored or real survival times are observed. We have employed accelerated failure time (AFT) model, see Sha *et al.*,[30] because of their attractive linear setting. The AFT model is given by

$$\log(T_i) = \alpha + \boldsymbol{x}_i'\boldsymbol{\beta} + \epsilon_i, \quad i = 1, \cdots, n \tag{9.6}$$

where T_i is the survival time, $\boldsymbol{x}_i$ is a $p \times 1$ covariate vector, $\boldsymbol{\beta}$ is a $p \times 1$ regression parameter vector and $\epsilon_i's$ are *iid* random variables which may have parametric distribution. For example, we obtain a log-normal model if ϵ_i follows a normal distribution, log-t model with a t distribution, and a Weibull model with an extreme value distribution. We focus on log-normal model throughout this chapter. Let c_i be the censored time which is independent of time t_i. We observe $t_i^* = min(t_i, c_i)$ and $\delta_i = I(t_i \leq c_i)$. Let's define $w_i = \log(t_i)$. i.e. $w_i = \log(t_i^*)$ if $\delta_i = 1$ and $w_i \geq \log(t_i^*)$ otherwise. Suppose that $\epsilon_i's$ are *iid* $\mathcal{N}(0, \sigma^2)$. Then

$$[\boldsymbol{W}|\boldsymbol{X}, \alpha, \boldsymbol{\beta}, \sigma^2] \sim \mathcal{N}(\alpha \boldsymbol{1}_n + \boldsymbol{X}\boldsymbol{\beta}, \sigma^2 \boldsymbol{I}_n)$$

where $\boldsymbol{1}_n = (1, \cdots, 1)'$, and $\boldsymbol{I}_n$ is identity matrix.

As for the previous modelling settings, Bayesian variable selection can be implemented by introducing a latent binary vector $\boldsymbol{\gamma}$ that induces mixture priors on the regression coefficients and conjugate priors can be specified on the model parameters α, $\boldsymbol{\beta}$ and σ^2.

9.2.4 Markov Chain Monte Carlo Algorithm

For posterior inference in the standard linear regression setting, fast and efficient Markov chain Monte Carlo (MCMC) can be implemented by integrating out $\boldsymbol{B}$ and Σ, so that $\boldsymbol{\gamma}$ becomes the only parameter that needs to be updated. In high-throughput experiments, the posterior space of $\boldsymbol{\gamma}$ is too big to be fully explored because there are many variables (genes) compared to the number of sample. For this, the following stochastic search inspired by MCMC is considered for exploration of a portion of the space that contains models with high posterior probability. Posterior samples for $\boldsymbol{\gamma}$ can be obtained via Gibbs sampling or a Metropolis algorithm. For instance, the latent vector can be updated using a Metropolis algorithm that generates at each iteration a new candidate $\boldsymbol{\gamma}^{new}$ by randomly choosing one of these transition moves:

1a. Add/Delete: randomly pick one of the indices in $\boldsymbol{\gamma}^{old}$ and change its value.

1b. Swap: draw independently and at random a 0 and a 1 in $\boldsymbol{\gamma}^{old}$ and switch their values.

The new candidate is accepted with probability

$$\min\left\{1, \frac{f(\boldsymbol{\gamma}^{new}|\boldsymbol{Y}, \boldsymbol{X})}{f(\boldsymbol{\gamma}^{old}|\boldsymbol{Y}, \boldsymbol{X})}\right\}. \tag{9.7}$$

The model fitting in the classification setting is a bit more intricate because the regression model is defined in terms of latent outcomes. The MCMC procedure needs to account for this and includes a step that updates the latent values Y from their full conditionals. In particular the following step is added to the MCMC:

2. Update Y from its marginal distribution $f(Y|X, Z, \gamma)$. For classification into nominal categories this distribution is a truncated matrix-variate t-distribution. For ordered classes it follows a multivariate truncated normal distribution. In this latter setting we also need to update the boundary parameters δ_j from their posterior densities, which are uniform on the interval

$$[\max\{\max\{Y_i : Z_i = j - 1\}, \delta_{j-1}\}, \min\{\min\{Y_i : Z_i = j\}, \delta_{j+1}\}]$$

with $j = 1, \cdots, K - 1$.

For log-normal AFT models, the additional step updates only the censored elements of W from their marginal distribution:

2. Update the censored elements, w_i with $\delta_i = 0$, from $f(w_i|W_{(-i)}, X, \gamma)$ as univariate truncated t-distributions.

9.2.5   Posterior Inference

The MCMC procedure results in a list of visited models, $\gamma^{(0)}, \cdots, \gamma^{(T)}$ and their corresponding posterior probabilities. Several criteria can be used for posterior inference on the selected variables. A selection rule based on the γ vectors with largest joint posterior probabilities among the visited models is given by

$$\hat{\gamma} = \underset{1 \leq t \leq T}{\mathrm{argmax}}\, p(\gamma^{(t)}|Y, X) \tag{9.8}$$

where T is the number of iterations. From a marginal point of view, one can estimate marginal posterior probabilities for inclusion of each γ_j by

$$p(\gamma_j = 1|X, Y) \approx \sum_{\gamma^{(t)}:\gamma_j=1} p(Y|X, \gamma^{(t)})p(\gamma^{(t)}) \tag{9.9}$$

and choose those γ_j's with marginals exceeding a given cut-off. In the example below we choose the cut-off point arbitrarily. Methods based on expected false discovery rates can also be employed, similarly to Newton et al.[25] In addition, the marginal posterior probabilities can also be estimated by simple frequencies, by counting how many times each gene is visited during the

MCMC sampler after burn-in. The probability of inclusion of gene j is given by

$$p(\gamma_j = 1|\boldsymbol{X}, \boldsymbol{Y}) = \frac{1}{T-b} \sum_{t=b+1}^{T} I(\gamma_j^{(t)} = 1) \qquad (9.10)$$

where b is the burn-in.

As for prediction of future observations, this can be achieved via least squares prediction based on a single "best" model or using Bayesian model averaging (BMA), which accounts for the uncertainty in the selection process by averaging over a set of *a posteriori* likely models to estimate $\boldsymbol{Y}_f$ as

$$\widehat{\boldsymbol{Y}}_f = \sum_{\gamma} \left(\boldsymbol{1}\tilde{\alpha}' + \boldsymbol{X}_{f(\gamma)} \tilde{\boldsymbol{B}}_{(\gamma)} \right) p(\boldsymbol{\gamma}|\boldsymbol{Y}, \boldsymbol{X}) \qquad (9.11)$$

where $\boldsymbol{X}_{f(\gamma)}$ consists of the covariates selected by $\boldsymbol{\gamma}$, $\tilde{\alpha} = \bar{\boldsymbol{Y}}$, $\tilde{\boldsymbol{B}}_\gamma = (\boldsymbol{X}'_\gamma \boldsymbol{X}_\gamma + \boldsymbol{H}_\gamma^{-1})^{-1} \boldsymbol{X}'_\gamma \boldsymbol{Y}$ and $\boldsymbol{H}$ is the prior row covariance matrix of $\boldsymbol{B}$.

Similar inference can be performed in the probit model setting, where the latent variables $\boldsymbol{Y}$ are first imputed using the corresponding MCMC averages, and then the corresponding predicted categorical value for a future observation can be computed via (9.4).

In survival, the censored failure times are imputed by the mean of their sampled values, $\widehat{\boldsymbol{W}} = \frac{1}{T} \sum_{t=1}^{T} \boldsymbol{W}^{(t)}$. The log-survival times are estimated by the posterior probabilities of the visited models:

$$\widehat{\boldsymbol{W}}_f = \sum_{\gamma} \boldsymbol{X}^*_{f(\gamma)} \hat{\boldsymbol{\beta}}^*_\gamma \, p(\boldsymbol{\gamma}|\boldsymbol{X}, \widehat{\boldsymbol{W}}). \qquad (9.12)$$

where $\hat{\boldsymbol{\beta}}^*_\gamma = \left(\boldsymbol{X}^{*\prime}_\gamma \boldsymbol{V} \boldsymbol{X}^*_\gamma + \Sigma^{*-1}_{0(\gamma)} \right)^{-1} \left(\boldsymbol{X}^{*\prime}_\gamma \boldsymbol{V} \widehat{\boldsymbol{W}} + \Sigma^{*-1}_{0(\gamma)} \boldsymbol{\beta}^*_{0\gamma} \right)$, $\boldsymbol{X}^*_\gamma = (\boldsymbol{1}, \boldsymbol{X}_\gamma)$, $\boldsymbol{\beta}^*_{0\gamma} = \binom{\alpha_0}{\beta_{0\gamma}}$ and $\Sigma^*_{0(\gamma)} = \mathrm{diag}(h_0, \Sigma_{0(\gamma)})$ and $\boldsymbol{V} = \boldsymbol{I}$ in the log-normal case.

Also, in the AFT model a sampling-based estimation approach can be implemented for estimation of the predictive survivor function. Let $\boldsymbol{x}_f$ be the covariate vector of a new subject. Then, we have

$$P(T > t|\boldsymbol{x}_f, \boldsymbol{X}, \widehat{\boldsymbol{W}}) = P(W > w|\boldsymbol{x}_f, \boldsymbol{X}, \widehat{\boldsymbol{W}})$$
$$= \int P(W > w|\boldsymbol{x}_f, \boldsymbol{X}, \widehat{\boldsymbol{W}}, \boldsymbol{\gamma}) \, p(\boldsymbol{\gamma}|\boldsymbol{x}_f, \boldsymbol{X}, \widehat{\boldsymbol{W}})$$
$$\approx \sum_{t=1}^{T} P(W > w|\boldsymbol{x}_f, \boldsymbol{X}, \widehat{\boldsymbol{W}}, \boldsymbol{\gamma}^{(t)}) \, p(\boldsymbol{\gamma}^{(t)}|\boldsymbol{X}, \widehat{\boldsymbol{W}}) \qquad (9.13)$$

where $\boldsymbol{\gamma}^{(t)}$ is the model visited at the t-th iteration and $p(\boldsymbol{\gamma}|\boldsymbol{X}, \widehat{\boldsymbol{W}})$ is used as importance sampling density for $p(\boldsymbol{\gamma}|\boldsymbol{x}_f, \boldsymbol{X}, \widehat{\boldsymbol{W}})$.

Applications of the methodologies described above to DNA microarray data and to mass spectrometry data can be found in Sha et al.,[30,31,32] Kwon et al.[20] and Vannucci et al.[39]

9.3 Model-Based Clustering

In recent years, there has been an increased interest in using DNA microarray technologies to uncover disease subtypes and identify discriminating genes. There is a consensus that the existing disease classes for various malignancies are too broad and need to be refined. Indeed, patients receiving the same diagnosis often follow significantly different clinical courses and respond differently to therapy. It is believed that gene expression profiles can capture disease heterogeneities better than currently used clinical and morphological diagnostics. The goal is to identify a subset of genes whose expression profiles can help stratify samples into more homogeneous groups. This is a problem of variable selection in the context of clustering samples.

From a statistical point of view, this is a more complicated problem than variable selection for linear models or classification, where the outcomes are observed. In clustering, the discriminating genes need to be selected and the different classes need to be uncovered simultaneously. The appropriate statistical approach therefore must identify genes with distinctive expression patterns, estimate the number of clusters and allocate samples to the different uncovered classes. We have proposed model-based methodologies that provide a unified approach to this problem. In Tadesse *et al.*[35] we formulate the clustering problem in terms of finite mixture models with an unknown number of components and use a reversible jump MCMC technique to allow creation and deletion of clusters. In Kim *et al.*[19] we propose an alternative approach that uses infinite mixture models with Dirichlet process priors. In both models, for the variable selection we introduce a latent binary vector $\boldsymbol{\gamma}$ and use stochastic search MCMC techniques to explore the space of variable subsets. The definition of this latent indicator and its inclusion into the models, however, are inherently different from the regression settings. In this section we present the basic intuitions behind the method. Readers are referred to the cited papers for details, particularly on the MCMC strategies.

9.3.1 Finite Mixture Models

In model-based clustering, the data are viewed as coming from a mixture of distributions:

$$f(\boldsymbol{x}_i|\boldsymbol{w}, \boldsymbol{\phi}) = \sum_{k=1}^{K} w_k f(\boldsymbol{x}_i|\boldsymbol{\phi}_k), \qquad (9.14)$$

where $f(\boldsymbol{x}_i|\boldsymbol{\phi}_k)$ is the density of sample $\boldsymbol{x}_i$ from group k and $\boldsymbol{w} = (w_1, \cdots, w_K)^T$ are the cluster weights ($\sum_k w_k = 1, w_k \geq 0$).[21] We assume that K is finite but unknown. In order to identify the cluster from which each observation is drawn, we introduce latent variables $\boldsymbol{c} = (c_1, \cdots, c_n)^T$, where

$c_i = k$ if the i-th sample comes from group k. The sample allocations, c_i, are assumed to be independently and identically distributed with probability mass function $p(c_i = k) = w_k$. We assume that the mixture distributions are multivariate normal with component parameters $\phi_k = (\mu_k, \Sigma_k)$. Thus, for sample i, we have

$$\boldsymbol{x}_i | c_i = k, \boldsymbol{w}, \boldsymbol{\phi} \sim \mathcal{N}(\boldsymbol{\mu}_k, \Sigma_k). \tag{9.15}$$

When dealing with high-dimensional data, it is often the case that a large number of collected variables provide no information about the group structure of the observations. The inclusion of too many unnecessary variables in the model could mask the true grouping of the samples. The discriminating variables therefore need to be identified in order to successfully uncover the clusters. For this, we introduce a latent binary vector $\boldsymbol{\gamma}$ to identify relevant variables

$$\begin{cases} \gamma_j = 1 \text{ if variable } j \text{ defines a mixture distribution,} \\ \gamma_j = 0 \text{ otherwise.} \end{cases} \tag{9.16}$$

In particular, $\boldsymbol{\gamma}$ is used to index the contribution of the variables to the likelihood. The set of variables indexed by a $\gamma_j = 1$, denoted $\boldsymbol{X}_{(\gamma)}$, define the mixture distribution, while the variables indexed by $\gamma_j = 0$, $\boldsymbol{X}_{(\gamma^c)}$, favor one multivariate normal distribution across all samples. The distribution of sample i is then given by

$$\begin{aligned} \boldsymbol{x}_{i(\gamma)} | c_i = k, \boldsymbol{w}, \boldsymbol{\phi}, \boldsymbol{\gamma} &\sim \mathcal{N}(\boldsymbol{\mu}_{k(\gamma)}, \Sigma_{k(\gamma)}) \\ \boldsymbol{x}_{i(\gamma^c)} | \boldsymbol{\psi}, \boldsymbol{\gamma} &\sim \mathcal{N}(\boldsymbol{\eta}_{(\gamma^c)}, \boldsymbol{\Omega}_{(\gamma^c)}), \end{aligned} \tag{9.17}$$

where $\boldsymbol{\psi} = (\boldsymbol{\eta}, \boldsymbol{\Omega})$. Notice that the use of the variable selection indicator here is different from the linear model context, where $\boldsymbol{\gamma}$ was used to induce mixture priors on the regression coefficients. In clustering, the outcome is not observed and the elements of the matrix $\boldsymbol{X}$ are viewed as random variables.

9.3.2 Dirichlet Process Mixture Models

Dirichlet process mixture (DPM) models have gained a lot of popularity in nonparametric Bayesian analysis and have particularly been successful in cluster analysis. Samples from a Dirichlet process are discrete with probability one and can therefore produce a number of ties, i.e., form clusters.

A general DPM model is written as

$$\begin{aligned} \boldsymbol{x}_i | \boldsymbol{\theta}_i &\sim F(\boldsymbol{\theta}_i) \\ \boldsymbol{\theta}_i | G &\sim G \\ G &\sim DP(G_0, \alpha), \end{aligned} \tag{9.18}$$

where $\boldsymbol{\theta}$ is a vector of sample-specific parameters and DP is the Dirichlet process with concentration parameter α and base distribution G_0. Due to properties of the Dirichlet process, some of the $\boldsymbol{\theta}_i$'s will be identical and can be set to $\boldsymbol{\theta}_i = \boldsymbol{\phi}_{c_i}$, where c_i represents the latent class associated with sample i, see Antoniak,[4] Ferguson.[12]

Mixture distributions with a countably infinite number of components can be defined in terms of finite mixture models by taking the limit as the number of components K goes to infinity

$$\boldsymbol{x}_i | c_i = k, \boldsymbol{\phi} \sim F(\boldsymbol{\phi}_k)$$
$$c_i | \boldsymbol{w} \sim \text{Discrete}(w_1, \cdots, w_K)$$
$$\boldsymbol{\phi}_k \sim G_0 \qquad\qquad (9.19)$$
$$\boldsymbol{w} \sim \text{Dirichlet}(\alpha/K, \cdots, \alpha/K).$$

As shown in Neal,[24] integrating over the mixing proportions $\boldsymbol{w}$ and letting K go to infinity leads to the following priors for the sample allocations:

$$p(c_i = c_l \text{ for some } l \neq i | \boldsymbol{c}_{-i}) = \frac{n_{-i,k}}{n-1+\alpha}$$
$$p(c_i \neq c_l \text{ for all } l \neq i | \boldsymbol{c}_{-i}) = \frac{\alpha}{n-1+\alpha}, \qquad (9.20)$$

where $\boldsymbol{c}_{-i}$ is the allocation vector $\boldsymbol{c}$ without the i-th element and $n_{-i,k}$ is the number of $c_l = k$ for $l \neq i$. Thus, sample i is assigned to an existing cluster with probability proportional to the cluster size and it is allocated to a new cluster with probability proportional to α. As in the finite mixture case, we assume that samples in group k arise from a multivariate normal distribution with component parameters $\boldsymbol{\phi}_k = (\boldsymbol{\mu}_k, \boldsymbol{\Sigma}_k)$ and we use the latent indicator $\boldsymbol{\gamma}$ to identify discriminating variables as in (9.16).

9.3.3 Prior Setting

The elements of $\boldsymbol{\gamma}$ can be taken to be independent Bernoulli random variables. Conjugate priors can be specified for the component parameters and the weights. In the finite mixture case, we impose a prior on the number of components, K, as a truncated Poisson or a discrete Uniform prior on $[1, \ldots, K_{\max}]$, where $K_{\max}$ is chosen arbitrarily large.

For inference, we are interested in estimating the variable selection vector $\boldsymbol{\gamma}$ and the sample allocations. The other model parameters can be integrated out from the likelihood leading to a more efficient MCMC algorithm. Some care is needed in the choice of the hyperparameters. In particular, the variance parameters need to be specified within the range of variability of the data. An extensive discussion on the prior specification and the MCMC procedure can be found in Tadesse *et al.*[35]

9.3.4 MCMC Implementation

In the finite mixture case the MCMC procedure iterates between the following steps:

1. Update γ using a Metropolis algorithm. The transition moves previously described for the linear setting can be used.

2. Update the component weights, w, from their full conditionals.

3. Update the sample allocation vector, c, from its full conditional.

4. Split/merge moves to create/delete clusters. We make a random choice between attempting to divide or combine clusters. The number of components may therefore increase or decrease by 1, and the necessary corresponding changes need to be made for the sample allocations and the component parameters. These moves require a sampler that jumps between different dimensional spaces, which is not a trivial task in the multivariate setting.

5. Birth/death of empty components.

In the infinite mixture case, the variable selection indicator γ and the cluster allocation vector c are updated using the following MCMC steps:

(i) Update γ using the Metropolis algorithm.

(ii) Update c via Gibbs sampling using the following conditional posterior probabilities to assign each sample to an existing cluster or to a newly created one:

$$p(c_i = c_l \text{ for } l \neq i | c_{-i}, x_i, \gamma) \propto \frac{n_{-i,k}}{n-1+\alpha} \int F(x_i; \phi_{k(\gamma)}) dH_{-i,k}(\phi_{k(\gamma)})$$

$$p(c_i \neq c_l \text{ for } l \neq i | c_{-i}, x_i, \gamma) \propto \frac{\alpha}{n-1+\alpha} \int F(x_i; \phi_{k(\gamma)}) dG_0(\phi_{k(\gamma)}) \quad (9.21)$$

where $H_{-i,k}$ is the posterior distribution of ϕ based on the prior G_0 and all observations x_l for which $l \neq i$ and $c_l = k$.

At each MCMC iteration, the number of clusters may decrease as components become empty, it may remain the same with possible changes in the sample allocations, or it may increase if new clusters are formed. The Gibbs sampler often exhibits poor mixing when dealing with mixture models. This can be improved by combining the MCMC algorithm with a split-merge method that essentially avoids local modes by separating or combining groups of observations based on a Metropolis-Hastings algorithm (see for example Jain & Neal[18]). In addition, parallel tempering can be used to further improve the performance of the sampler. Details on the MCMC implementation can be found in Kim *et al.*[19]

9.3.5 Posterior Inference

The MCMC output can then be used to draw posterior inference for the sample allocations and the variable selection. Posterior inference for the cluster structure is complicated by the varying number of components. A simple approach for estimating the most probable cluster structure c is to use the maximum *a posteriori* (MAP) configuration, which corresponds to the vector with highest conditional posterior probability among those visited by the MCMC sampler. An alternative is to estimate the number of clusters, K, by the value most frequently visited by the MCMC sampler then draw inference conditional on $\widehat{K}$. With this approach, we first need to address the label switching problem using for instance Stephens' relabeling algorithm.[34] The sample allocation vector, c, can then be estimated by the mode of the marginal posterior probabilities given by

$$\hat{c}_i = \operatorname*{argmax}_{1 \leq k \leq K} \{ p(c_i = k | \boldsymbol{X}, \widehat{K}) \}. \tag{9.22}$$

Several other inferential approaches are possible. An alternative estimator, for example, is the one proposed by Dahl,[7] which relies on the posterior pairwise probabilities of allocating samples to the same cluster.

For the variable selection, inference can be based on the $\boldsymbol{\gamma}$ vectors with highest posterior probability among the visited models or on the γ_j's with largest marginal posterior probabilities.

9.4 Breast Cancer DNA Microarrays

Here we illustrate the Bayesian method for classification using a breast cancer DNA microarray dataset from collaborators at the M.D. Anderson Cancer Center in Houston, Texas.

9.4.1 Experimental Design

Breast cancers are characterized by their active "receptors": The estrogen receptor, the progesterone receptor, and the human epidermal growth factor receptor 2. Approximately 15% of breast cancers lack all three of these receptors, and are thus called triple negative. Triple negative breast cancers do not generally respond to endocrine therapies that target one of these receptors, although they do respond to chemotherapy. Compared to other breast cancers, triple negative breast cancers can be particularly aggressive, and are more likely to recur. As a group, patients with triple negative breast cancer have much lower five year survival rates than patients with other subtypes. Some patients, however, respond particularly well to therapy with no pathologically detectable tumor remaining, while in others the tumor does not shrink or even grows.

The variable of interest in this study is a binary response variable that describes the patient response to the treatment as pathologic complete response (pCR) and residual disease (RD). There is also an interest in finding markers

TABLE 9.1:   MDACC285 and USO88 Datasets

	MDACC294				USO88		
	pCR	RD	NA	Total	pCR	RD	Total
TN	17	54	2	73	12	13	25
Others	52	162	7	221	23	40	63
	69	216	9	294	35	53	88

for the subgroup of the triple negative breast cancers. We have available data from two different sources. One set of data (MDACC294 in the sequel) was collected at the MD Anderson Cancer Center. 294 breast cancer patients were treated with neoadjuvant chemotherapy. The gene expression profiling of these patients (fine needle aspiration speciments) was performed using Affymetrix U133A GeneChips. Patients were recruited from MDACC, Lyndon B Johnson Hospital in Houston, Mexico, and Peru. For the other dataset (USO88) US Oncology provided pretreatment fine needle aspiration (FNA) specimens to the MDACC breast cancer pharmacogenomic laboratory. The gene expression profiling of all these samples was performed at MD Anderson Cancer Center using Affymetrix U133A GeneChips. Table 9.1 shows the details on the two datasets. Excluding the 9 samples in the MDACC294 dataset for which we have no information available, there is a total of 285 samples for analysis.

9.4.2   Pre-Processing

Quality control (QC) checks were performed at three levels: (1) total RNA yield had to be at least 0.8 μg and OD260/280 ratio at least 1.8, (2) cRNA yield had to be at least 10 μg, and (3) after completion of the hybridization procedure, every array had to have at least 30% of present calls, scaling factors within 3-fold of each other, HSAC07 and GAPDH 3'/5' signal ratios less than 3, less than 15% array outliers and less than 5% single outliers. These metrics were calculated in Bioconductor and dChip V1.3. All arrays passed QC were then normalized to one standard array (University of Texas MD Anderson Cancer Center. Bioinformatics: publicly available data. http://bioinformatics.mdanderson.org/pubdata.html) in dChip V1.3. The model-based gene expression data were exported from dChip and log-transformed for further analysis. The triple negative (TN) breast cancer was defined based on mRNA expression values for ESR1 and HER2/neu receptor probesets (205225-at and 216836-s-at), and immunohistochemistry for PR.

Modern DNA microarrays obtain thousands of quantitative gene expression measurements simultaneously. The U133-A array, for example, provides 22283 gene expression measurements (probesets) for each sample. However, many of these genes are not active in the tissue of interest, so their gene expression measurements have very low values across nearly all samples, and are almost all noise. At best, the presence of these variables merely increases the computation needed to analyze an experiment, but, more likely, the noise

TABLE 9.2:   MDACC285: Training and Test

	Training			Test		
	pCR	RD	Total	pCR	RD	Total
TN	11	36	47	6	18	24
Others	34	108	142	18	54	72
	45	144	189	24	72	96

in thousands of such variables will increase the error rate of any model. A gene screening or filtering procedure should therefore be applied before any sophisticated statistical model. Here, we used a two step procedure to exclude uninteresting probesets. We first selected only those probesets whose value was in its sample's top quartile of expression levels in at least one quarter of the MDACC294 samples and in at least one quarter of the USO88 samples. This first step selected 6612 probesets. We then logged the data and selected the 4000 probesets with the largest variance.

9.4.3   Classification via Bayesian Variable Selection

We first focused on the MDACC285 data only and divided the samples into a training and a test set, using a ratio 2 to 1, as shown in Table 9.2.

We applied our Bayesian variable selection procedure for multinomial probit models with $K = 2$. In order to induce parsimonious models we set the prior expected number of discriminating genes to 10. We chose $\Sigma \sim IW(\delta, Q)$ with $Q = cI$. Sha et al.[32] provide insights on the choice of the hyperparameter c. Here we report the results for $c = 1$ and $\delta = 3$. Two MCMC chains were run for 100,000 iterations and the first 50,000 iterations were used as burn-in. Inference was done by calculating posterior probabilities based on the pooled set of visited models by the two chains. Results on the "best" model, i.e. the model with the highest posterior probability, and on the marginal model are provided below.

By marginal frequency five genes were selected. These are reported in Table 9.3, together with some summary statistics. Least squares prediction results calculated on the test set are given in Table 9.4 as percentages of correctly classified samples in the two groups of samples, the TN and the "other" subgroups. Prediction performances in the two subgroups are comparable. By best model five genes were selected. These are reported in Table 9.5 and least squares prediction results calculated on the test set are given in Table 9.4. The prediction results we obtained are similar to those reported by other investigators in the biomedical literature. Genomic predictors show overall accuracies between 75-85% in most reports, see for example Andre et al.[2] However, almost all previous studies generated chemotherapy response predictors for all breast cancer patients considered together without any attempt to separate the estrogen receptor-positive cases from the receptor-negative cases. Because hormone receptor status itself is associated with response (i.e. estrogen re-

TABLE 9.3: MDACC285: Selected Genes by Marginal Model

Gene ID/Probeset ID	RD (mean/var)	pCR (mean/var)	Description
ERBB2/216836-s-at	2.8506/0.2085	3.1583/0.4192	v-erb-b2 erythroblastic leukemia viral oncogene homolog 2, neuro/ glioblastoma derived oncogene homolog (avian)
IGFBP4/201508-at	2.6701/0.1109	2.4455/0.0710	insulin-like growth factor binding protein 4
TIMP1/201666-at	3.1172/0.0920	2.9001/0.0884	TIMP metallopeptidase inhibitor 1
CD44/212063-at	3.1096/0.0728	2.9257/0.0858	CD44 molecule (Indian blood group)
MAPT/203928-x-at	2.5863/0.0440	2.4271/0.0250	microtubule-associated protein tau

ceptor negative cancers have a higher chance of pCR), the clinically most useful predictors will need to adjust for this known association. Our attempt to develop predictors separately for these two types of breast cancers is an important step in this direction.

We also considered a second scenario, where we used the entire MDACC285 data as training set and the USO88 samples for validation. This is a particularly challenging problem because subtle and sometimes seemingly trivial variations in processing, such as the batch of chemicals used, the particular habits of different technicians and the precise calibrations of the processing machines, can have effects that are far more easily detected than the biological condition in which we are interested.

We used the same hyperparameter setting as before. Table 9.6 presents some statistic values for six genes selected by the marginal model with a

TABLE 9.4: MDACC285: LS Prediction (% Correct)

Model	TN	Others	Total
Marginal	70%	76%	75%
Best	83%	76%	78%

TABLE 9.5:   MDACC285: Selected Genes by Best Model

Gene ID/Probeset ID	RD (mean/var)	pCR (mean/var)	Description
IGFBP4/201508-at	2.6701/0.1109	2.4455/0.0710	insulin-like growth factor binding protein 4
RAB31/217764-s-at	2.7671/0.0782	2.5813/0.0427	RAB31, member RAS oncogene family
LISCH7/208190-s-at	2.5158/0.0424	2.5886/0.0445	lipolysis stimulated lipoprotein receptor
FXYD5/218084-x-at	2.7902/0.0399	2.7757/0.0391	FXYD domain containing ion transport regulator 5
HNRPH3/208990-s-at	2.5134/0.0313	2.4551/0.0260	heterogeneous nuclear ribonucleoprotein H3 (2H9)

cutoff probability of .1. Least squares predictions are reported in Table 9.7. By best model five genes were selected. These are reported in Table 9.8 and the least squares prediction results calculated on the test set are given in Table 9.7. As expected, the prediction performances worsen considerably.

It was reassuring to see that our method identified two of the most important previously known predictors of response to chemotherapy. Increased expression of the ERBB2 gene, also called HER2, is associated with increased rates of pCR to chemotherapy in both Estrogen receptor-positive and estrogen receptor negative cancers.[3] High expression of HER2 is almost always caused by amplification of the chromosome region where this gene is located. The gene GRB7 that is listed in Table 9.6 is located adjacent to the HER2 gene on chromosome 17, they are invariably amplified together, and it is considered to be a marker of HER2 amplification status. Similarly, low MAPT expression was discovered from gene expression data as an important predictor and mediator of increased sensitivity to paclitaxel chemotherapy (one of the components of the therapy that these patients received. Experiments in cell lines showed that MAPT protects microtubules from the deleterious effects of paclitaxel.[29] The other genes have not previously been linked to chemotherapy response. However, CD44 is a membrane protein that regulates growth factor signaling and RAB31 is also a component of the signal transduction pathways that

TABLE 9.6: MDACC285: Selected Genes by Marginal Model

Gene ID/Probeset ID	RD (mean/var)	pCR (mean/var)	Description
IGKC/216401-x-at	2.4236/0.2692	2.6944/0.2934	immunoglobulin kappa variable 1-37
ERBB2/216836-s-at	2.8506/0.2085	3.1583/0.4192	v-erb-b2 erythroblastic leukemia viral oncogene homolog 2, neuro/ glioblastoma derived oncogene homolog (avian)
TPBG/203476-at	2.9093/0.1008	2.7329/0.0843	trophoblast glycoprotein
GRB7/210761-s-at	2.5596/0.0789	2.7649/0.1346	growth factor receptor-bound protein 7
RAB31/217762-s-at	2.6454/0.0718	2.4634/0.0408	RAB31, member RAS oncogene family
NFBL/203961-at	2.6354/0.0547	2.4887/0.0458	nebulette

TABLE 9.7: US088: LS
Prediction (% Correct)

Model	TN	Others	Total
Marginal	48%	71%	60%
Best	52%	68%	63%

induce cells to proliferate. A mechanistic link between these 2 molecules and increased proliferation and increased sensitivity to chemotherapy is plausible.

9.5 Conclusion

Bayesian variable selection techniques for gene selection with high-throughput data have been briefly reviewed in this chapter. In particular, we have described methods to integrate the variable selection task into models for classification and clustering. The methodologies were briefly illustrated on a DNA microarray data example.

TABLE 9.8: MDACC285: Selected Genes by Best Model

Gene ID/Probeset ID	RD (mean/var)	pCR (mean/var)	Description
ERBB2/216836-s-at	2.8506/0.2085	3.1583/0.4192	v-erb-b2 erythroblastic leukemia viral oncogene homolog 2, neuro/ glioblastoma derived oncogene homolog (avian)
—/217281-x-at	2.4475/0.1870	2.6742/0.2528	
TPBG/203476-at	2.9093/0.1008	2.7329/0.0843	trophoblast glycoprotein
AZIN1/212461-at	2.8639/0.0448	2.8297/0.0381	antizyme inhibitor 1
MCM2/202107-s-at	2.5611/0.0297	2.6383/0.0338	minichromosome maintenance complex component 2

The Bayesian approaches we have described offer a coherent framework in which variable selection and clustering or classification of the samples are performed simultaneously. Bayesian variable selection techniques can cope with a large number of regressors and have the flexibility of allowing us to handle data where the number of covariates is larger than the sample size. In addition, these methods allow the evaluation of the joint effect of sets of variables and the use of stochastic search techniques to explore the high-dimensional variable space. They provide joint posterior probabilities of sets of variables, as well as marginal posterior probabilities for the inclusion of single variables.

Acknowledgments

Sha, Tadesse, and Vannucci are supported by NIH/NHGRI grant R01HG003319. Vannucci is also supported by NSF award DMS-0605001.

References

[1] Albert, J.H. and Chib, S. (1993) Bayesian analysis of binary and polychotomous response data, *Journal of the American Statistical Association*, **88**, 669–679.

[2] Andre, F., Mazouni, C., Hortobagyi, G.N. and Pusztai, L. (2006) DNA arrays as predictors of efficacy of adjuvant/neoadjuvant chemotherapy in breast cancer patients: Current data and issues on study design. *Biochim Biophys Acta*, **1766**, 197–204.

[3] Andre, F., Mazouni, C., Liedtke, C., Kau, S.W., Frye, D., Green, M., Gonzalez-Angulo, A.M., Symmans, W.F., Hortobagyi, G.N. and Pusztai, L. (2008) HER2 expression and efficacy of preoperative paclitaxel/FAC chemotherapy in breast cancer, *Breast Cancer Res Treat*, **108(2)**, 183–190.

[4] Antoniak, C.E. (1974) Mixtures of Dirichlet processes with applications to Bayesian nonparametric problems, *Annals of Statistics*, **2**, 1152–1174.

[5] Benjamini, Y. and Hochberg, Y. (1995) Controlling the false discovery rate: a practical and powerful approach to multiple testing, *Journal of the American Statistical Association*, **57**, 289–300.

[6] Brown, P.J., Vannucci, M. and Fern, T. (1998) Multivariate Bayesian variable selection and prediction, *Journal of the Royal Statistical Society, Series B*, **60**, 627–641.

[7] Dahl, D.B. (2006) Model-based clustering for expression data via a Dirichlet process mixture model. In *Bayesian Inference for Gene Expression and Proteomics*, K. Anh-Do, P. Mueller and M. Vannucci (Eds). Cambridge University Press (this book).

[8] Dempster, A., Laird, N. and Rubin, D. (1977) Maximum likelihood for incomplete data via the EM algorithm (with discussion), *Journal of the Royal Statistical Society, Series B*, **39**, 1–38.

[9] Diebolt, J. and Robert, C.P. (1994) Estimation of finite mixture distributions through Bayesian sampling, *Journal of the Royal Statistical Society, Series B*, **56**, 363–375.

[10] Dudoit, S., Yang, Y.H., Callow, M.J. and Speed, T.P. (2002) Statistical methods for identifying differentially expressed genes in replicated cDNA microarray experiments, *Statistica Sinica*, **12**, 111–139.

[11] Dudoit, S., Fridlyand, J. and Speed, T.P. (2002) Comparison of discrimination methods for the classification of tumors using gene expression data, *Journal of the American Statistical Association*, **97**, 77–87.

[12] Ferguson, T.S. (1983) Bayesian density estimation by mixtures of normal distributions. In *Recent Advances in Statistics*, H. Rizvi and J. Rustagi (Eds). New York: Academic Press.

[13] Fowlkes, E.B., Gnanadesikan, R. and Kettering, J.R. (1988) Variable selection in clustering, *Journal of Classification*, **5**, 205–228.

[14] Friedman, J.H. and Meulman, J.J. (2004) Clustering objects on subsets of attributes (with discussion), *Journal of the Royal Statistical Society, Series B*, **66**, 815–849.

[15] George, E.I. and McCulloch, R.E. (1993) Variable selection via Gibbs sampling, *Journal of the American Statistical Association*, **88**, 881–889.

[16] Golub, T.R., Slonim, D.K., Tamayo, P., Huard, C., Gassenbeek, M., Mesirov, J.P., Coller, H., Loh, M.L., Downing, J.R., Caligiuri, M.A., Bloomfield, C.D. and Lander, E.S. (1999) Molecular classification of cancer: class discovery and class prediction by gene expression monitoring, *Science*, **286**, 531–537.

[17] Jain, A. and Dubes, R. (1988) *Algorithms for Clustering Data*, Englewood Cliffs: Prentice-Hall.

[18] Jain, S., and Neal, R.M. (2004) A split-merge Markov chain Monte Carlo procedure for the Dirichlet process mixture model, *Journal of Computational and Graphical Statistics*, **13**, 158–182.

[19] Kim, S., Tadesse, M.G. and Vannucci, M. (2005) Variable selection in clustering via Dirichlet process mixture models, *Biometrika*, **93(4)**, 877–893.

[20] Kwon, D., Tadesse, M.G., Sha, N., Pfeiffer, R.M. and Vannucci, M. (2007) Identifying biomarkers from mass spectrometry data with ordinal outcome, *Cancer Informatics*, **3**, 19–28.

[21] McLachlan, G. and Basford, K. (1988) *Mixture Models: Inference and Applications to Clustering*, New York: Marcel Dekker.

[22] Miller, A. (1990) *Subset Selection in Regression*, London:Chapman & Hall.

[23] Mitchell, T.J. and Beauchamp, J.J (1988) Bayesian variable selection in linear regression, *Journal of the American Statistical Association*, **83**, 1023–1036.

[24] Neal, R.M. (2000) Markov chain sampling methods for Dirichlet process mixture models, *Journal of Computational and Graphical Statistics*, **9**, 249–265.

[25] Newton, M.A., Noueiry, A., Sarkar, D. and Ahlquist, P. (2004) Detecting differential gene expression with a semiparametric hierarchical mixture model, *Biostatistics*, **5(2)**, 155–176.

[26] Nguyen, D. and Rocke, D. (2002) Tumor classification by partial least squares using microarray gene expression data, *Bioinformatics*, **18**, 39–50.

[27] Pounds, M. and Morris, S.W. (2003) Estimating the occurrence of false positives and false negatives in microarray studies by approximating and partitioning the empirical distribution of p-values, *Bioinformatics*, **19**, 1236–1242.

[28] Richardson, S. and Green, P.J. (1997) Bayesian analysis of mixtures with an unknown number of components (with discussion), *Journal of the Royal Statistical Society, Series B*, **59**, 731–792.

[29] Rouzier, R., Rajan, R., Wagner, P., Hess, K.R., Gold, D.L., Stec, J., Ayers, M., Ross, J.S., Zhang, P., Buchholz, T.A., Kuerer, H., Green, M., Arun, B., Hortobagyi, G.N., Symmans, W.F. and Pusztai, L. (2005) Microtubule-associated protein tau: A marker of paclitaxel sensitivity in breast cancer, *Proc Natl Acad Sci USA*, **102(23)**, 8315–8320.

[30] Sha, N., Tadesse, M.G. and Vannucci, M. (2006). Bayesian variable selection for the analysis of microarray data with censored outcome, *Bioinformatics*, **22(18)**, 2262 2268.

[31] Sha, N., Vannucci, M., Brown, P.J., Trower, M.K., Amphlett, G. and Falciani, F. (2003) Gene selection in arthritis classification with large-scale microarray expression profiles, *Comparative and Functional Genomics*, **4**, 171–181.

[32] Sha, N., Vannucci, M., Tadesse, M.G., Brown, P.J., Dragoni, I., Davies, N., Roberts, T. C., Contestabile, A., Salmon, N., Buckley, C. and Falciani, F. (2004) Bayesian variable selection in multinomial probit models to identify molecular signatures of disease stage, *Biometrics*, **60**, 812–819.

[33] Stephens, M. (2000a) Bayesian analysis of mixture models with an unknown number of components – an alternative to reversible jump methods, *Annals of Statistics*, **28**, 40–74.

[34] Stephens, M. (2000b) Dealing with label switching in mixture models, *Journal of the Royal Statistical Society, Series B*, **62**, 795–809.

[35] Tadesse, M.G., Sha, N. and Vannucci, M. (2005) Bayesian variable selection in clustering high-dimensional data, *Journal of the American Statistical Association*, **100**, 602–617.

[36] Tibshirani, R. (1996) Regression shrinkage and selection via the lasso, *Journal of the Royal Statistical Society, series B*, **58**, 267–288.

[37] Tibshirani, R., Hastie, T., Narashiman, B. and Chu, G. (2002) Diagnosis of multiple cancer types by shrunken centroids of gene expression, *Proceedings of the National Academy of Sciences*, **99**, 6567–6572.

[38] Tusher, V.G., Tibshirani, R. and Chu, G. (2003) Significance analysis of microarrays applied to the ionizing radiation response, *Proceedings of the National Academy of Sciences*, **98**, 5116–5121.

[39] Vannucci, M., Sha, N. and Brown, P.J. (2005) NIR and mass spectra classification: Bayesian methods for wavelet-based feature selection, *Chemometrics and Intelligent Laboratory Systems*, **77**, 139–148.

[40] Vapnik, V. (1998) *Statistical Learning Theory*, New York: Wiley-Interscience.

[41] West, M. (2003) Bayesian factor regression models in large p small n paradigm. In *Bayesian Statistics 7*, J.M. Bernardo, M.J. Bayarri, J.O. Berger, A.P. David, D. Heckerman, A.F.M. Smith and M. West (Eds). Oxford: Oxford University Press.

Chapter 10

Sparsity Priors for Protein–Protein Interaction Predictions

Inyoung Kim[1], Yin Liu[2], and Hongyu Zhao[3*]

1 Department of Statistics, Virginia Tech., 2 Department of Neurobiology and Anatomy, University of Texas Medical School at Houston, 3* Department of Epidemiology and Public Health, Yale University School of Medicine

Protein–protein interactions play important roles in most fundamental cellular processes. It is important to develop effective statistical approaches to predicting protein interactions based on recently available high throughput experimental data. Since protein domains are the functional units of proteins and protein–protein interactions are mostly achieved through domain-domain interactions, the modeling and analysis of protein interactions at the domain level may be more informative. However, due to the large number of domains, the number of parameters to be estimated is very large, whereas the number of observed protein–protein interactions is relatively small. Hence the amount of information for statistical inference is quite limited. In this chapter we describe a Bayesian method for simultaneously estimating domain-domain interaction probabilities, the false positive rate, and the false negative rate of high-throughput data through integrating data from several organisms. Since we expect the domain and protein interaction networks to be sparse, a point-mass mixture prior is applied to incorporate network sparsity. We compare the prediction results between models with and without sparsity prior using high throughput yeast two-hybrid data from different organisms. Our results clearly demonstrate the advantages of the Bayesian approach with a sparsity prior in modeling and predicting protein interaction data.

Keywords: Bayesian method; Domain-domain interaction; Expectation Maximization algorithm; Point-mass mixture prior; Protein–protein interaction.

10.1   Introduction

Because protein–protein interactions are critical in many biological processes, various high-throughput experimental approaches have been developed and enormous amounts of data have been generated to identify interacting proteins. The protein interaction data used in our study are generated by genome-wide yeast two-hybrid assays. In this method, one protein is fused to a DNA-binding domain, and the other is fused to a transcription activation domain. The interaction between the protein pair can be detected by the formation of a transcription activator that activates a reporter construct (Uetz *et al.*, 2000). However, this experimental approach suffers from high false negative and false positive rates due to a number of limitations in this approach (Mrowka *et al.*, 2001; von Mering *et al.*, 2002). For example, a self-activating protein being tested in the experiment can lead to a false positive result, and a protein that cannot be targeted to the yeast nucleus may not yield positive results though it may potentially interact with other proteins, which leads to false negative results. It is reported that the false negative rate of the yeast two-hybrid assay used to construct *S. cerevisiae* interaction maps is larger than 70% (Deng *et al.*, 2002).

Complementary to experimental approaches, a number of computational methods have been proposed to predict protein–protein interactions (Enright *et al.*, 1999; Marcotte *et al.*, 2001; Pazos *et al.*, 2001; Tsoka *et al.*, 2000; Goh *et al.*, 2002; Ramani *et al.*, 2003; Jansen *et al.*, 2003; Lu *et al.*, 2003; Aloy *et al.*, 2004). However, most methods do not consider the fact that domains are the functional units of proteins and protein–protein interactions (PPIs) are achieved mostly through domain-domain interactions (DDIs). Protein domains are the basic modules of the overall protein structure and are mostly conserved during evolution. Some proteins consist of only a single domain, but many proteins contain more than one domain to perform multiple functions. For example, the protein DNA-directed RNA polymerase II subunit 9 is a multi-domain protein that contains two domains, the TFIIS domain for DNA binding and the RNA polymerase M domain for RNA synthesis. Protein domains serve as the units for PPIs and the specificity of PPIs is achieved from the binding of a modular domain to another in proteins (Pawson *et al.*, 2003). Therefore, the modeling and analysis of protein interactions at the domain level may be more informative and insightful.

Several methods have been proposed for PPI predictions based on protein domains (Sprinzak and Margalit, 2001; Gomez *et al.*, 2001, 2003; Deng *et al.*, 2002). The likelihood based approach (Deng *et al.*, 2002) has been compared with three other methods (Sprinzak and Margalit, 2001; Gomez *et al.*, 2001, 2003) and was shown to be among the best performing methods (Liu *et al.*, 2005). Liu *et al.* (2005) further extended the likelihood based approach to improve PPI predictions by pooling information from three organisms:

S. cerevisiae, C. elegans and *D. melanogaster.* In the likelihood based approach, all DDIs and PPIs are treated as missing data. For a given specified set of false negative and false positive rates, DDI probabilities are estimated using the Expectation Maximization (EM) algorithm (Dempster *et al.*, 1977) and the estimated DDI probabilities are used to infer PPI probabilities.

However, in general, the false negative rate (f_n) and the false positive rate (f_p) of PPI data are not known and may depend on many factors, e.g., data-set specific. Therefore it is more appropriate to estimate f_p and f_n together with DDI probabilities and PPI probabilities. In addition, the number of domains in many proteins is more than one. When we infer PPIs using DDIs, it is important to take into account the varying numbers of domains across different proteins. Furthermore, the number of parameters is very large but the number of observed PPIs is relatively small, limiting the amount of information for statistical inference. Hence, Kim *et al.* (2007) proposed a full Bayesian method as well as a semi-Bayesian method for simultaneously estimating f_n, f_p, and DDI probabilities from high-throughput yeast two hybrid data from several organisms. A new model was also proposed to associate PPI probabilities with DDI probabilities that reflects the number of domains in each protein. It was found that the Bayesian approaches yielded more accurate predictions than the likelihood-based approach.

We note that the number of parameters to be estimated is very large in this context. For example, if there are more than 2000 annotated domains within an organism, the number of domain pairs is in the order of millions. However the number of protein–protein interactions to be observed is relatively very small. On the other hand, we do expect the true DDI and PPI networks are sparse. From real data analysis, we did observe that many of our estimated DDI probabilities were very close to zero. But this underlying sparsity was not explicitly modeled in our previous work. The sparsity assumption also complements the statistical view of parsimony in modelling - that is, we aim to use as few parameters as possible to adequately represent the observed patterns in data set. Several groups have used this approach in functional expression genomics studies (Ishwaran *et al.*, 2003; Ishwara and Rao, 2005, Muller *et al.*, 2005, Lucas *et al.*, 2006) where a point mass mixture prior was used to model the sparsity. This line of thinking has recently been extended with new hierarchical specifications for such "point-mass mixture" priors (Lucas *et al.*, 2006).

This chapter is organized as follows. In Section 10.2, we introduce a model to relate PPI probabilities with DDI probabilities that reflects the number of domains in each protein and describe the modeling of PPI using DDI information proposed by Kim *et al.* (2007). In Section 10.3, we discuss the sparsity specification based on a point-mass mixture and the Bayesian methods for PPI predictions. In Section 10.4, we apply our approaches to large-scale PPI data obtained from high throughput yeast two-hybrid experiments analyzed by Liu *et al.* (2005) and compare the prediction results of a Beta prior with that of a point-mass Beta mixture prior. Section 10.5 concludes this chapter.

10.2   Model

This section describes our model to estimate PPI probabilities using DDI probabilities (Kim *et al.*, 2007). Before we explain our model in detail, we define some notations and briefly describe the model proposed by Deng *et al.* (2002) and Liu *et al.* (2005). The model proposed by Deng *et al.* (2002) has two assumptions: A1) DDIs are independent, so whether two domains interact or not does not depend on the interactions among other domain pairs; A2) two proteins P_i and P_j interact if and only if at least one domain pair from the two proteins interact. To pool information across different organism, Liu *et al.* (2005) made another assumption: A3) the probability that two domains interact is the same among all organisms. This assumption is biologically meaningful because domains are evolutionarily conserved across different organisms (Shin *et al.*, 2005; Napetschning *et al.*, 2009). This assumption allows the integration of large-scale PPI data from different organisms to estimate DDI probabilities. By considering each protein as a collection of domains, we can then estimate PPI probabilities in any organism based on the inferred DDI probabilities. More specifically, let λ_{mn} represent the probability that domain D_m interacts with domain D_n. Define $D_{mn}^{(ij)} = 1$ if D_m and D_n interact in protein pair P_i and P_j and $D_{mn}^{(ij)} = 0$ otherwise. Let $(D_{mn}^{(ij)} \in P_{ijk})$ denote all pairs of domains from protein pair P_i and P_j in organism k, where $k = 1, ..., K$ and K is the number of organisms. Let P_{ijk} represent the interaction event between P_i and P_j in organism k, with $P_{ijk} = 1$ if they interact in organism k and $P_{ijk} = 0$ otherwise. Further let $O_{ijk} = 1$ if P_i and P_j are observed to interact in organism k, and $O_{ijk} = 0$ otherwise. In our example, we focus on $K = 3$ organisms, where $k = 1, 2, 3$ represents *S. cerevisiae*, *C. elegans*, and *D. melanogaster*, respectively.

The false negative (f_n) and false positive rates (f_p) are defined as

$$f_p = Pr(O_{ijk} = 1 | P_{ijk} = 0),$$
$$f_n = Pr(O_{ijk} = 0 | P_{ijk} = 1).$$

We further define $O = \{O_{ijk} = o_{ijk}, \ \forall i \leq j\}$, $\Lambda = \{\lambda_{mn}; \ D_{mn}^{(ij)} \in P_{ijk}, \forall m \leq n, \ \forall i \leq j\}$. With the above assumptions and notation, we have

$$Pr(P_{ijk} = 1) \quad = \quad 1 - \prod_{(D_{mn}^{(ij)} \in P_{ijk})} (1 - \lambda_{mn})$$
$$=_{\text{def.}} h_{ij}^1(\Lambda) \tag{10.1}$$

and

$$Pr(O_{ijk} = 1) = Pr(P_{ijk} = 1)(1 - f_n) + \{1 - Pr(P_{ijk} = 1)\} f_p. \tag{10.2}$$

The likelihood for the observed PPI data across all K organisms is then

$$L(f_n, f_p, \Lambda | O) = \prod_{ijk} Pr(O_{ijk} = 1)^{O_{ijk}} \{1 - Pr(O_{ijk} = 1)\}^{1-O_{ijk}},$$

which is a function of (λ_{mn}, f_n, f_p). Deng *et al.* (2002) and Liu *et al.* (2005) specified values for f_n and f_p, and then λ_{mn} were estimated using the EM algorithm by treating all DDIs and PPIs as missing data.

Although these methods have seen successes, f_n and f_p are generally unknown and they are also likely organism/dataset dependent. In our earlier work (Kim *et al.*, 2007), we allowed different organisms to have different f_n and f_p values. That is, we have organism specific rates f_{n_k} and f_{p_k}, $k = 1, ..., K$. In this case, equation 10.2 is replaced by

$$Pr(O_{ijk} = 1) = Pr(P_{ijk} = 1)(1 - f_{n_k}) + \{1 - Pr(P_{ijk} = 1)\} f_{p_k}.$$

We also extended equation (10.1) to incorporate varying numbers of domains across different proteins. This extension is motivated from observing that the value of equation (10.1) increases as the number of domains increases. For example, if all domain pairs have $1/2$ chance to interact, the PPI probability approaches 1 when the number of domain pairs is large. Therefore, we formulate function $h_{ij}(\Lambda)$, where $h_{ij}(\Lambda) = Pr(P_{ijk} = 1)$, to satisfy the following four conditions:

C1: If $\lambda_{mn} = 1$ for at least one domain pair, $h_{ij}(\Lambda) = 1$;

C2: If $\lambda_{mn} = 0$ for all domain pairs, $h_{ij}(\Lambda) = 0$;

C3: If $\lambda_{mn} = 1/2$ for all domain pairs, $h_{ij}(\Lambda) = 1/2$;

C4: (Strictly increasing condition) If $\lambda_{mn} < \lambda_{m'n'}$ and all other λs are the same, $h_{ij}(\Lambda) < h_{ij}(\Lambda')$.

We note that $h_{ij}^1(\Lambda)$ in equation (10.1) does not satisfy **C3**. In this paper, we consider one possible function for $h_{ij}(\Lambda)$ in the following form:

$$h_{ij}^a(\Lambda) = 1 - \prod_{(D_{mn}^{(ij)} \in P_{ijk})} (1 - \lambda_{mn}^a),$$

where a can be derived from condition **C3** as,

$$a = \frac{\log\{1 - (1/2)^{\frac{1}{M_{ij}}}\}}{\log(1/2)}$$

and M_{ij} represents the total number of domain pairs between P_i and P_j. It is easy to see that $h_{ij}^a(\Lambda)$ is the same as $h_{ij}^1(\Lambda)$ when P_i and P_j each has only one domain, i.e, $M_{ij} = 1$.

10.2.1   Example

Suppose there are 5 proteins, $P_1, \ldots, P_5$ in organism *S. cerevisiae*, and the protein interaction data are generated by genome-wide yeast two-hybrid assays. If two proteins are interacted, the observed PPI is 1 and otherwise 0.

TABLE 10.1: Example of Observed Protein–Protein Interaction Obtained from Fenome-Wide Yeast Two-Hybrid Assays and Protein Domain Annotation Obtained from PFAM and SMART

P_i	P_j	O_{ij}	Protein domain annotation
P_1	P_2	1	$P_1 = \{D_1, D_2, D_3\}$
P_1	P_3	0	$P_2 = \{D_2, D_5\}$
P_2	P_3	0	$P_3 = \{D_1, D_2, D_4\}$
P_2	P_4	1	$P_4 = \{D_1, D_3\}$

The domain annotation for each protein is obtained from PFAM and SMART. For protein P_1, the domain annotations are D_1, D_2, and D_3, We denote these information as $P_1 = \{D_1, D_2, D_3\}$, $P_2 = \{D_2, D_5\}$, $P_3 = \{D_1, D_2, D_4\}$, $P_4 = \{D_1, D_3\}$. The data can be summarized as the Table 10.1.

We note that there are two missing data, O_{14} and O_{34}. Since the probability of PPI is the probability of that at least one domain pair from the two proteins interact based on A2), the probability of interaction between two proteins P_1 and P_2 is calculated as

$$Pr(P_{12} = 1) = 1 - (1 - \lambda_{12})(1 - \lambda_{15})(1 - \lambda_{22})(1 - \lambda_{25})(1 - \lambda_{32})(1 - \lambda_{35}),$$

where $\lambda_{mn} = Pr(D_{mn} = 1) = \lambda_{nm}$ is the probability of interaction between two domains, D_m and D_n, which can be estimated using the Bayesian approach explained in Section 10.3, $m = 1, 2, 3$ and $n = 2, 5$.

The PPI probabilites, $Pr(P_{13} = 1)$, $Pr(P_{14} = 1)$, $Pr(P_{23} = 1)$, $Pr(P_{24} = 1)$, and $Pr(P_{34} = 1)$ can be calculated by the similar way.

10.3   A Bayesian Approach

In this section, we discuss the Bayesian approaches to inferring DDIs and PPIs. Define (P_{ij}^k) to be all pairs of proteins in organism k, where $k = 1, ..., K$. We assume that λ_{mn} has a prior distribution: $\lambda_{mn} \sim \pi\delta_0(\lambda_{mn}) + (1 - \pi)\text{Beta}(\alpha, \beta)$, where $\delta_0(\cdot)$ is a point mass at zero. We treat f_{n_k} and f_{p_k} as unknown but are within a reasonable range that can be established from prior knowledge. We also assume that $f_{n_k} \sim \text{unif}[u_{n_k}, v_{n_k}]$ and $f_{p_k} \sim \text{unif}[u_{p_k}, v_{p_k}]$. The hyperprior parameters are chosen to be proper but vague; We set a $f_n > 0.5$ and $f_p < 0.5$ because Deng *et al.* (2002) was reported that the false negative rate of the yeast two-hybrid assay used to construct *S. cerevisiae* (yeast) interaction maps is larger than 70%. We choose $u_{n_k} \sim \text{Unif}(0, 0.3)$ and $v_{n_k} \sim \text{Unif}(0.5, 1)$ or we can choose them to be mean equal to 0.7 and prior standard deviation equal to 0.2. The α and β can be chosen to be mean equal to 0.5 and prior standard deviation 0.2.

We have varied (α, β) in our analysis without appreciable changes in results. The full conditional distributions of λ_{mn} are proportional to

$$
\begin{aligned}
[\Lambda|rest] &\propto L(O|f_{n_k}, f_{p_k}, \Lambda) f(\Lambda|f_{n_k}, f_{p_k}) \\
&\propto \prod_{ijk} [h_{ij}(\Lambda)(1 - f_{n_k}) + \{1 - h_{ij}(\Lambda)\} f_{p_k}]^{o_{ijk}} \\
&\quad [1 - \{h_{ij}(\Lambda)(1 - f_{n_k}) + \{1 - h_{ij}(\Lambda)\} f_{p_k}\}]^{1 - o_{ijk}} \\
&\quad f(\Lambda|f_{n_k}, f_{p_k}).
\end{aligned}
$$

The full conditional distributions of f_{n_k} and f_{p_k} are proportional to

$$
\begin{aligned}
[f_{n_k}|rest] &\propto L(O|f_{n_k}, f_{p_k}, \Lambda) f(f_{n_k}|\Lambda, f_{p_k}) f(f_{p_k}|\Lambda) \\
&\propto \prod_{(P_{ij}^k)} [h_{ij}(\Lambda)(1 - f_{n_k}) + \{1 - h_{ij}(\Lambda)\} f_{p_k}]^{o_{ijk}} \\
&\quad [1 - \{h_{ij}(\Lambda)(1 - f_{n_k}) + (1 - h_{ij}(\Lambda)) f_{p_k}\}]^{1 - o_{ijk}} \\
&\quad f(f_{n_k}|\Lambda, f_{p_k}); \\
[f_{p_k}|rest] &\propto L(O|f_{n_k}, f_{p_k}, \Lambda) f(f_{p_k}|\Lambda, f_{n_k}) f(f_{n_k}|\Lambda) \\
&\propto \prod_{(P_{ij}^k)} [h_{ij}(\Lambda)(1 - f_{n_k}) + \{1 - h_{ij}(\Lambda)\} f_{p_k}]^{o_{ijk}} \\
&\quad [1 - \{h_{ij}(\Lambda)(1 - f_{n_k}) + (1 - h_{ij}(\Lambda)) f_{p_k}\}]^{1 - o_{ijk}} \\
&\quad f(f_{p_k}|\Lambda, f_{n_k}).
\end{aligned}
$$

The full conditional distributions of λ_{mn}, f_{n_k}, and f_{p_k} have the following properties:

(P.1) Under $h_{ij}^1(\Lambda)$, the full conditional distribution of λ_{mn} is a log-concave function.

(P.2) Under $h_{ij}^a(\Lambda)$, the full conditional distribution of λ_{mn} is not a log-concave function but the behavior of the mean and tails is the same as that of $h_{ij}^1(\Lambda)$.

(P.3) Under both $h_{ij}^1(\Lambda)$ and $h_{ij}^a(\Lambda)$, the full conditional distributions of f_{n_k} and f_{p_k} are log-concave functions.

The proofs of these properties were given in Kim *et al.* (2007). In order to generate the posterior samples, we use the adaptive rejection Metropolis sampling (Gilks *et. al*, 1995) methods to generate sample from the posterior distributions.

10.4   Real Data Analysis

We used large-scale PPI data from three organisms, *S. cerevisiae*, *C. elegans*, and *D. melanogaster*, obtained from high throughput yeast two-hybrid experiments to infer DDI probabilities. For *S. cerevisiae*, we used

5,295 interactions from two independent studies (Ito *et al.*, 2001; Utez *et al.*, 2000). For *C. elegans* and *D. melanogaster*, 4,714 and 20,349 interactions were extracted from yeast two-hybrid experiments (Li *et al.*, 2004; Giot *et al.*, 2003), respectively. The domain annotation for each protein in *S. cerevisiae*, *C. elegans*, and *D. melanogaster* was obtained from PFAM (Bateman *et al.*, 2004) and SMART (Letunic *et al.*, 2004). Using these data sets, we first estimated DDI probabilities. We considered the following six cases of prior and model specification:

- Case 1: Use only one organism, *S. cerevisiae* ($K = 1$) and $h_{ij}(\Lambda) = h_{ij}^1(\Lambda)$, with prior distributions $f_{n_k} \sim \text{Unif}(u_k, 1)$, $f_{p_k} \sim \text{Unif}(0, v_k)$, $\lambda_{mn} \sim \text{Beta}(2,2)$, $u_k \sim \text{Unif}(0, 0.3)$, and $v_k \sim \text{Unif}(0.5, 1)$.

- Case 2: The same setting as Case 1 except that we considered all three organisms ($K = 3$).

- Case 3: The same setting as Case 2 except $h_{ij}(\Lambda) = h_{ij}^a(\Lambda)$.

- Case 4: The same setting as Case 1 except $\lambda_{mn} \sim \pi\delta_0(\lambda_{mn}) + (1 - \pi)\text{Beta}(2,2)$.

- Case 5: The same setting as Case 4 except that we considered all three organisms ($K = 3$).

- Case 6: The same setting as Case 5 except $h_{ij}(\Lambda) = h_{ij}^a(\Lambda)$.

For sparsity priors, we varied the value of π between 0.5 and 0.9 with 0.1 increments. Based on Case 4 with $\pi = 0.5$, the estimated false negative rate for the three organisms was about 0.60, 0.80, and 0.56, respectively. The estimated false positive rate was about 0.00018, 0.00012, and 0.00008, respectively. These estimated rates were almost same when $0.5 \leq \pi \leq 0.7$. With $0.8 \leq \pi \leq 0.9$, the estimated false negative rate for the three organisms was about 0.60, 0.75, and 0.55, respectively, and the estimated false positive rate was about 0.00017, 0.00014, and 0.00008, respectively. We observed that these estimated rates for *S. cerevisiae* and *D. melanogaster* were not sensitive to the value of π. For Cases 5 and 6 with $\pi = 0.5$, the results are shown in Table 10.2. We note that the results based on Case 5 were similar to those of Case 4. However these values were different from those of Case 2 and Case 3 which did not use sparsity prior. We also note that Deng *et al.* (2002) and Liu *et al.* (2005) fixed $f_n = 0.8$ and $f_p = 0.0003$ for all organisms.

We compared the ROC curves among the Bayesian methods as well as the likelihood approach based on the above six cases. For the ROC curves, we used the 3,543 experimentally verified physical interactions in *S. cerevisiae* at the Munich Information Center for Protein Sequences (MIPS) (http://mips.gsf.de/genre/proj/yeast/) as positive examples and the other 6,895,215 pairs as negative examples. We calculated the true positive rate and the false positive rate for different thresholds. The true positive rate was calculated as the

TABLE 10.2: The Estimated False Negative Rate and False Positive Rate Using the Bayesian Method Based on Three Organisms Data for the Large-Scale Protein–Protein Interaction Data

Method	k	Organism	$\widehat{f_{n_k}}$	$\widehat{f_{p_k}}$
	1	*S. cerevisiae*	0.912	0.000026
Bayesian with case 2	2	*C. elegans*	0.950	0.000018
	3	*D. melanogaster*	0.931	0.000031
	1	*S. cerevisiae*	0.901	0.000027
Bayesian with case 3	2	*C. elegans*	0.946	0.000020
	3	*D. melanogaster*	0.929	0.000033
	1	*S. cerevisiae*	0.592	0.00018
Bayesian with case 5	2	*C. elegans*	0.800	0.00012
	3	*D. melanogaster*	0.556	0.00008
	1	*S. cerevisiae*	0.590	0.00019
Bayesian with case 6	2	*C. elegans*	0.800	0.00015
	3	*D. melanogaster*	0.553	0.00008

These are obtained from high throughput yeast two-hybrid experiments. Cases 2-3 are based on non-sparsity prior Beta$(2, 2)$. The estimated values of Cases 5-6 are based on sparsity prior, $\pi\delta_0(\lambda) + (1 - \pi)$Beta$(2, 2)$, with $\pi = 0.5$.

number of predicted protein pairs that were included in the positive examples divided by 3,543 (the total number of positives) and the false positive rate was calculated as the number of predicted protein pairs that were included in the negative examples divided by 6,895,215 (the total number of negatives). The ROC curves are shown in Figure 10.1.

Based on a thorough comparison of true positive rate and false negative rate in Figure 10.1, we first notice that the Bayesian approaches without the sparsity prior are more effective in borrowing information from multiple organisms than the likelihood-based approach by Liu *et al.* (2005). Second, we note that the Bayesian approach based on Case 2 and Case 3 can further improve PPI predictions. This is likely achieved by allowing different organisms having different false negative and false positive rates and using our model $h_{ij}^a(\Lambda)$ to relate PPI probabilities with DDI probabilities. Third, the Bayesian approaches with sparsity prior yielded more accurate predictions than those without sparsity prior to predict PPI probabilities. However, the ROC curves based on sparsity prior using data from all three organisms are only slightly higher than that based on this prior using data from *S. cerevisiae* alone. Hence the performance of the Bayesian approach with sparsity and all three organisms did not show a significant improvement. Lastly, we found that the ROC curves almost did not change with different values of π although ROC curve with $\pi = 0.8$ gave the best performance among them (data not shown).

We further examined that the correlation between the inferred PPI probabilities based on the sparsity priors for *S. cerevisiae* alone and for all three organisms is 0.74, which is larger than the correlation, 0.69, between the sparsity prior for all three organisms and the non-sparsity for all three organisms.

10.5 Conclusion and Discussion

In this chapter, we have described a Bayesian method for simultaneously estimating DDI probabilities, false positive rates, and false negative rates, from high-throughput yeast two hybrid data. Sparsity priors for protein interaction prediction are incorporated in the analysis through a Bayesian method. We have compared the prediction results between with and without sparsity prior.

Compared to previous methods by Deng *et al.* (2002) and Liu *et al.* (2005), our methods using non-sparsity prior may be more efficient in dealing with a large number of parameters using some prior information, more effective to allow for different false positive and false negative rates across different data sets, and more appropriate when the proportion of proteins having more than three domains increases. Compared to the prediction results between with and without sparsity prior, the results of Bayesian approaches with sparsity prior give much more improvement than those without sparsity prior to predict PPI probabilities. However we found that the performance of the Bayesian approach with sparsity and all three organisms did not show a significant improvement compared to that with only one organism *S. cerevisiae*, although the ROC curve based on sparsity prior and all three organism is slightly higher than that of the other case.

We note that the interacting protein pairs we use as gold standard are incomplete because the number of protein interactions verified by experiments is limited. As the number of annotated interactions increases the values of the false positive rate and false negative rate will certainly change.

In our study, we have assumed that the domains are independent of each other. However, domains may be classified into different super families based on structural or functional evidence for a common evolutionary ancestor (Gough and Chothis, 2002). Domains within the same super family would have similar interaction profiles due to the similarity of their structures. We can further develop our Bayesian approach to handle this dependence structure. In addition, in our example considered in this chapter, we have only used yeast two-hybrid data to make PPI predictions. We may further improve our predictions by integrating multiple data sources, e.g., gene expression data and Gene Ontology information (Lee *et al.*, 2004; Lee *et al.*, 2006).

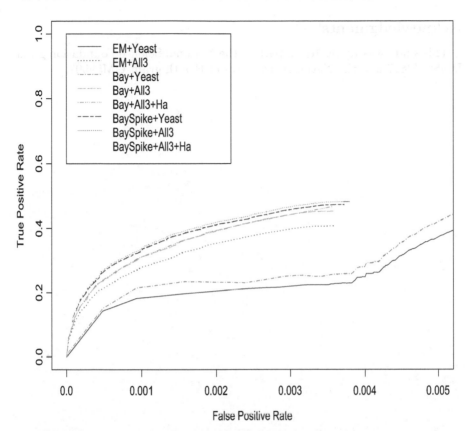

FIGURE 10.1: ROC curves of the Bayesian methods. We use MIPS proteins interactions as a golden standard: EM+Yeast(**S. cerevisiae**)= the likelihood based approach using **S. cerevisiae** organism and using $f_n = 0.8$ and $f_p = 0.0003$; EM+All3= the likelihood based approach using all three organisms and using $f_n = 0.8$ and $f_p = 0.0003$; Bay+Yeast(**S. cerevisiae**)= the Bayesian method with the estimated $\hat{f}_{n_k}$ and $\hat{f}_{p_k}$ using **S. cerevisiae** organism alone; Bay+All3= the Bayesian method with the estimated $\hat{f}_{n_k}$ and $\hat{f}_{p_k}$ using all three organisms; Bay+All3+Ha= the Bayesian method with the estimated $\hat{f}_{n_k}$ and $\hat{f}_{p_k}$ using all three organisms with $h_{ij}^a(\Lambda)$; BaySpike+Yeast(**S. cerevisiae**)= the Bayesian method with sparsity prior, the estimated $\hat{f}_{n_k}$ and $\hat{f}_{p_k}$ using **S. cerevisiae** organism alone; BaySpike+All3= the Bayesian method with sparsity prior, the estimated $\hat{f}_{n_k}$ and $\hat{f}_{p_k}$ using all three organisms; BaySpike+All3+Ha= the Bayesian method with sparsity prior, the estimated $\hat{f}_{n_k}$ and $\hat{f}_{p_k}$ using all three organisms with $h_{ij}^a(\Lambda)$.

Acknowledgments

This study was supported in part by the National Science Foundation grant DMS0714817 and the National Institutes of Health grant GM59507.

References

Aloy, P., Bottcher, B., Ceulemans, H., Leutwein, C., Mellwig, C., Fischer, S., Gavin, A. C., Bork, P., Superti-Furga, G., Serrano, L., and Russell, R. B. 2004. Structure-based assembly of protein complexes in yeast. *Science* 302: 2026-2029.

Bateman, A., Coin, L., Durbin, R., Finn, R. D., Hollich, V., Griffiths-Jones, S., Khanna, A., Marshall, M., Moxon, S., Sonnhammer, E. L., Syudholme, D. J., Yeasts, C., and Eddy, S. R. 2004. The Pfam protein families database. *Nucleic Acids Research* 32: D138-141.

Dempster, A. P., Laird, N. M., and Rubin, D. B. 1977. Maximum likeliood from incomplete data via the EM algorithm. *Journal of the Royal Statistical Society, Ser. B* 39: 1-38.

Deng, M., Mehta, S., Sun, F. and Chen, T. 2002. Inferring domain-domain interactions from protein–protein interactions. *Genome Research* 12: 1504-1508.

Enright, A. J., Iliopoulos, I., Kyrpides, N.C., and Quzounis, C. A. 1999. Protein interaction maps for complete genomes based on gene fusion events *Nature* 402: 86-90.

Giot, L. Bader, J. S., Brouwer, C., Chaudhuri, A., Kuang, B., Li, Y., Hao, Y. L., Ooi, C. E., Godwin, B., Vitols, E., et al. 2003. A protein interaction map of Drosophila melanogaster. *Science* 302: 1727-1736.

Goh, C. S. and Cohen, F. E. 2002. Co-evolutionary analysis reveals insights into protein protein interactions. *Journal of Molecular Biology* 324: 177-179.

Gomez, S. M., Lo, S. H., and Rzhetsky, A. 2001. Probabilistic prediction of unknown metabolic and signal-transduction networks. *Genetics* 159: 1291-1298.

Gomez, S. M., Noble, W. S., and Rzhetsky, A. 2003. Learning to predict protein–protein interactions from protein sequences. *Bioinformatics* 19: 1875-1881.

Gough, J. and Chothis, C. 2002. SUPERFAMILY: HMMs representing all proteins of known structure. SCOP sequence searches, alignments and genome assignments. *Nucleic Acids Research* 30: 268-72.

Gilks, W. R., Best, N. G., and Tan, K. K. C. 1995. Adaptive rejection Metropolis sampling. *Applied Statistics* 44: 455-472.

Ishwaran, H. and Rao, J. 2003. Detecting differentially expressed genes in microarray using Bayesian model selection. *Journal of the American Statistical Association* 98: 438-455.

Ishwaran, H. and Rao, J. 2005. Spike and slab gene selection for multigroup microarray data. *Journal of the American Statistical Association* 100: 764-780.

Ito, T., Chiba, T., Ozawa, R., Yoshidam, M., Hattori, M. *et al.* 2001. A comprehensive two-hybrid analysis to explore the yeast protein interactome. *Proceedings of the National Academy of Sciences* 98: 4569-4574.

Jansen, R., Yu, H., Greenbaum, D., Kluger, Y., Krogan, N. J., Chung, S., Emili, A., Snyder, M., Greenblatt, J. F., and Gerstein, M. 2003. A Bayesian networks approach for predicting protein–protein interactions from genomic data. *Science* 302: 449-453.

Lee, H., Deng, M., Sun, F., and Chen, T. 2006. An integrated approach to the prediction of domain-domain interactions. *BMC Bioinformatics* 7: 269.

Lee, I., Date, S., Adai, A., and Marcotte, E. 2004. A probabilistic functional network of yeast genes. *Science* 306: 1555-1558.

Letunic, I., Copley, R. R., Schmidt, S., Ciccarelli, F. D., Doerks, T., Schultz, J., Ponting, C. P., and Bork, P. 2004. SMART 4.0: *Nucleic Acids Research* 32: D142-144.

Li, S., Armstrong, C. M., Bertin, N., Ge, H., Milstein, S., Boxem, M., Vidalain, P. O., Han, J. D., Chesneau, A., Hao, T. *et al.* 2004. A map of the interactome network of the metazoan C. elegans. *Science* 303: 540-543.

Lin, N., Wu, B., Jansen, R., Gerstein, M., and Zhao, H. 2004. Information assessment on predicting protein–protein interactions. *BMC Bioinformatics* 5: 154.

Liu, Y., Liu, N., and Zhao, H. 2005. Inferring protein–protein interactions through high-throughput interaction data from diverse organisms. *Bioinformatics* 21: 3279-3285.

Lu, L., Arakaki, A. K., Lu, H., and Skolnic, J. 2003. Multimeric threading-based prediction of protein–protein interactions on a genomic scale: application to the *Saccharomyces cerevisiae* proteome. *Genome Research* 13: 1146-1154.

Lucas, J., Carvalho, C., Wang, A. B, Nevins, J. R., and West, M. 2006. Sparse statistical modelling in gene expression genomics. Bayesian Inference for Gene Expression and Proteomics, 155-176, ed. Kim-Anh Do, Peter Mueller, Marina Vannucci, 155-176, Cambridge, Cambridge University Press.

Kim, I., Liu, Y., and Zhao, H. 2007. Bayesian methods for predicting interacting protein pairs using domain information, *Biometrics* 63: 824-833.

Marcotte, E. M., Xenarios, I., and Eisenberg, D. 2001. Mining literature for protein–protein interactions. *Bioinformatics* 17: 357-363.

Mrowka, R., Patzak, A., and Herzel, H. 2001. Is there a bias in proteome research ? *Genome Research* 11: 1971-1973.

Muller, P., Do, K. A., and Tang, F. 2005. A Bayesian mixture model for differential gene expression. *Journal of the Royal Statistical Society, Ser. C (Applied Statistics)* 54: 627-644.

Napetschnig, J., Kassube, S. A., Debler, E. W., Wong, R. W., Blobel, G., Hoelz, A. 2009. Structural and functional analysis of the interaction between the nucleoporin Nup214 and the DEAD-box helicase Ddx19. *Proceedings of the National Academy of Sciences* 106: 3089-3094.

Pazos, F. and Valencia, A. 2001. Similarity of phylogenetic trees as indicator of protein–protein interaction. *Protein Engineering* 14: 609-614.

Pawson, T. and Nash, P. 2003. Assembly of cell regulatory systems through protein interaction domains. *Science* 300: 445-452.

Ramani, A. K. and Marcotte, E. M. 2003. Exploiting the co-evolution of inter-acting proteins to discover interaction specificity. *Journal of Molecular Biology* 327: 273-284.

Shin, O. H., Han, W., Wang, Y., Südhof, T. C. 2005. Evolutionarily conserved multiple C2 domain proteins with two transmembrane regions (MCTPs) and unusual Ca2+ binding properties. *The Journal of Biological Chemistry* 280: 1641-1651.

Sprinzak, E. and Margalit, H. 2001. Correlated sequence-signatures as markers of protein–protein interaction. *Journal of Molecular Biology* 311: 681-692.

Tsoka, S., and Ouzounis, C. A. 2000. Prediction of protein interactions: metabolic enzymes are frequently involved in gene fusion. *Nature Genetics* 26: 141-142.

Uetz, P., Giot, L., Cagney, G., Mansfield, T. A., Judson, R. S., Knight, J. R., et al. 2000. A comprehensive analysis of protein–protein interactiona in *Saccharomyces cerevisiae*. *Nature* 403: 623-627.

von Mering, C., Krause, R., Snel, B., Cornell, M., Oliver, S. G., Fields, S., and Bork, P. 2002. Comparative assessment of large-scale data sets of protein–protein interactions. *Nature* 417: 399-403.

Symbol Description

Symbol	Description		
α	To solve the generator maintenance scheduling, in the past, several mathematical techniques have been applied.		algorithms have also been tested.
		$\theta\sqrt{abc}$	This paper presents a survey of the literature
σ^2	These include integer programming, integer linear programming, dynamic programming, branch and bound etc.	ζ	over the past fifteen years in the generator
		∂	maintenance scheduling. The objective is to
		sdf	present a clear picture of the available recent literature
$\sum$	Several heuristic search algorithms have also been developed. In recent years expert systems,	ewq	of the problem, the constraints and the other aspects of
abc	fuzzy approaches, simulated annealing and genetic	bvcn	the generator maintenance schedule.

Chapter 11

Learning Bayesian Networks for Gene Expression Data

Faming Liang

Department of Statistics, Texas A&M University, College Station, TX 77845, USA

11.1 Introduction

DNA microarray experiments simultaneously measure the expression levels of thousands of genes. An important problem in computational biology is to discover, from such measurements, gene interaction networks and key biological features of cellular systems. One of the most used approaches for this problem is to learn a Bayesian network (Pearl, 1988; Cowell, 1999) from the gene expression data. The Bayesian network is a powerful knowledge representation and reasoning tool under conditions of uncertainty that is typical of real-life applications. The Bayesian network, as illustrated by Figure 11.1, is a directed acyclic graph (DAG) in which the nodes represent the variables in the domain and the edges correspond to direct probabilistic dependencies between them. In a Bayesian network of gene interaction, the nodes represent different genes.

During the past decade, many approaches have been developed for learning of Bayesian networks. These approaches can be grouped into three categories: the approaches based on conditional independence tests, the approaches based on an optimization procedure, and the approaches based on MCMC simulations.

The approaches in the first category perform a qualitative study of dependence relationships between the nodes, and generate a network that repre-

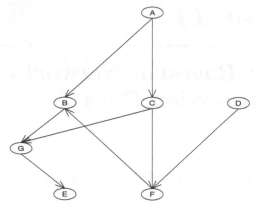

FIGURE 11.1: An example of Bayesian networks.

sents most of the relationships. The approaches described in Spirtes *et al.* (1993), Wermuth and Lauritzen (1983) and de Campos and Huete (2000) belong to this category. The networks constructed by these approaches are usually asymptotically correct, but as pointed out by Cooper and Herskovits (1992) that the conditional independence tests with large condition-sets may be unreliable unless the volume of data is enormous. Due to limited research resources, the sample size of gene expression data is usually small, ranging from 3 to 20, and thus the approaches included in this category are usually not suitable for gene expression data.

The approaches in the second category attempt to find a network that optimizes a selected scoring function, which evaluates the fitness of each feasible network to the data. The scoring functions can be formulated based on different principles, such as entropy (Herskovits and Cooper, 1990), the minimum description length (Lam and Bacchus, 1994), and Bayesian scores (Cooper and Herskovits, 1992; Heckerman *et al.*, 1995). The optimization procedures are usually heuristic, such as Tabu search (Bouckaert, 1995) and evolutionary computation (de Campos and Huete, 2000; Neil and Korb, 1999). Unfortunately, the task of finding a network structure that optimizes the scoring function is a combinatorial optimization problem, and is known to be NP-hard (Chickering, 1996). Hence, the optimization process often stops at a local optimal structure. On the other hand, for most gene expression datasets, the sample size is usually small relative to the network size. In this case, there are usually a large number of high scoring networks with quite different structures, so the inference of network features, such as presence/absence of a particular edge, based on a single network is of high risk.

The approaches in the third category work by simulating a Markov chain over the space of feasible network structures, whose stationary distribution is the posterior distribution of the network. A non-exhaustive list of the work in this category include Madigan and Raftery (1994), Madigan and York

(1995), and Giudici and Green (1999), where the simulation is done using the Metropolis-Hastings (MH) algorithm (Metropolis *et al.*, 1953; Hastings, 1970), and the network features are inferred by averaging over a large number of networks simulated from the posterior distribution. Averaging over different networks significantly can reduce the risk suffered by the single model-based inference procedure. Although the approaches seem attractive, they can only work well for the problems with a very small number of variables. This is because the energy landscape of the Bayesian network can be quite rugged, with a multitude of local energy minima separated by high energy barriers, especially when the network size is large. Here, the energy function refers to the negative log-posterior distribution function of the Bayesian network. As known by many researchers, the MH algorithm is prone to get trapped into a local energy minimum indefinitely in simulations from a system for which the energy landscape is rugged. To alleviate this difficulty, Friedman and Koller (2003) introduce a two-stage algorithm: use the MH algorithm to sample a temporal order of the nodes, and then sample a network structure compatible with the given order. As discussed in Friedman and Koller (2003), for any Bayesian network, there exists a temporal order of the nodes such that for any two nodes X and Y, if there is an edge from X and Y, then X must be preceding to Y in the order. For example, for the network shown in Figure 11.1, a temporal order compatible with the network is ACDFBGE. The two-stage algorithm does improve the mixing over the space of network structures, however, the structures sampled by it does not follow the correct posterior distribution, because the temporal order does not induce a partition of the space of network structures. A network may be compatible with more than one order. For example, the network shown in Figure 11.1 is compatible with at least three orders ACDFBGE, ADCFBGE, and DACFBGE. Refer to Ellis and Wong (2008) for more discussions on this issue.

Based on Liang and Zhang (2009), we describe in this chapter how to learn Bayesian networks using the stochastic approximation Monte Carlo (SAMC) algorithm (Liang *et al.*, 2007). A remarkable feature of the SAMC algorithm is that it possesses the self-adjusting mechanism and is thus not trapped by local energy minima. This is very important for learning of Bayesian networks. In addition, SAMC belongs to the class of dynamic weighting algorithms (Wong and Liang, 1997; Liu *et al.*, 2001), and the samples generated by it can be used to evaluate the network features via a dynamically weighted estimator. Like Bayesian model averaging estimators, the dynamically weighted estimator can have a much lower risk than the single model-based estimator.

The remainder of this chapter is organized as follows. In Section 11.2, we give the formulation of Bayesian networks. In Section 11.3, we first give a brief review of the SAMC algorithm and then describe its implementation for Bayesian networks. In Section 11.4, we present the numerical results on a simulated example and a real gene expression data example. In Section 11.5, we conclude the chapter with a brief discussion.

11.2 Bayesian Networks

A Bayesian network model can be defined as a pair $B = (\mathcal{G}, \boldsymbol{\rho})$. Here, $\mathcal{G} = (\mathcal{V}, \mathcal{E})$ is a directed acyclic graph that represents the structure of the network, $\mathcal{V}$ denotes the set of nodes, and $\mathcal{E}$ denotes the set of edges. For a node $V \in \mathcal{V}$, a parent of V is a node from which there is a directed link to V. The set of parents of V is denoted by $pa(V)$. In this article, we study only the discrete case where V is a categorical variable taking values in a finite set $\{v_1, \ldots, v_{r_i}\}$. There are $q_i = \prod_{V_j \in pa(V_i)} r_j$ possible values for the joint state of the parents of V_i. Each element of the parameter vector $\boldsymbol{\rho}$ represents a conditional probability. For example, ρ_{ijk} is the probability of variable V_i in state j conditioned on that $pa(V_i)$ is in state k. Naturally, $\boldsymbol{\rho}$ is restricted by the constraints $\rho_{ijk} \geq 0$ and $\sum_{j=1}^{r_i} \rho_{ijk} = 1$. The joint distribution of the variables $\boldsymbol{V} = \{V_1, \ldots, V_d\}$ can be specified by the decomposition

$$P(\boldsymbol{V}) = \prod_{i=1}^{d} P(V_i | pa(V_i)). \qquad (11.1)$$

Let $\mathcal{D} = \{\boldsymbol{V}_1, \ldots, \boldsymbol{V}_N\}$ denote a set of independently and identically distributed (*iid*) samples drawn from (11.1). Let n_{ijk} denote the number of samples for which V_i is in state j and $pa(V_i)$ is in state k. Then, the counts $(n_{i1k}, \ldots, n_{ir_ik})$ follow a multinomial distribution, i.e.,

$$(n_{i1k}, \ldots, n_{ir_ik}) \sim \text{Multinomial}(\sum_{j=1}^{r_i} n_{ijk}, \boldsymbol{\rho}_{ik}), \qquad (11.2)$$

where $\boldsymbol{\rho}_{ik} = (\rho_{i1k}, \ldots, \rho_{ir_ik})$. Thus, the likelihood function of the Bayesian network model can be written as

$$P(\mathcal{D}|\mathcal{G}, \boldsymbol{\rho}) = \prod_{i=1}^{d} \prod_{k=1}^{q_i} \binom{\sum_{j=1}^{r_i} n_{ijk}}{n_{i1k}, \ldots, n_{ir_ik}} \rho_{i1k}^{n_{i1k}} \cdots \rho_{ir_ik}^{n_{ir_ik}}. \qquad (11.3)$$

To carry out a Bayesian analysis for the model, we have the following prior specification for the network structure and parameters. Since a network with a large number of edges is often less interpretable and there is a risk of overfitting, it is important to use priors over the network space that encourage sparsity. For this reason, we let $\mathcal{G}$ be subject to the prior:

$$P(\mathcal{G}|\beta) \propto \left(\frac{\beta}{1-\beta}\right)^{\sum_{i=1}^{d} |pa(V_i)|}, \qquad (11.4)$$

where $0 < \beta < 1$ is a user-specified parameter. To force the network to be as simple as possible, a small value of β is usually specified in practice. In this article, we set $\beta = 0.1$ for all examples. If the network size is not a big concern and one wants to avoid an explicit specification for the value of β, a hierarchical prior can also be specified here for β, say, assuming β follows a uniform distribution on (0,1). The parameters ρ is subject to a product Dirichlet distribution

$$P(\rho|\mathcal{G}) = \prod_{i=1}^{d}\prod_{k=1}^{q_i} \frac{\Gamma(\sum_{j=1}^{q_i}\alpha_{ijk})}{\Gamma(\alpha_{i1k})\cdots\Gamma(\alpha_{ir_ik})}\rho_{i1k}^{\alpha_{i1k}-1}\cdots\rho_{ir_ik}^{\alpha_{ir_ik}-1}, \qquad (11.5)$$

where $\alpha_{ijk} = 1/(r_iq_i)$ as suggested by Ellis and Wong (2008). Combining with the likelihood function and the prior distributions, and integrating out ρ, we get the posterior distribution (up to a multiplicative constant):

$$P(\mathcal{G}|\mathcal{D}) \propto \prod_{i=1}^{d}\left(\frac{\beta}{1-\beta}\right)^{|pa(V_i)|} \prod_{k=1}^{q_i} \frac{\Gamma(\sum_{j=1}^{r_i}\alpha_{ijk})}{\Gamma(\sum_{j=1}^{r_i}(\alpha_{ijk}+n_{ijk}))} \prod_{j=1}^{r_i}\frac{\Gamma(\alpha_{ijk}+n_{ijk})}{\Gamma(\alpha_{ijk})},$$
$$(11.6)$$

which contains all the information of the network structure provided by the data.

We note that the Bayesian network is conceptually different from the causal Bayesian network. In the causal Bayesian network, each edge can be interpreted as a direct causal relation between a parent node and a child node, relative to the other nodes in the network (Pearl, 1988). The formulation of Bayesian networks, as described above, is also not sufficient for causal inference. To learn a causal Bayesian network, one needs a dataset obtained through experimental interventions. In general, one can not learn a causal Bayesian network from the observational data alone. Refer to Cooper and Yoo (1999) and Ellis and Wong (2008) for more discussions on this issue.

11.3 Learning Bayesian Networks Using SAMC

11.3.1 A Review of the SAMC Algorithm

Suppose that we are working with the following Boltzmann distribution,

$$f(x) = \frac{1}{Z}\exp\{-U(x)/\tau\}, \quad x \in \mathcal{X}, \qquad (11.7)$$

where Z is the normalizing constant, τ is the temperature, $\mathcal{X}$ is the sample space, and $U(x)$ is called the energy function in terms of physics. In the context of Bayesian networks, $U(x)$ corresponds to $-\log P(\mathcal{G}|\mathcal{D})$, the negative logarithm of the posterior distribution (11.6), and the sample space $\mathcal{X}$ is finite and thus compact. Furthermore, we suppose that the sample space has been partitioned according to the energy function into m disjoint subregions denoted by $E_1 = \{x : U(x) \leq u_1\}$, $E_2 = \{x : u_1 < U(x) \leq u_2\}$,

..., $E_{m-1} = \{x : u_{m-2} < U(x) \le u_{m-1}\}$, and $E_m = \{x : U(x) > u_{m-1}\}$, where $u_1, \ldots, u_{m-1}$ are real numbers specified by the user. Let $\psi(x)$ be a non-negative function defined on the sample space with $0 < \int_{\mathcal{X}} \psi(x)dx < \infty$, and $\theta_i = \log(\int_{E_i} \psi(x)dx)$. In practice, we often set $\psi(x) = \exp\{-U(x)/\tau\}$.

SAMC seeks to draw samples from each of the subregions with a pre-specified frequency. If this goal can be achieved, then the local-trap problem can be avoided essentially. Let $x^{(t+1)}$ denote a sample drawn from a MH kernel $K_{\theta^{(t)}}(x^{(t)}, \cdot)$ with the proposal distribution $q(x^{(t)}, \cdot)$ and the stationary distribution

$$f_{\theta^{(t)}}(x) \propto \sum_{i=1}^{m} \frac{\psi(x)}{e^{\theta_i^{(t)}}} I(x \in E_i), \qquad (11.8)$$

where $\theta^{(t)} = (\theta_1^{(t)}, \ldots, \theta_m^{(t)})$ is an m-vector in a space Θ.

Let $\boldsymbol{\pi} = (\pi_1, \ldots, \pi_m)$ be an m-vector with $0 < \pi_i < 1$ and $\sum_{i=1}^{m} \pi_i = 1$, which defines a desired sampling frequency for the subregions. Henceforth, $\boldsymbol{\pi}$ will be called the desired sampling distribution. Define $H(\theta^{(t)}, x^{(t+1)}) = (e^{(t+1)} - \boldsymbol{\pi})$, where $e^{(t+1)} = (e_1^{(t+1)}, \ldots, e_m^{(t+1)})$ and $e_i^{(t+1)} = 1$ if $x^{(t+1)} \in E_i$ and 0 otherwise. Let $\{\gamma_t\}$ be a positive, non-decreasing sequence satisfying the conditions,

$$(i) \ \sum_{t=0}^{\infty} \gamma_t = \infty, \qquad (ii) \ \sum_{t=0}^{\infty} \gamma_t^{\delta} < \infty, \qquad (11.9)$$

for some $\delta \in (1, 2)$. In the context of stochastic approximation (Robbins and Monro, 1951), $\{\gamma_t\}$ is called the gain factor sequence. In this article, we set

$$\gamma_t = \frac{t_0}{\max(t_0, t)}, \quad t = 0, 1, 2, \ldots \qquad (11.10)$$

for a pre-specified value of $t_0 > 1$. A large value of t_0 will allow the sampler to reach all subregions very quickly even for a large system. Let $J(x)$ denote the index of the subregion that the sample x belongs to. With above notations, one iteration of SAMC can be described as follows.

The SAMC algorithm:

(i) Generate $x^{(t+1)} \sim K_{\theta^{(t)}}(x^{(t)}, \cdot)$ with a single MH step.

(ii) Set $\theta^* = \theta^{(t)} + \gamma_t H(\theta^{(t)}, x^{(t+1)})$.

(iii) If $\theta^* \in \Theta$, set $\theta^{(t+1)} = \theta^*$; otherwise, set $\theta^{(t+1)} = \theta^* + c^*$, where $c^* = (c^*, \ldots, c^*)$ and c^* is chosen such that $\theta^* + c^* \in \Theta$.

The self-adjusting mechanism of the SAMC algorithm is obvious: If a proposal is rejected, the weight of the subregion that the current sample belongs to will be adjusted to a larger value, and thus the proposal of jumping out from the current subregion will less likely be rejected in the next iteration. This mechanism warrants that the algorithm will not be trapped by local minima. The SAMC algorithm represents a significant advance in simulations of complex systems for which the energy landscape is rugged.

The parameter space Θ is set to $[-B_\Theta, B_\Theta]^m$ with B_Θ being a huge number, e.g., 10^{100}, which, as a practical matter, is equivalent to setting $\Theta = \mathcal{R}^m$. In theory, this is also fine. As implied by Theorem 5.4 of Andrieu *et al.* (2005), the varying truncation of θ^* can only occur a finite number of times, and thus $\{\theta^{(t)}\}$ can be kept in a compact space during simulations. Note that $f_{\theta^{(t)}}(x)$ is invariant with respect to a location transformation of $\theta^{(t)}$—that is, adding to or subtracting a constant vector from $\theta^{(t)}$ will not change $f_{\theta^{(t)}}(x)$.

The proposal distribution used in the Metropolis-Hastings moves satisfies the following condition: For every $x \in \mathcal{X}$, there exist $\epsilon_1 > 0$ and $\epsilon_2 > 0$ such that

$$|x - y| \leq \epsilon_1 \Longrightarrow q(x, y) \geq \epsilon_2, \qquad (11.11)$$

where $q(x, y)$ is the proposal distribution, and $|x-y|$ denotes a certain distance measure between x and y. This is a natural condition in study of MCMC theory (Roberts and Tweedie, 1996). In practice, this kind of proposals can be easily designed for both discrete and continuum systems, as discussed in Liang *et al.* (2007). Since, for Bayesian networks both Θ and $\mathcal{X}$ are compact, it is easy to verify that the proposal distributions described in §11.3.2 satisfies the condition (11.11).

SAMC falls into the category of stochastic approximation algorithms (Robbins and Monro, 1951; Benveniste, Métivier and Priouret, 1990; Andrieu, Moulines and Priouret, 2005). Under conditions (11.9) and (11.11), Liang *et al.* (2007) showed that

$$\theta_i^{(t)} \to \begin{cases} C + \log\left(\int_{E_i} \psi(x)dx\right) - \log(\pi_i + \pi_0), & \text{if } E_i \neq \emptyset, \\ -\infty. & \text{if } E_i = \emptyset, \end{cases} \qquad (11.12)$$

as $t \to \infty$, where C is an arbitrary constant, $\pi_0 = \sum_{j \in \{i: E_i = \emptyset\}} \pi_j/(m - m_0)$, and $m_0 = \#\{i : E_i = \emptyset\}$ is the number of empty subregions. A subregion E_i is called empty if $\int_{E_i} \psi(x)dx = 0$. In SAMC, the sample space partition can be made blindly by simply specifying some values of $u_1, \ldots, u_{m-1}$. This may lead to some empty subregions. The constant C can be determined by imposing a constraint on $\theta^{(t)}$, say, $\sum_{i=1}^m e^{\theta_i^{(t)}}$ is equal to a known number.

Let $\hat{\pi}_i^{(t)} = P(x^{(t)} \in E_i)$ be the probability of sampling from the subregion E_i at iteration t. Equation (11.12) implies that as $t \to \infty$, $\hat{\pi}_i^{(t)}$ will converge to $\pi_i + \pi_0$ if $E_i \neq \emptyset$ and 0 otherwise. With an appropriate specification of $\boldsymbol{\pi}$, sampling can be biased to the low energy subregions to increase the chance of locating the global energy optimizer.

Let $(x_1, \theta_1), \ldots, (x_n, \theta_n)$ denote a set of samples generated by SAMC. Let $y_1, \ldots, y_{n'}$ denote the distinct samples among $x_1, \ldots, x_n$. Generate a random variable/vector Y such that

$$P(Y = y_i) = \frac{\sum_{t=1}^n e^{\theta_{t J(x_t)}} I(x_t = y_i)}{\sum_{t=1}^n e^{\theta_{t J(x_t)}}}, \quad i = 1, \ldots, n', \qquad (11.13)$$

where $I(\cdot)$ is the indicator function, and $J(x_t)$ denote the index of the sub-region to which the sample x_t belongs. Since Θ has been restricted to a compact set, $\theta_{tJ(x_t)}$ is finite. By calling some results from the literature of non-homogeneous Markov chains, Liang (2009) showed that the random variable/vector Y generated in (11.13) is asymptotically distributed as $f(\cdot)$. Note that the samples $(x_1, \theta_1), \ldots, (x_n, \theta_n)$ form a non-homogeneous Markov chain. Therefore, for an integrable function $h(x)$, the expectation $E_f h(x)$ can be estimated by

$$\widehat{E_f h(x)} = \frac{\sum_{t=1}^{n} e^{\theta_{tJ(x_t)}} h(x_t)}{\sum_{t=1}^{n} e^{\theta_{tJ(x_t)}}}. \tag{11.14}$$

As $n \to \infty$, $\widehat{E_f h(x)} \to E_f h(x)$ for the same reason that the usual importance sampling estimate converges (Geweke, 1989).

11.3.2   Learning Bayesian Networks Using SAMC

In this subsection, we first describe how to make the MH moves over the space of feasible Bayesian networks, and then discuss some practical issues on implementation of SAMC. Let $\mathcal{G}$ denote a feasible Bayesian network for the data $\mathcal{D}$. At each iteration of SAMC, the MH moves can be performed as follows:

(a) Uniformly randomly choose between the following possible changes to the current network $\mathcal{G}^{(t)}$ producing $\mathcal{G}'$:

 (a.1) Temporal order change: Swap the order of two neighboring models. If there is an edge between them, reverse its direction.

 (a.2) Skeletal change: Add (or delete) an edge between a pair of randomly selected nodes.

 (a.3) Double skeletal change: Randomly choose two different pairs of nodes, and add (or delete) edges between each pair of the nodes.

(b) Calculate the ratio

$$r = e^{\theta_{J(\mathcal{G}')}^{(t)} - \theta_{J(\mathcal{G}^{(t)})}^{(t)}} \frac{\psi(\mathcal{G}')}{\psi(\mathcal{G}^{(t)})} \frac{T(\mathcal{G}' \to T(\mathcal{G}^{(t)}))}{T(\mathcal{G}^{(t)} \to \mathcal{G}')},$$

where $\psi(\mathcal{G})$ is defined as the right hand side of (11.6), and the ratio of the proposal probabilities $T(\mathcal{G}' \to T(\mathcal{G}^{(t)}))/T(\mathcal{G}^{(t)} \to \mathcal{G}') = 1$ for all of the three types of the changes. Accept the new network structure $\mathcal{G}'$ with probability $\min(1, r)$. If it is accepted, set $\mathcal{G}^{(t+1)} = \mathcal{G}'$; otherwise, set $\mathcal{G}^{(t+1)} = \mathcal{G}^{(t)}$.

It is easy to see that this proposal satisfies the condition (11.11). We note that a similar proposal has also been used by Wallace and Korb (1999) in a Metropolis sampling process of Bayesian networks. Later we will show by numerical examples that the SAMC sampling process can mix much faster over the space of Bayesian networks than the Metropolis sampling process. Note that the double changes are not necessary for the algorithm to work, but are included to help accelerate the sampling process.

Let $h(\mathcal{G})$ denote a quantity of interest for a Bayesian network, such as the presence/absence of an edge or a future observation. It follows from (11.14) that $E_p h(\mathcal{G})$, the expectation of $h(\mathcal{G})$ with respect to the posterior (11.6), can be estimated by

$$\widehat{E_P h(\mathcal{G})} = \frac{\sum_{k=n_0+1}^{n} h(\mathcal{G}_k) e^{\theta_{J(\mathcal{G}_k)}^{(k)}}}{\sum_{k=n_0+1}^{n} e^{\theta_{J(\mathcal{G}_k)}^{(k)}}}, \tag{11.15}$$

where $(\mathcal{G}_{n_0+1}, \theta_{J(\mathcal{G}_{n_0+1})}^{(n_0+1)}), \ldots, (\mathcal{G}_n, \theta_{J(\mathcal{G}_n)}^{(n)})$ denotes a set of samples generated by SAMC, and n_0 denotes the number of burn-in iterations.

For an effective implementation of SAMC, several issues need to be considered.

- Sample space partitioning. For learning Bayesian networks, the sample space is usually partitioned according to the energy function. The maximum energy difference in each subregion should be bounded by a reasonable number, say, 2, which ensures that the MH moves within the same subregion have a reasonable acceptance rate. Note that within the same subregion, the SAMC move is reduced to the conventional MH move.

- Choice of the desired sampling distribution π. Since π controls the sampling frequency of each subregion, intuitively, one may choose it to bias sampling to the low energy subregions to increase the chance of finding the global energy minima. In this article, we set in all computations.

$$\pi_i \propto (m - i + 1)^2, \quad i = 1, 2, \ldots, m, \tag{11.16}$$

- Choice of t_0 and N, where N denotes the total number of iterations. Since a large value of t_0 will force the sampler to reach all subregions quickly, even in the presence of multiple local energy minima, t_0 should be set to a large value for a complex problem. The appropriateness of the choice of t_0 and N can be diagnosed by checking the convergence of multiple runs (starting with different points) through an examination for the variation of $\widehat{\theta}$ or $\widehat{\pi}$, where $\widehat{\theta}$ and $\widehat{\pi}$ denote, respectively, the estimates of θ and π obtained at the end of a run. A rough examination for $\widehat{\theta}$ is to see visually whether the $\widehat{\theta}$ vectors produced in the multiple runs follow the same pattern or not. Existence of different patterns implies that the gain factor is still large at the end of the runs or some parts of the sample space are not visited in all runs.

TABLE 11.1: The Definition of the Distribution P_1

X	Z	$P(Z\|X)$	X	Z	$P(Z\|X)$
0	0	0.8	0	1	0.2
1	0	0.2	1	1	0.8

An examination for $\widehat{\pi}$ can base on the following statistic

$$\epsilon_f(E_i) = \begin{cases} \frac{\widehat{\pi}_i - (\pi_i + \widehat{d})}{\pi_i + \widehat{d}} \times 100\%, & \text{if } E_i \text{ is visited,} \\ 0, & \text{otherwise,} \end{cases} \tag{11.17}$$

which measures the deviation of the realized sampling distribution from the desired one, where $\widehat{d} = \sum_{j \in \{i : E_i \text{ is not visited}\}} \pi_j / (m - m'_0)$ and m'_0 is the number of subregions not visited in the run. It is said $\{\epsilon_f(E_i)\}$, output from all runs and for all subregions, matches well if the following two conditions are satisfied: (i) there does not exist such a subregion which is visited in some runs but not in others, and (ii) $\max_{i=1}^{m} |\epsilon_f(E_i)|$ is less than a threshold value, say, 10%, for all runs. A group of $\{\epsilon_f(E_i)\}$ which does not match well implies that some parts of the sample space are not visited in all runs, t_0 is too small, or the number of iterations is too small.

If a run is diagnosed as unconverged, SAMC should be re-run with a large value of N or a larger value of t_0.

11.4 Numerical Examples

11.4.1 A Simulated Example

Consider the Bayesian network shown in Figure 11.1 again. Suppose that a dataset consisting of 500 independent observations has been generated from the network according to the following distributions: $V_A \sim$ Bernoulli(0.7), $V_D \sim$ Bernoulli(0.5), $V_C|V_A \sim P_1$, $V_F|V_C, V_D \sim P_2$, $V_B|V_A, V_F \sim P_2$, $V_G|V_B$, $V_C \sim P_2$, and $V_E|V_G \sim P_1$, where P_1 and P_2 are defined as in Tables 11.1 and 11.2, respectively.

SAMC was first applied to this example, where we partitioned the sample space into 501 subregions with an equal energy bandwidth, $E_1 = \{x : U(x) \leq 2000\}$, $E_2 = \{x : 2000 < U(x) \leq 2001\}$, ..., $E_{500} = \{x : 2498 < U(x) \leq 2499\}$, and $E_{501} = \{x : U(x) > 2499\}$, and set other parameters as follows: $\psi(x) = e^{-U(x)}$ and $t_0 = 5000$. SAMC was run for 10^6 iterations, and 10,000 samples were collected at equally spaced time points. Each run costs about 90

TABLE 11.2: The Definition of the Distribution P_2

X	Y	Z	$P(Z\|X,Y)$	X	Y	Z	$P(Z\|X,Y)$
0	0	0	0.9	0	0	1	0.1
0	1	0	0.5	0	1	1	0.5
1	0	0	0.5	1	0	1	0.5
1	1	0	0.1	1	1	1	0.9

seconds CPU time on a 2.8GHZ personal computer (all computations reported in this article were done on the same computer). The overall acceptance rate of the SAMC moves is about 0.18. For comparison, MH was also applied to this example. MH was run for 10^6 iterations with the same proposal as used by SAMC, and 10,000 samples were collected at equally spaced time points. Each run costs about 81s CPU time. The overall acceptance rate of the MH moves is 0.006, which is extremely low. Figures 11.2 (a)& (b) show, respectively, the sample paths (in the space of energy) produced by SAMC and MH. The sample paths indicate that SAMC can move very fast over the space of Bayesian networks, while MH tends to get trapped in a local energy minimum. The minimum energy values found by SAMC and MH are 2052.88 and 2099.19, respectively. The corresponding network structures are shown in Figure 11.3. It is easy to see that the network produced by SAMC has the same skeleton with the true network, but with the directions of three edges reversed; while the network structure produced by MH is quite different from the true one. We call the network produced by SAMC the putative maximum a posteriori (MAP) network. We note that the energy value of the true network is 2106.6, which is much higher than that of the MAP network.

Regarding the edge direction of Bayesian networks, we note again that the direction of an edge, in our formulation, does not necessarily imply the causal relation between the parent node and the child node. Therefore, for Bayesian networks, inference of the direction of an edge is far less important than inference of presence/absence of the edge.

Later, SAMC was re-run 5 times, but each run was lengthened to 2.5×10^6 iterations. In each of the five runs, the MAP network shown in Figure 11.3(a) was re-located, but no networks with lower energy values were found. For the purpose of estimation, in each of the five runs, we discarded the first 5×10^5 iterations for the burn-in process, and retained the remaining iterations for the network inference. Table 11.3 shows the estimates of the presence probabilities of all possible edges of the Bayesian network, which are calculated using (11.15) based on five independent runs. Table 11.3 can be viewed as a random graph. Further inference of the Bayesian network, for example, prediction for a future observation, can then be made from the random graph based on its Markov property.

For comparison, MH was also run 5 times with each run consisting of $2.5 \times$

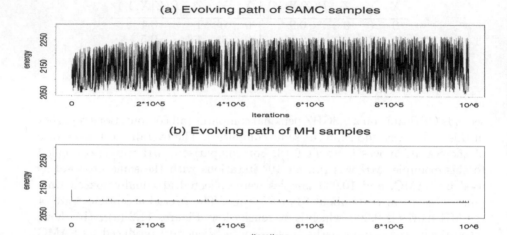

FIGURE 11.2:   The sample paths (in the space of energy) produced by SAMC (upper panel) and MH (lower panel).

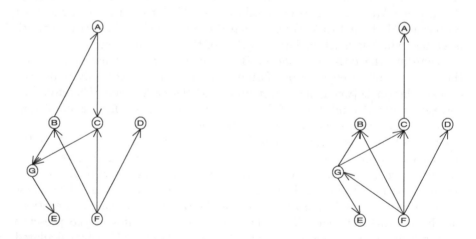

FIGURE 11.3:   The maximum a posteriori network structures learned by SAMC (left panel) and MH (right panel) for the simulated example.

TABLE 11.3: Estimates of the Presence Probabilities of the Edges for the Bayesian Network Shown in Figure 11.1

	A	B	C	D	E	F	G
A	—	0 (0)	0.9997 (0.0001)	0 (0)	0 (0)	0 (0)	0 (0)
B	1 (0)	—	0 (0)	0 (0)	0.0046 (0.0009)	0.4313 (0.0552)	1 (0)
C	0.0003 (0.0001)	0 (0)	—	0 (0)	0 (0)	0 (0)	0.9843 (0.0036)
D	0 (0)	0 (0)	0 (0)	—	0.0002 (0)	0.0476 (0.0233)	0 (0)
E	0 (0)	0 (0)	0 (0)	0 (0)	—	0 (0)	0.0044 (0.0009)
F	0 (0)	0.5687 (0.0552)	1 (0)	0.9524 (0.0233)	0.1638 (0.0184)	—	0 (0)
G	0 (0)	0 (0)	0.0003 (0.0001)	0 (0)	0.9956 (0.0009)	0 (0)	—

Note: The numbers in parentheses show the standard errors of the estimates.

10^6 iterations. Figure 11.4 compares the progression paths of minimum energy values produced by SAMC and MH. It indicates again that SAMC is superior to MH for learning of Bayesian networks. SAMC can locate the MAP network in each of the five runs, while the MH algorithm failed to locate the MAP network in all of the five runs.

For this example, SAMC costs a little more time than MH in each iteration. For larger networks, e.g., the network studied in the next example, SAMC and MH will cost about the same CPU time, because in this case the CPU time cost by each algorithm is dominated by the part used for energy evaluation, and the part used for weight updating in SAMC is ignorable.

11.4.2 The Yeast Cell Cycle Data

The yeast cell cycle dataset (Cho *et al.*, 1998) shows the fluctuation of expression levels of approximately 6000 genes over two cell cycles (17 time points). Cho *el al.* (1998) identified 384 genes which have only one phase and peak at different time points. The normalized expression data of the 384 one-phase genes are available at http://faculty.washington.edu/kayee/model. The data have been used by many authors to test various clustering methods. In this paper, we use a subset of the data as one example to illustrate learning of Bayesian networks for gene expression data. The sub-dataset was generated as follows: Randomly select 50 genes from the set of 384 genes, and then

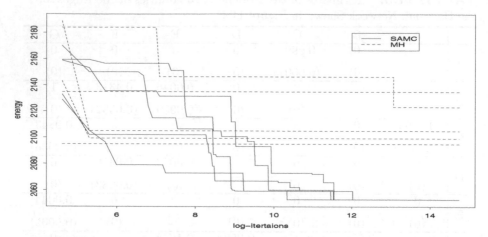

FIGURE 11.4: Progression paths of minimum energy values produced in the five runs of SAMC (solid lines) and in the five runs of MH (dashed lines) for the simulated example.

discretize the data into three categories by setting

$$V_{ij} = \begin{cases} 1, & \text{if } X_{ij} < Q_{i,1/3}, \\ 2, & \text{if } Q_{i,1/3} \leq X_{ij} < Q_{i,2/3}, \\ 3, & \text{if } X_{ij} \geq Q_{i,2/3}, \end{cases} \qquad (11.18)$$

for $i = 1, \ldots, 50$ and $j = 1, \ldots, 17$, where X_{ij} denotes the expression level of gene i at time j, and $Q_{i,1/3}$ and $Q_{i,2/3}$ denote, respectively, the 1/3 and 2/3 quantiles of the expression values of gene i. The names of the genes included in the sub-dataset are given in Table 11.4.

SAMC was applied to this example. We partitioned the sample space into 301 subregions with an equal energy bandwidth, $E_1 = \{x : U(x) \leq 900\}$, $E_2 = \{x : 900 < U(x) \leq 901\}$, ..., $E_{300} = \{x : 1198 < U(x) \leq 1199\}$, and $E_{301} = \{x : U(x) > 1200\}$, and set other parameters as follows: $\psi(x) = e^{-U(x)}$ and $t_0 = 5000$. In addition, to make the sample space smaller, we restricted the number of parent nodes of each node to be no more than 5. SAMC was run 5 times. Each run consisted of 2×10^7 iterations, and cost about 29 minutes CPU time. For comparison, MH was also run 5 times for this example with the same proposal as used by SAMC. Each run consisted of 2×10^7 iterations, and cost about 27 minutes CPU time. Figure 11.5 compares the progression paths of minimum energy values produced by SAMC and MH in the respective runs. It indicates that SAMC outperforms MH for this example; the minimum energy value produced by SAMC in any of the five runs is lower than that produced by MH in all of the five runs.

Later, a run of SAMC was lengthened to 5×10^7 iterations, and a network

TABLE 11.4:   Genes Included in the Sub-Dataset for Bayesian
Network Learning

No.	Name	No.	Name	No.	Name	No.	Name
1	YGR044c	2	YLR273c	3	YCR005c	4	YBR067c
5	YPL058c	6	YGL055w	7	YMR254c	8	YDR511w
9	YBR231c	10	YJL187c	11	YLR313c	12	YJL173c
13	YDL164c	14	YNL082w	15	YML021C	16	YJL196c
17	YHR153c	18	YLR121c	19	YHR110w	20	YPL014w
21	YHR039c	22	YOR144c	23	YDR113c	24	YEL061c
25	YGR140w	26	YBL063w	27	YDR356w	28	YMR190c
29	YDL093w	30	YER017c	31	YBR276c	32	YFR038w
33	YOR188w	34	YJL099w	35	YOR274w	36	YDR464w
37	YCR085w	38	YMR003w	39	YOR073w	40	YJR110w
41	YOL012c	42	YBL032w	43	YGL021w	44	YCR042c
45	YDR146c	46	YNL053w	47	YIL162w	48	YDL138W
49	YPL186c	50	YPR157w					

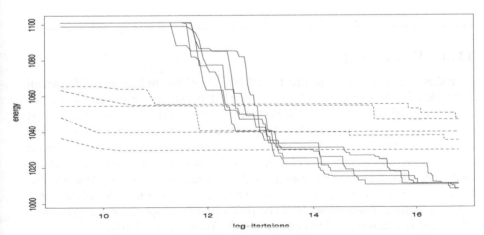

FIGURE 11.5:   Progression paths of minimum energy values produced in
the five runs of SAMC (solid lines) and in the five runs of MH (dashed lines)
for the yeast cell cycle example.

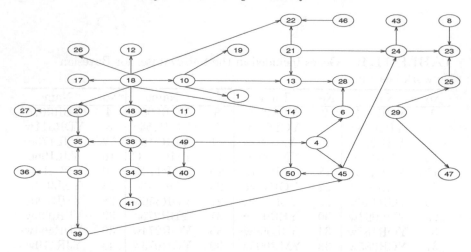

FIGURE 11.6: The maximum a posteriori Bayesian network learned by SAMC for a subset of the yeast cell cycle data.

with the energy value 997.68 was found. Figure 11.6 shows the major part of the network, which consists of 37 genes. The other genes not shown in Figure 11.6 form a number of small networks disconnected from the major part of the network.

11.5 Discussion

In this article, we have applied the SAMC algorithm to learning of Bayesian networks. The numerical results indicate that SAMC can outperform significantly the MH algorithm for this problem. In our study of the yeast cell cycle data, the data are discretized into three categories. We note that the resulting networks may depend on the discretization scheme. In general, discretization with a small number of categories can lead to poor accuracy, while discretization with a large number of states can lead to excessive computation efforts. Recently, some researchers seek to discretize the data non-uniformly according to a certain criterion, such as the conditional entropy (Fayyad and Irani, 1993) and the minimum description length (Wang *et al.*, 2006).

In this article, the inference of Bayesian networks is done via the dynamic importance estimator (11.15) for which the weights of many terms are very low. Alternatively, the inference can be done using the Occam's window approach as described in Madigan and Raftery (1994). That is, estimating

$E_P h(\mathcal{G})$ by

$$\widetilde{E_p h(\mathcal{G})} = \frac{\sum_{\mathcal{G} \in \mathcal{A} \smallsetminus \mathcal{B}} h(\mathcal{G}) P(\mathcal{G}|\mathcal{D})}{\sum_{\mathcal{G} \in \mathcal{A} \smallsetminus \mathcal{B}} P(\mathcal{G}|\mathcal{D})}, \tag{11.19}$$

where

$$\mathcal{A} = \left\{ \mathcal{G}_k : \frac{\max_l P(\mathcal{G}_l|\mathcal{D})}{P(\mathcal{G}_k|\mathcal{D})} \leq c \right\},$$

for some constant c, say, a value between 10 and 100 as suggested by Jeffreys (1961, app. B); and

$$\mathcal{B} = \left\{ \mathcal{G}_k : \exists \mathcal{G}_l \in \mathcal{A}', \mathcal{G}_l \subset \mathcal{G}_k, \frac{P(\mathcal{G}_l|\mathcal{D})}{P(\mathcal{G}_k|\mathcal{D})} > 1 \right\}.$$

Two basic principles underly the estimator (11.19). First, if a model predicts the data far less well than the best model in the class, it should be discarded, so the models not belonging to the set $\mathcal{A}$ should be excluded from estimation. Second, appealing to Occam's razor, a model receiving less support from the data than any of its simpler sub-models should no longer be considered, so the models belonging to $\mathcal{B}$ should also be excluded. The estimator (11.19) is potentially less variable than the estimator (11.15).

Acknowledgment

The author's research was partially supported by grants from the National Science Foundation (DMS-0607755) and the National Cancer Institute (CA104620).

References

Andrieu, C., Moulines, É., and Priouret, P. (2005). Stability of stochastic approximation under verifiable conditions. *SIAM J. Control and Optimization*, **44**, 283-312.

Benveniste, A., Métivier, M., and Priouret, P. (1990). *Adaptive Algorithms and Stochastic Approximations*, New York: Springer-Verlag.

Bouckaert, R.R. (1995). Bayesian belief networks: from construction to inference. Ph.D. Thesis. University of Utrecht.

Chickering, D.M. (1996). Learning Bayesian networks is NP-complete. In *Learning from Data: Artificial Intelligence and Statistics V*. New York: Springer-Verlag.

Cho, R.J., Campbell, M.J., Winzeler, E.A., Steinmetz, L., Wodicka, L., Wolfsberg, T.G., Gabrielian, A.E., Landsman, D., Lockhart, D.J., and Davis, R.W. (1998). A genome wide transcriptional analysis of the mitotic cell cycle. *Mol. Cell*, **2**, 65-73.

Cooper, G.F. and Herskovits, E. (1992). A Bayesian method for the induction of probabilistic networks from data. *Machine Learning*, **9**, 309-347.

Cooper, G.F. and Yoo, C. (1999). Causal discovery from a mixture of experimental and observational data. In *Proceedings of the 15th Conference on Uncertainty in Artificial Intelligence*, pp.116-125. Morgan Kaufmann Publishers.

Cowell, R. (1999). Introduction to inference for Bayesian networks. In *Learning in Graphical Models*, M.I. Jordon (eds), pp.9-26. Cambridge: The MIT Press.

de Campos, L.M. and Huete, J.F. (2000). A new approach for learning belief networks using independence criteria. *Int. J. Approx. Reason*, **24**, 11-37.

Ellis, B. and Wong, W.H. (2008). Learning causal Bayesian network structures from experimental data. *J. Amer. Statist. Assoc.*, **103**, 778-789.

Fayyad, U. and Irani, K. (1993). Multi-interval discretization of continuous-valued attributes for classification learning. In *Proceedings of International Joint Conference on Artificial Intelligence*, Chambery, France, pp.1022-1027.

Friedman, N. and Koller, D. (2003). Being Bayesian about network structure. A Bayesian approach to structure discovery in Bayesian networks. *Machine Learning*, **50**, 95-125.

Geweke, J. (1989). Bayesian inference in econometric models using Monte Carlo integration. *Econometrica*, **57**, 1317-1339.

Giudici, P. and Green, P. (1999). Decomposable graphical Gaussian model determination. *Biometrika*, **86**, 785-801.

Hastings, W.K. (1970). Monte Carlo sampling methods using Markov chains and their applications. *Biometrika*, **57**, 97-109.

Heckerman, D., Geiger, D. and Chickering, D.M. (1995). Learning Bayesian networks: The combination of knowledge and statistical data. *Machine Learning*, **20**, 197-243.

Herskovits, E. and Cooper, G.F. (1990). Kutató: an entropy-driven system for the construction of probabilistic expert systems from datasets. In: Bonissone P. (eds) Proceedings of the Sixth Conference on Uncertainty in Artificial Intelligence, Cambridge, pp. 54-62.

Jeffreys, H. (1961). *Theory of probability* (3rd ed.), Oxford: Oxford University Press.

Lam, W. and Bacchus, F. (1994). Learning Bayesian belief networks: an approach based on the MDL principle. *Comput. Intell.*, **10**, 269-293.

Liang, F. (2009). On the use of stochastic approximation Monte Carlo for Monte Carlo Integration. *Statistics & Probability Letters*, **79**, 581-587.

Liang, F., Liu, C. and Carroll, R.J. (2007). Stochastic approximation in Monte Carlo computation. *J. Amer. Statist. Assoc.*, **102**, 305-320.

Liang, F. and Zhang, J. (2009). Learning Bayesian networks for discrete data. *Computational Statistics and Data Analysis*, **53**, 865-876.

Liu, J.S., Liang, F., and Wong, W.H. (2001). A theory for dynamic weighting in Monte Carlo. *J. Amer. Statist. Assoc.*, **96**, 561-573.

Madigan, D. and Raftery, E. (1994). Model selection and accounting for model uncertainty in graphical models using Occam's Window. *J. Amer. Statist. Assoc.*, **89**, 1535-1546.

Madigan, D. and York, J. (1995). Bayesian graphical models for discrete data. *International Statistical Review*, **63**, 215-232.

Metropolis, N., Rosenbluth, A.W., Rosenbluth, M.N., Teller, A.H., and Teller, E. (1953). Equation of state calculations by fast computing machines. *Journal of Chemical Physics*, **21**, 1087-1091.

Neil, J.R. and Korb, K.B. (1999). The evolution of causal models. In *Third Pacific Asia Conference on Knowledge Discovery and Data Mining*, N. Zhong and L. Zhous, Eds., Spinger-Verlag, pp.432-437.

Pearl, J. (1988). *Probabilistic Reasoning in Intelligent Systems: Networks of Plausible Inference*. San Mateo: Morgan Kaufmann.

Robbins, H. and Monro, S. (1951). A stochastic approximation method. *Annals of Mathematical Statistics*, **22**, 400-407.

Roberts, G.O. and Tweedie, R.L. (1996). Geometric convergence and central limit theorems for multidimensional Hastings and Metropolis algorithms. *Biometrika*, **83**, 95-110.

Spirtes, P., Glymour, C., Scheines, R. (1993). *Causation, Prediction and Search*, Lecture Notes in Statistics 81. New York: Springer-Verlag.

Wallace, C.S. and Korb, K.B. (1999). Learning linear causal models by MML sampling. In *Causal Models and Intelligent Data Management*, A. Gammerman, (Eds). Heidelberg: Springer-Verlag.

Wang, S., Li, X., and Tang, H. (2006). Learning Bayesian networks structure with continuous variables. In *Lecture Notes in Computer Science*, Vol. 4093, X. Li, O.R. Zaiane, and Z. Li (Eds), pp.448-456. Heidelberg: Springer-Verlag.

Wermuth, N. and Lauritzen, S. (1983). Graphical and recursive models for contingency tables. *Biometrika*, **72**, 537-552.

Wong, W.H. and Liang, F. (1997). Dynamic weighting in Monte Carlo and optimization, *Proc. Natl. Acad. Sci. USA*, **94**, 14220-14224.

Symbol Description

i	An index for the list of samples from a microarray experiment (e.g. a particular microarray).
g	An index for the list of probe IDs – i.e., genes – on the expression array.
$x_{g,i}$	(log base 2) Gene expression for gene g and sample i.
x_i	p–vector of the $x_{g,i}$.
X	$p \times n$ matrix whose columns are the x_i.
H	Transpose of the $n \times r$ design matrix in a designed experiment.
h_i	The ith column of H corresponding to the ith sample.
B	A $p \times r$ matrix of regression coefficients, where r is the number of elements in the design matrix and p is the number of genes – or, more properly, probe sets – on the array.
Λ	$k \times n$ matrix of latent factors.
λ_i	The ith column Λ.
A	A $p \times k$ matrix of factor loadings, where k is the number of latent factors.
ψ_g	Residual variance of the expression measurement error for gene g.
$\pi_{g,j}$	Sparsity probability governing non-zero values of (g, j) elements of A and B.
$\circ^*$	For any quantity $\circ$, the MCMC-based approximation to the posterior mean $E(\circ|X)$.

Chapter 12

In-Vitro to In-Vivo Factor Profiling in Expression Genomics

Joseph E. Lucas[1], Carlos M. Carvalho[2], Daniel Merl[1] and Mike West[1]

[1] *Duke University*, [2] *University of Chicago*

12.1 Introduction

The widespread use of DNA microarray gene expression technology offers the potential to substantially increase our understanding of cellular pathways and apply that knowledge in clinical as well as basic biological studies. In human biology, there are two common types of study performed with expression arrays. The first involves samples derived from cloned cells in a highly controlled environment, subject to some designed experimental intervention, such as up/down regulation of a particular target gene (Huang, West and Nevis , 2003; Huang, et al., 2003; Bild, et al. , 2006) or manipulation of the pH level of the growing environment (Chen, et al. , 2008). Such experiments are useful in that they are designed to minimize all sources of gene expression variability except that of the experimental intervention. One result generated by this type of experiment is the identification of a subset of genes that are differentially expressed when subject to a given intervention. A list of such genes, together with the magnitudes by which those genes are differentially over/under-expressed and additional numerical summaries are said to comprise the *signature* of the intervention. The second, common type of study involves samples derived from tissue removed from living subjects, such as tumor tissue obtained via biopsy (Perou, et al., 2000; West, et al. , 2001; Huang, et al., 2002, 2003; Pittman, et al. , 2004; Seo, et al., 2004; Miller, et al. , 2005; Riech, et al., 2005). A central goal of this second type of study is to find statistical associations between the measured expression levels in a tissue sample and clinical variables associated with the subject, e.g. survival time or metastasis.

Much of our interest in recent and current studies lies in relating these two types of experiments. We concentrate in this chapter on studies in cancer genomics, though the general questions of cross-study projection of genomic markers, and the more general questions of trans-study and trans-platform integration of genomic data and analysis results, are central to all human disease studies with genomic methodologies (e.g., Seo, et al., 2007). In cancer genomics, for example, a designed *in vitro* experiment in cultured cells may identify a list of genes that are differentially expressed in the presence of an up-regulated oncogene; we might then focus on data on the same genes in a collection of tumor tissue samples to determine the extent to which each tumor exhibits signs of de-regulation of that oncogene. One way to accomplish this is to make use of the oncogene signature derived from the controlled study to compute an appropriately weighted average of the relevant gene expression levels observed in the tumor tissue samples. This technique has been shown to be useful for quantifying pathway activity in a variety of different settings (e.g., Bild, et al. , 2006).

However, there are at least two difficulties with this approach. First, it is exceedingly rare to find that the exact patterns of differential expression discovered in the designed experiment are conserved in the tissue samples. Second, the highly controlled setting of the designed experiment results in a very limited representation of the various sources of variability in gene expression that are likely to affect expression in tissues grown *in vivo*. Returning to the example of up-regulation of a particular gene, consider an example of the MYC gene (myelocytomatosis viral oncogene homolog) in mammalian cell developmental processes. MYC is a transcription factor that is well-known to be involved in programmed cell death, cellular differentiation, proliferation, and potentially a number of other biological pathways. MYC is not, however, the only gene associated with each of these pathways, and therefore a tissue sample which exhibits high levels of expression of MYC may not exhibit high levels of activity of all of its associated downstream pathways due to the activity of other genes, especially other transcription factors. For the same reason, there will be genes in the MYC pathways that are not discovered by the simple MYC upregulation experiment due to low levels of expression of co-regulators. In short, the designed experiment lacks information pertaining to possible synergistic or antagonistic activity between MYC and other genes of the sort likely to be encountered *in vivo*.

Translation of patterns of differential gene expression derived from *in vitro* experiments to *in vivo* tissue studies therefore requires a methodology for decomposing an experimental signature into functional sub-units, each reflecting modular but interacting components of a larger cellular process. This chapter overviews and exemplifies our approach to this general problem area using sparse factor models, an approach that has been used effectively in a number of applications in cancer genomics and other contexts. Here we focus on the application to trans-study analyses connecting *in vitro* expression discovery to *in vivo* contexts in cancer. Based on formal Bayesian models for both contexts, we describe methodology for projection and then evolving experimentally derived signatures into sets of interacting sub-signatures, as represented by multiple latent factors in formal latent factor models of observational human cancer data. We demonstrate how latent factors can be linked to components of known cellular pathways, as well as how latent factors can be predictively related to clinical outcomes. Issues of data calibration and other practicalities are also addressed in this formal factor model framework for *in vitro* to *in vivo* translation of biological markers. A flow chart showing the overall process for the use of sparse factor modeling in the context of microarrays is shown in Figure 12.9.

12.2   Modeling Gene Expression

In general, for experiments performed on cloned cells under strictly controlled conditions, our models utilize multivariate regression and analysis of variance with a fixed, known design to describe observed expression patterns. We use the notation of (Carvalho, et al., 2008). Suppose gene expression assay has p genes (probe sets on a DNA microarray) and we perform the experiment on n samples, and let X be the $p \times n$ data matrix of observed expression values with elements $x_{g,i}$. Let the $r \times n$ matrix H be the transpose of the design matrix; so H has elements $h_{j,i}$ running over treatments/factors $j = 1 : r$ (rows) and expression array samples $i = 1 : n$ (columns). We model the measured expression values (on a log base 2, or fold-change scale) as:

$$x_{g,i} = \mu_g + \sum_{j=1}^{r} \beta_{g,j} h_{j,i} + \nu_{g,i}, \qquad \nu_{g,i} \perp\!\!\!\perp N(\nu_{g,i}|0, \psi_g)$$

or, in vector form,

$$x_i = \mu + B h_i + \nu_i, \qquad (i = 1 : n),$$

where μ is the $p-$vector of baseline average expression levels μ_g for all genes, B is the $p \times r$ matrix of regression coefficients with elements $\beta_{y,j}$, h_i is the $r-$vector column i of H, ν_i is the $p-$vector of gene-specific experimental noise terms $\nu_{g,i}$ with individual normal error variances ψ_g and $\Psi = \text{diag}(\psi_{1:g})$. In matrix form,

$$X = \mu 1' + BH + N$$

where 1 is the $n-$vector of 1s and N the $p \times n$ error matrix with columns ν_i.

Depending on the experimental context, the rows of H will include 0/1 entries reflecting main effects and interactions of treatment factors, genetic or environmental interventions and so forth, with the corresponding $\beta_{g,j}$ values representing the changes in average expression for each gene g as a function of *design factor j*. The design matrix may also include measured values of regressor variables. In most of our studies, we generally include values of sample-specific control information constructed from housekeeping probe sets on the array; these covariates carry *assay artifact* information reflecting experimental bias and errors that are often reflected at least sporadically, if not in some cases substantially, across the samples. Ignoring the potential for such assay artifacts to corrupt expression on a gene-sample specific level can lead to false discovery and/or obscure biological variation, and our approach to gene-sample specific normalisation using housekeeping gene data is generally applicable to at least partially address this in the automatic analysis (Lucas et al., 2006; Carvalho, et al., 2008; Wang, et al. , 2007). These references also describe model fitting and evaluation under specified priors, and describe – with a number of examples – the MCMC implementation in the Bayesian Factor Regression Models (BFRM) software, available at the Duke Department of Statistical Science software web page.

12.2.1   Sparsity Priors

In the case of very high-dimensional genomic assays, we expect that most genes will not show differential expression in association with any particular design factor - biology is complex, but not that complex. That is, any column j of B will be sparse, with typically many, many zeros. Our studies over the last several years have demonstrated the importance of sparsity prior modeling utilising the class of prior distributions (Lucas et al., 2006) that reflect this inherent biological view that, in parallel for all $j = 1 : r$, *many genes g have zero probability of a non-zero* $\beta_{g,j}$. That is, the priors utilise the standard Bayesian zero point-mass/mixture form in which each $\beta_{g,j}$ has an individual probability $\pi_{g,j} = Pr(\beta_{g,j} \neq 0)$, but now the extension allows some (typically many) of the $\pi_{g,j}$ themselves to be zero, inducing a more aggressive shrinkage of the $\beta_{g,j}$ to zero. Specifically, the hierarchical sparsity prior model is (see (Lucas et al., 2006))

$$\beta_{g,j} \sim (1 - \pi_{g,j})\delta_0(\beta_{g,j}) + \pi_{g,j}N(\beta_{g,j}|0, \tau_j),$$
$$\pi_{g,j} \sim (1 - \rho_j)\delta_0(\pi_{g,j}) + \rho_j Be(\pi_{g,j}|a_j m_j, a_j(1 - m_j)),$$

with hyper-priors on the elements ρ_j and τ_j. We refer to such distributions interchangeably as sparsity priors and variable selection priors. Our assumption is that the probability of any particular probe associating with a given design factor is quite low, so that we assign the priors $Be(\rho_j|st, s(1-t))$ with s is large and t is small. The priors on the τ_j, conditionally conjugate inverse gamma priors, are specified in context of the design and, critically, the know range of (log base 2) expression data (see many details and examples in (Lucas et al., 2006; Carvalho, et al., 2008; Wang, et al. , 2007) and the BFRM software web site linked to the latter paper, under the Duke Department of Statistical Science software web page).

12.2.2   Signature Scores

Assume a model has been fitted to a given experimental data set, generating posterior summaries from the MCMC computations. We can explore, characterize and summarise significant aspects of changes in gene expression expression with respect to design factors in a number of ways. Of particular interest in studies that aim to connect the results of such *in vitro* studies to data from human observational studies is the definition and evaluation of a summary numerical *signature score* for any set of genes that are taken as defining a signature of one of the design factors.

Suppose that for the design variable j there are a number of genes with very high values of $\pi_{g,j}^*$; one typical approach is to threshold the probabilities to identify a set of most highly significant genes, and then define a weighted average of expression on those genes as a signature score. In a number of studies we have used this approach with the weightings defined as follows.

Consider a potential future sample x_{new} with a fixed and known value h_{new}. For each g, j and conditional on all model parameters, the likelihood ratio $p(x_{g,new}|\beta_{g,j} \neq 0)/p(x_{g,new}|\beta_{g,j} = 0)$ depends on $x_{g,new}$ monotonically and only through the term

$$s_{g,j}x_{g,new} \quad \text{with} \quad s_{g,j} = \psi_g^{-1}\beta_{g,j}.$$

Hence the future observation will be more consistent with non-zero effects on design factor j for larger values of the sum of these terms over all genes g, the sum being implied by the conditional independence of the $x_{g,\cdot}$ given all model parameters. This leads to the obvious *signature score* for design factor j as

$$sig_{j,new} = \sum_{g=1}^{p} s_{g,j}x_{g,new} = s_j'x_{new}$$

for the signature j weight vector $s_j = (s_{1,j}, \ldots, s_{p,j})'$. In practice, we can estimate the signature weight vectors s_j by posterior estimates based on the MCMC samples of the observed data, and use these to evaluate the score on any future samples – i.e., to project the signature measure of the activity levels of genes related to design factor j onto the new sample, predictively. This can be done within the MCMC analysis by saving and summarizing posterior samples of the set of s_j, or as an approximation after the analysis by substituting posterior estimates for the parameters in the expression for $s_{g,j}$, such as by plugging-in posterior means $\pi_{g,j}^* = Pr(\beta_{g,j} \neq 0|X)$, posterior estimates of residual variance ψ_g^* for each gene g, and posterior estimates of effects

$B^* = (\beta^*_{g,j})$ where $\beta^*_{g,j} = E(\beta_{g,j}|X, \beta_{g,j} \neq 0)$. This latter approach leads to the defined signature scores $sig^*_{j,new}$ used here, in which $s_{g,j}$ is estimated by

$$s^*_{g,j} = \pi^*_{g,j}\beta^*_{g,j}/\psi^*_g.$$

Further, we may modify this definition to sum over just a selected set of signature genes for which $\pi^*_{g,j}$ exceeds some specified, high threshold. This is important for communication and transfer to other studies, as it generates a more manageable subset of signature genes rather than requiring the full set.

12.2.3 Sparse, Semi-Parametric Latent Factor Models

Some of our major applications in recent studies of observational data sets in cancer, as in other areas, have utilized large-scale Bayesian latent factor models (Lopes and West , 2003; West , 2003), either alone or as components of more elaborate *latent factor regression models* (Lucas et al., 2006; Carvalho, et al., 2008). The value of latent factor models to reflect multiple intersecting patterns underlying observed variations in gene expression data across samples of, for example, lung cancers is clear. It is also relevant in some experimental contexts. The model as described above assumes that all sources of variation in expression are known and represented by the fixed design, though imprecision in the design specification and other aspects of biological activity may lead to influences on expression reflected in common patterns in multiple subsets of genes that may be captured by latent factors.

The extension of the model to include latent factors is simply

$$X = \mu 1' + BH + A\Lambda + N$$

where the $k \times n$ matrix Λ represents the realized values of k latent factors across the n samples, having elements $\lambda_{j,i}$ for factor $j = 1:k$ on sample $i = 1:n$. Λ is a now uncertain analogue to the known matrix H from the design. The columns of the $p \times k$ factor loadings matrix $A = \alpha_{g,j}$ are the weights or coefficients of genes on factors (analogous to the regression coefficients in B). In observational cancer studies with known covariates, we will generally use the full model above; in other studies, either of the design or factor component may be absent, depending on context.

The elements of A are given sparsity priors as described for $\beta_{g,j}$. Our models for latent factor space utilize non-parametric components to allow for what may be quite non-Gaussian patterns in these common, underlying patterns that reflect underlying "gene pathway" activities influencing gene expression. It is common to observe genes in subsets that are unexpressed in some samples but clearly expressed at possibly various sets of levels across others, for example; subtle examples include expression in the presence/absence of a genetic mutation, in diseased versus non-diseased patients, or before and after treatment. In these situations, and others, a standard normal latent factor model would be lacking. Our models assume a Dirichlet process component for this. With λ_i representing the $k-$vector column of Λ, we have

$$\lambda_i \sim F(\cdot),$$
$$F \sim DP(\alpha F_0),$$
$$F_0(\lambda) = N(\lambda|0, I)$$

with total mass $\alpha > 0$. This encourages latent factor-space clustering of the samples with relation to gene expression patterns and can lead to cleaner separation when expression differences are due to binary or categorical phenotypes when appropriate, as well as allowing for adaptation to multiple other non-Gaussian aspects. The model and MCMC computational extensions are routine and implemented in BFRM (Wang, et al. , 2007).

Consider now a potential future observation, x_{new} at a known design vector h_{new}, and set $z_{new} = x_{new} - BH$. Conditional on all model parameters and any realized Λ values (the full set of conditioning values being simply denoted by $\circ$ below) it then routinely follows that

$$(\lambda_{new}|z_{new}, \circ) \sim c_{new} N(\lambda_{new}|d, D) + \sum_{i=1}^{n} c_i \delta_{\lambda_i}(\lambda_{new}),$$

where $\delta_e(\cdot)$ is the Dirac delta function representing a point mass at e; the c_q are probabilities given by

$$c_{new} = \alpha CN(z_{new}|0, AA' + \Psi),$$
$$c_i = CN(z_{new}|A\lambda_i, \Psi), \quad i = 1 : n$$

where C is a constant of normalization; and

$$d = DA'\Psi^{-1}z_{new} \quad \text{and} \quad D^{-1} = I + A'\Psi^{-1}A.$$

We know, incidentally, that the λ_i cluster into subsets of common values, and hence the sum above collapses to a simpler sum over the distinct values; that fact and its implications is central to efficient MCMC computation though not immediately relevant for this specific theoretical discussion.

As with *in vitro* defined signature projection, we are often interested in predicting factor variability in a new sample, and the above distributions show how this is done. Given full posterior samples of the model parameters – now including Λ – based on the observed data X, we can at least approximately evaluate the terms defining the predictive distribution for λ_{new}. Plugging-in posterior estimates of model quantities as a practicable approximation leads to the ability to simulate λ_{new}, and also to generate point estimates such as those used in our examples below, viz.

$$\lambda_{new}^* = c_{new}^* d^* + \sum_{i=1}^{n} c_i^* \lambda_i^* \tag{12.1}$$

where λ_q^* are the MCMC posterior means of the λ_i and the the terms c^* and d^* are as theoretically defined above but now evaluated at posterior estimates of the model parameters and latent factors set at posterior means.

12.3 Signature Factor Profiling Analysis

Signature Factor Profiling Analysis (SFPA) refers to the overall enterprise of evaluating *in vitro* defined gene expression signatures of biological or environmental interventions, projecting those signatures into observational data sets – such as a sample of arrays from breast cancer tissues – and then using latent factor models to explore and evaluate the expression patterns of signature genes, and other related genes, in the cancer data set. The complexity of *in vivo* biology compared to controlled experiments on cultured cells inevitably means that projected signatures involve genes that are simply not expressed in the cancer samples, and others that show far more complex and intricate patterns in the cancer samples. Thus factor analysis of a signature gene set can generate multiple (k) factors that reflect this increased complexity; this represents the $k-$dimensional *in vivo profile* of the one-dimensional *in vitro* signature.

Our factor modeling uses the evolutionary computational method for model search and gene inclusion that was introduced and described in (Carvalho, et al., 2008). Beginning with a small set of signature genes, MCMC analysis is used to fit an initial factor model, evolving the model in terms of the number of latent factors included. At the next step the model predicts outside that initial set of genes to evaluate all other (thousands of) genes on the array to assess concordance in expression of each of these genes with estimated latent factors in the "current" model. Of those genes showing high concordance, we can then select a subset to add to the initial set of genes and refit the factor model, potentially increasing the number of factors. Many examples are given in the references. In the current context, the biological focus is perfectly reflected in this evolutionary analysis: we are initally concerned with the "simple" expression signature of intervention on a single biological pathway – the design factor in question elicits that pathway response *in vitro*. When moving into observational cancer samples, biological action involving multiple other, intersecting pathways will generally be evident in the tumor data set. Thus (i) we will be interested in identifying other genes that seem to show patterns of related expression; and (ii) as we add in such additional genes, we are moving out of the initial signature pathway into intersecting pathways that will lead to the need for additional latent factors to reflect the increasingly rich and complex patterns observed. Operationally, this evolutionary process is allowed to continue until a certain number of factors has been found, until a certain number of genes have been added, or until no more factors or genes can be added based on thresholds on estimated posterior gene-factor loading probabilities (for details, see (Carvalho, et al., 2008)). By restricting the termination requirements, we can control how closely the final list of factors remains to the initial set of probes, or how rich it can in principle become by exploring the biological neighborhood of the original *in vitro* response.

12.4   The E2F3 Signature in Ovarian, Lung, and Breast Cancer

The E2F family of transcription factors is central to cell cycle regulation and synthesis of DNA in mammalian cells, playing roles in multiple aspects of cell growth, proliferation and fate (Harbour and Dean , 2000; Zhu, et al., 2004). Deregulation of E2F family members is a feature of many human cancers (Nevis, et al., 2001). Two of the family members, E2F1 and E2F3, are among the genes investigated in experiments defining *in vitro* signatures of oncogene deregulation in human mammary epithelial cells (HMEC) (Bild, et al. , 2006). In these oncogene experiments, expression profiles were collected from cloned HMECs with each experimental group consisting of replicate observations from the same cloned HMECs after transfection with a viral plasmid containing a known oncogene. A series of control samples were also generated; see (Lucas et al., 2006) for additional insights and examples using one-way Anova models under sparsity priors, as well as for discussion of the extension of this analysis to include assay artifact control regression variables. The nine oncogenes transfected were MYC, Src, Akt, P110, E2F1, E2F3, Ras, β-catenin, and p63. From among these, we focus here on E2F3 and explore the E2F3 signature projection to human cancers to exemplify signature factor profiling. The analysis generates biological connections in the E2F pathways consonant with known biology, as well as new and novel biological insights relevant to the oncogenic role of E2F3 when deregulated.

To define our initial signature gene set for evolutionary factor analysis, we reanalyzed the full set of oncogene expression data (Affymetrix u133) in a one-way Anova design under sparsity priors. For design factor $j = E2F3$, the $\beta_{g,E2F3}$ then represent the average changes in expression on genes g due to upregulation of E2F3. We identify those probe sets g for which $\pi^*_{g,E2F3} > 0.99$ in this analysis.

Our interest here is in projecting the E2F3 signature to three different human cancer data sets: lung cancer (Bild, et al. , 2006), ovarian cancer (Dressman, et al. , 2007), and breast cancer (Miller, et al. , 2005), also all on Affymetrix arrays. Of the initially screened genes, we first identified all of those probsets on the Affymetrix arrays used in each of the cancer studies;

from among these, we then selected the 100 genes with the largest positive expression changes $\beta^*_{g,E2F3}$ in the experimental context. This defined our *in vitro* E2F3 signature gene set and the corresponding signature weight vector s^*_{E2F}.

Evolutionary factor analysis was applied to each of the three cancer data sets separately, initializing based on this signature gene set. The analyses evolved through a series of iterations, at each stage bringing in at most 25 additional genes most highly related to the "current" estimates of latent factors, and then exploring whether or not to add additional latent factors re-fitting the model and evaluating the estimated gene-factor loadings $\pi^*_{g,j}$ for all genes g and factors j in the model. Thresholding these probabilities at 0.75 was used for both new gene inclusion and adding factors; an additional factor was added to the model only when at least 15 genes showed $\pi^*_{g,j} > 0.75$. The evolutionary analysis was run in the context of allowing expansion to no more than 1000 genes and 20 factors. The analyses terminated with ($p = 1000, k = 15$) for lung, ($p = 1000, k = 14$) for ovarian and ($p = 1000, k = 13$) for breast cancer. Consistent with our strong belief that factors represent small collections of genes, the prior on ρ_j was a beta distribution with mean .001 and prior observations 1000. Both τ_j and Ψ_j were given diffuse inverse gamma priors.

In the remaining discussion, we describe some aspects of the resulting factor analyses of the three data sets and explore how these signature factor profiles relate across cancers and also relate to some underlying biological pathway interconnections. We refer to posterior means $\lambda^*_{j,i}$ as, simply, factor j or, the value of factor j, on sample i in any data set. Referring to the full vector of factors on sample i implicitly denotes the approximate posterior mean λ^*_i. Further, in discussing genes related to a specific factor we refer to genes being involved in the factor, or significantly loaded on the factor, based on $\pi^*_{g,j} > 0.99$ in the corresponding analysis.

12.4.1   Indications of E2F3 Co-Regulation with NF-Y

To begin, Figure 12.1 shows the high correlation between the first factor discovered in the ovarian cancer data (x-axis) compared to projections into the ovarian data of the first factors from each of the lung and breast analyses. This is done, according to equation (12.1), by projection of the estimated factor loadings of lung and breast factor 1 into the ovarian data set.

The first factor can be expected to represent the *in vitro* instantiation of key aspects of the projected E2F3 signature, suggesting that at least some, key aspects of E2F3 pathway activation across the cancer types is reflective of that seen in the experimental context. The consonance of the structure across the three cancers suggests a strong and meaningful biological underlying function is picked up in this component of the projected signature factor profile. This can be explored by simply looking at the names of genes identified as being highly loaded on factor 1 in each analysis.

If we look more closely at the three factors 1 in the different cancer types

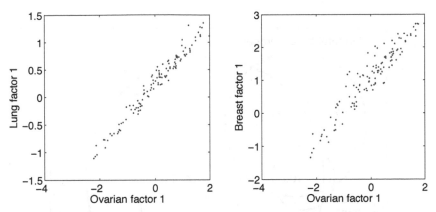

FIGURE 12.1: Scatter plot of estimated values $\lambda^*_{1,i}$ in the ovarian cancer data set (x-axis) against scores on the ovarian tumors of the lung and breast cancer factors 1; these two sets of scores are constructed by projection of the estimated factor loadings of lung and breast factor 1 into the ovarian data set, as described in Section 12.2.3.

we see that, while the lung cancer data set yields a factor that highly loads substantially more genes than those discovered in breast and ovarian cancer, there is substantial overlap in highly loaded genes (Figure 12.2). If we look at the standardized expression levels of the genes in the intersection of these three factors in each of the three data sets, we see that there is a clear, shared pattern of expression across cancer types (Figure 12.3).

With the strongly conserved nature of the first factor across all three data sets, it is natural to ask whether there is a coherent biological structure one might attach to the associated genes. We used GATHER (Chang and Nevins , 2006) to search for significant association between the 109 probes (representing 83 genes) in the intersection of the three factors and other known gene lists. All E2F's are involved in cell cycle regulation, so it is of no surprise that this factor shows a strong association with the cell cycle pathway (38 of the 83 genes, GATHER Bayes' factor of 65). More remarkable is the fact that 50 of the 83 genes are in a list of 178 genes known to have binding sites for the transcription factor NF-Y in their promoter region (Mantonavi , 1998). There is strong evidence for the association between NF-Y and E2F; there is known to be a high correspondence between the presence of E2F3 and NF-Y binding sites in the promoter regions of many genes (Zhu, et al., 2005). These two genes are also known to work in concert to regulate Cyclin B and the beginning of G_2/M phase of cell proliferation (Zhu, et al., 2004; Hu, et al., 2006). Thus there is strong evidence that this factor represents co-regulation of the cell cycle by E2F3 and NF-Y.

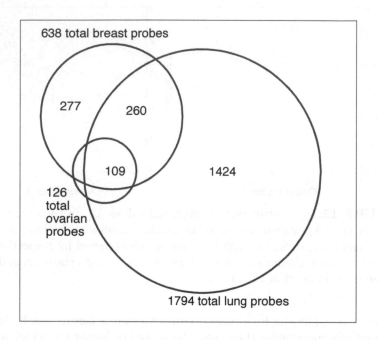

638 total breast probes

277   260

109   1424

126
total
ovarian
probes

1794 total lung probes

FIGURE 12.2:   A Venn diagram depicting intersections of the gene lists of factor 1 from breast, lung, and ovarian cancers (total number of genes in each factor listed outside the circles, the number of genes in each intersection listed inside). Although the ovarian cancer factor 1 involves a significantly smaller number of genes than the other cancers, almost all of these (109 of 124) are common with those in the lung and breast factors.

12.4.2   Adenocarcinoma versus Squamous Cell Carcinoma

The lung cancer data set (from (Bild, et al. , 2006)) is comprised of some tissues from patients with squamous cell carcinoma, and some from patients with adenocarcinoma. These are significantly different types of non-small cell lung carcinoma (as defined histologically) and we expect that many pathways would show differential expression between these two groups. To explore this, we generated binary regression models with the lung cancer sub-type as response and the 15 estimate factors λ_i^* in lung cancer as candidate predictors. Bayesian model fitting and exploration of subset regressions on these factors used Shotgun Stochastic Search (SSS) of (Hans, et al., 2007a; Hans, et al. , 2007b) – following previous uses of SSS methodology in "large p" regression in genomics (Riech, et al., 2005; Blackwell, et al., 2006; Hans and West , 2006) – to search over models involving factors which distinguish between squamous

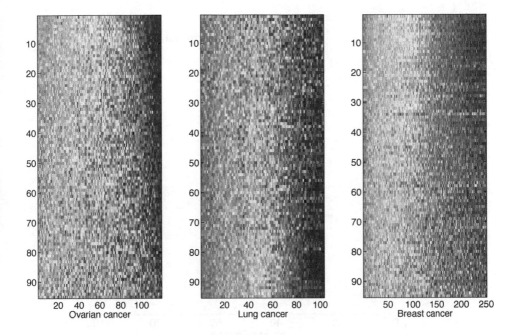

FIGURE 12.3: Patterns of expression of the 109 genes in the intersection group in Figure 12.2. The genes are ordered the same in the three heat-map images, and the patterns of expression across samples are evidently preserved across genes in this set.

cell and adenocarcinoma.

A model involving lung cancer E2F3 profiled factors 4 and 5 shows a clear ability to distinguish between the expression patterns of squamous cell carcinoma and adenocarcinoma (Figure 12.4). Given that there is no such distinction among the patients in the ovarian or breast cancer data sets, it is gratifying to note that the expression pattern exhibited by these genes is not conserved across the other two cancer data sets (Figure 12.5). This is a nice example of the use of factor profiling to identify clinically relevant predictive factor structure, one of the key reasons for interest in these models and the overall approach and is being used in other applications (e.g., Chen, et al. , 2008; Seo, et al , 2007; Carvalho, et al., 2008).

12.4.3   Survival and the E2F3 Pathway

The same general strategy was explored to assess whether the E2F3 profiled factors relate at all to tumor malignancy and aggressiveness, proxied by survival data available in all three cancer studies. Again we used SSS for re-

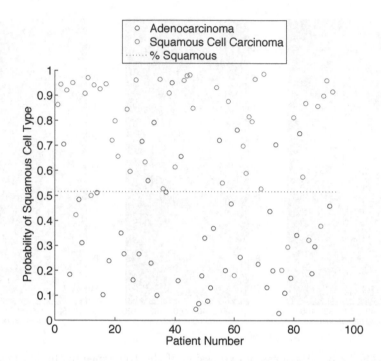

FIGURE 12.4: Factors 4 and 5 from the E2F3 signature factor profile in the lung cancer data set show a clear ability to distinguish between adenocarcinoma and squamous cell carcinoma. The figure displays probabilities of squamous versus adenocarcinoma from binary regression models identified using SSS.

gression model evaluation, this time using Weibull survival models that draw on the set of estimated factors as candidate predictors (Hans, et al., 2007a; Hans, et al. , 2007b). From each of the three separate analyses, we then identified those factors with the highest posterior probabilities of being included in the survival regression across many models explored and evaluated on subsets of factors. To exemplify the immediate clinical relevance of these "top survival factors" and their connections back to the underlying E2F3 pathway related biology, we mention here one of the top factors from each of the three analyses.

First, it is a measure of the importance of the E2F3/NF-Y pathway that the long term survival of breast cancer patients depends significantly on the E2F/NF-Y factor (Figure 12.6), the factor earlier discussed as being highly consonant across cancer types. Interestingly, from the survival regressions and follow-on exploration of survival curves, this pathway factor appears to be

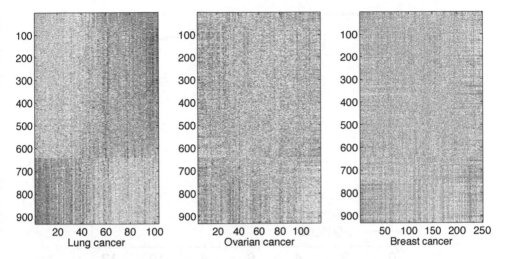

FIGURE 12.5: Patterns of expression of the genes in factor 4 in lung cancer. This factor shows differential expression of 350 genes that distinguish between squamous cell carcinoma and adenocarcinoma. It is therefore appropriate that the pattern of expression is clearly not conserved in the ovarian and breast cancers.

only mildly important in connection with ovarian cancer survival and really unimportant in lung cancer survival.

Survival in ovarian cancer is strongly and most heavily associated with ovarian factor 5 (Figure 12.7), though 1, 5, and 12 in combination produced the most probable individual survival model candidate in SSS. Factor 5 significantly loads on 129 probe ID's representing 113 genes. We again utilized GATHER to identify that 47 of these genes contain known binding sites for Transcription Factor DP-2 in their promoter region. While this factor does not share a strong correlation with any factor discovered in the breast or lung tissues, it is true that the expression pattern of the genes in the factor are conserved in the other data sets. Transcription Factor DP-2 is a known dimerization partner of the E2F family (Rogers, et al., 1996), and is important in cell cycle regulation via this interaction and perhaps other roles, and so this suggests a plausible interpretation of factor 5 as related to the interactions of E2Fs with DP-2. Again, this factor naturally emerges in the factor analysis as reflecting key aspects of variation of genes in the E2F3 related pathways, and this very strong linkage to survival outcomes is novel and a starting point for further biological investigation.

The most probable Weibull survival model in the lung cancer analysis involved lung cancer factor 9, and some of the strength of association of factor 9 is illustrated in Figure 12.8. This factor contains significant 80 probes rep-

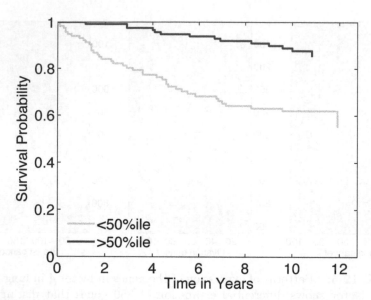

FIGURE 12.6:   Kaplan–Meyer curves representing survival in breast cancer patients with above/below median scores for the E2F/NF-Y factor. Suppression of the E2F/NF-Y pathway appears to be associated with the decreased probability of long-term survival.

resenting 66 genes. Of these, 44 contain a known binding motif for E74-like factor 1 (ELF1) (Nishiyama, et al., 2000), a gene known to be important in the interleukin-2 immune response pathway. Additionally, ELF1 binds to the key Retinoblastoma protein (Wang, et al., 1993). Rb is a central checkpoint in the Rb/E2F network controlling cell proliferation, and is also bound by the members of the E2F family (Harbour and Dean , 2000; Nevis, et al., 2001). This factor seems specific to lung cancer, and the genes it involves do not appear to show consistent expression in either breast or ovarian samples. A possible explanation of this is that the lung tumor tissue may have a higher level of invasion by immune cells than the other tissues. This possibility is corroborated by factor 8, which shows a high degree of relatedness (GATHER Bayes factor >18) to the biologically annotated pathways representing *response to biotic stimulus, defense response* and *immune response* as calculated by GATHER. If we examine the expression patterns of lung factor 8 genes in the other two tumor tissue types, we find that the signature is visibly preserved, but the pattern is weaker and contains a higher noise level than in lung cancers.

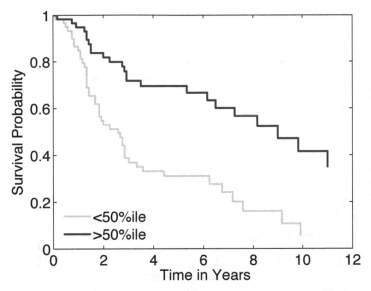

FIGURE 12.7: Kaplan–Meyer curves representing survival in ovarian cancer patients with above/below median scores for the E2F/TFDP2 factor. Suppression of the E2F/TFDP2 pathway appears to be associated with decreased probability of long-term survival.

12.5 Closing Comments

Much recent and current expression genomics research has been devoted to the discovery of lists of genes showing differential expression across some known phenotypes, or that demonstrate an ability to stratify patients into different risk groups. While studies of this type have been shown to be useful in many applications, they have typically relied on the use of clustering and other simple methods to make the transition between *in vitro* and *in vivo* studies. The overall framework of Bayesian factor regression modeling utilizing sparsity priors, as implemented in the BFRM software and exemplified in a range of recent studies, provides a formal, encompassing framework for such trans-study analysis.

The *in vivo* factor model profiling of *in vitro* defined expression signatures of controlled biological perturbations or environmental changes is a powerful and statistically sound approach that we have further explored and exemplified in the current chapter. Multiple current studies in cancer genomics, and other areas of both basic and human disease related biology, are using this approach (Lucas et al., 2009a,b; Merl et al., 2010a,b). The example of the E2F3 signature factor profiling across three distinct cancer types and

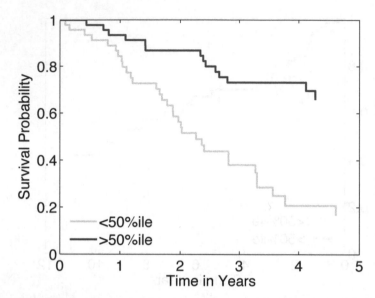

FIGURE 12.8:   Kaplan–Meyer curves representing survival in lung cancer patients with above/below median scores for the E2F/ELF1 factor. Suppression of the E2F/ELF1 pathway appears to be associated with decreased probability of long-term survival.

sample data sets is a vivid illustration of the kinds of results that can be expected: factor model refinements of *in vitro* signatures define multiple factors that relate to underlying biological pathway interconnections, generating suggestions for interpretation of factors, biologically interesting contrasts across cancer types, and suggesting novel directions for functional evaluation as well as identifying clinically useful predictive factors as potential biomarkers of cancer subtypes, survival, and other outcomes. Critically, the methodological framework of Bayesian regression factor modeling under sparsity priors for both designed experiments and observational samples enables and drives the overall strategy.

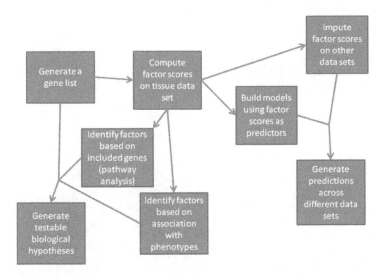

FIGURE 12.9: This flowchart shows the overall strategy for analyzing microarray expression data.

Acknowledgments

We are grateful to Ashley Chi and Joe Nevins of Duke University for useful discussions. We acknowledge support of the National Science Foundation (grants DMS-0102227 and DMS-0342172) and the National Institutes of Health (grants NHLBI P01-HL-73042-02 and NCI U54-CA-112952-01). Any opinions, findings and conclusions or recommendations expressed in this work are those of the authors and do not necessarily reflect the views of the NSF or NIH.

References

A.H. Bild, G. Yao, J.T. Chang, Q. Wang, A. Potti, D. Chasse, M. Joshi, D. Harpole, J.M. Lancaster, A. Berchuck, J.A. Olson, J.R. Marks, H.K. Dressman, M. West and J.R. Nevins (2006). Oncogenic pathway signatures in human cancers as a guide to targeted therapies. *Nature*, **439**, 353–357.

C.M. Carvalho, J. Lucas, Q. Wang, J. Chang, J.R. Nevins and M. West (2008). High-dimensional sparse factor modeling - Applications in gene expression genomics. *Journal of the American Statistical Association*, **103**, 1438–1456.

J.T. Chang and J.R. Nevins (2006). GATHER: A systems approach to interpreting genomic signatures. *Bioinformatics*, **22**, 2926–2933.

J.L. Chen, J.E. Lucas, T. Schroeder, S. Mori, J.R. Nevins, M. Dewhirst, M. West J.T. Chi (2008). The genomic analysis of latic acidiosis and acidiosis response in human cancers. *PLoS Genetics* **4(12):e1000293**.

H.K. Dressman, A. Berchuck, G. Chan, J. Zhai, A. Bild, R. Sayer, J. Cragun, J.P. Clarke, R. Whitaker, L.H. Li, J. Gray, J. Marks, G.S. Ginsburg, A. Potti, M. West, J.R. Nevins and J.M. Lancaster (2007). An integrated genomic-based approach to individualized treatment of patients with advanced-stage ovarian cancer. *Journal of Clinical Oncology*, **25**, 517–525.

H.K. Dressman, C. Hans, A. Bild, J. Olsen, E. Rosen, P.K. Marcom, V. Liotcheva, E. Jones, Z. Vujaskovic, J.R. Marks, M.W. Dewhirst, M. West, J.R. Nevins and K. Blackwell (2006). Gene expression profiles of multiple breast cancer phenotypes and response to neoadjuvant therapy. *Clinical Cancer Research*, **12**, 819–216.

C. Hans and M. West (2006). High-dimensional regression in cancer genomics. *Bulletin of the International Society for Bayesian Analysis*, **13**, 2–3.

C. Hans, A. Dobra and M. West (2007). Shotgun stochastic search in regression with many predictors. *Journal of the American Statistical Association*, **102**, 507–516.

C. Hans, Q. Wang, A. Dobra and M. West (2007). SSS: High-dimensional Bayesian regression model search. *Bulletin of the International Society for Bayesian Analysis*, **14**, 8-9.

J.W. Harbour and D.C. Dean (2000). The Rb/E2F pathway: Expanding roles and emerging paradigms. *Genes and Development* **14(19)**, 2393–2409.

Q. Hu, J-F. Lu, R. Luo, S. Sen and S. Maity (2006). Inhibition of CBF/NF-Y mediated transcription activation arrests cells at G_2/M phase and suppresses expression of genes activated at G_2/M phase of the cell cycle. *Nucleic Acids Research*, **34**, 6272–6285.

E. Huang, M. West and J.R. Nevins (2002). Gene expression profiles and predicting clinical characteristics of breast cancer. *Hormone Research*, **58**, 55–73.

E. Huang, M. West and J.R. Nevins (2003). Gene expression phenotypes of oncogenic pathways. *Cell Cycle*, **2**, 415–417.

E. Huang, S. Chen, H. Dressman, J. Pittman, M.H. Tsou, C.F. Horng, A. Bild, E.S. Iversen, M. Liao, C.M. Chen, M. West, J.R. Nevins and A.T. Huang (2003). Gene expression predictors of breast cancer outcomes. *The Lancet*, **361**, 1590–1596.

E. Huang, S. Ishida, J. Pittman, H. Dressman, A. Bild, M. D'Amico, R. Pestell, M. West and J.R. Nevins (2003). Gene expression phenotypic models that predict the activity of oncogenic pathways. *Nature Genetics*, **34**, 226–230.

H. Lopes and M. West (2003). Bayesian model assessment in factor analysis. *Statistica Sinica*, **14**, 41–67.

J.E. Lucas, C.M. Carvalho, Q. Wang, A. Bild, J.R. Nevins and M. West (2006). Sparse statistical modeling in gene expression genomics. In *Bayesian Inference for Gene Expression and Proteomics*, (eds: M. Vannucci *et al*), Cambridge University Press, 155–176.

J.E. Lucas, C.M. Carvalho, M. West (2009). A Bayesian analysis strategy for cross-study translation of gene expression biomarkers. *Statistical Applications in Genetics and Molecular Biology*, **81**, article 11.

J.E. Lucas, C.M. Carvalho, J.L. Chen, J.T.A. Chi and M. West (2009). Cross-study projection og genomic markers: An evaluation in cancer genomics. *PLoS ONE*, **4:e4523**

R. Mantovani (1998). A survey of 178 NF-Y binding CCAAT boxes. *Nucleic Acids Research*, **26**(5), 1135–1143.

D. Merl, J.L. Chen, J.T. Chi and M. West (2010). Integrative analysis of cancer gene expression studies using Bayesian latent factor modeling. *Annals of Applied Statistics*, in press.

D. Merl, J.E. Lucas, J.R. Nevins, H. Shen and M. West (2010). Trans-study projection of genomic biomarkers in analysis of oncogene deregulation and brest cancer. In *The Handbokk of Applied Bayesian Analysis* (A. O'Hagan and M. West, eds.), Oxford University Press.

L.D. Miller, J. Smeds, J. George, V.B. Vega, L. Vergara, A. Ploner, Y. Pawitan, P. Hall, S. Klaar, E.T. Liu and J. Bergh (2005). An expression signature for p53 status in human breast cancer predicts mutation status, transcriptional effects, and patient survival. *Proceedings of the National Academy of Sciences*, **102**, 13550–13555.

J.R. Nevins (2001). The Rb/E2F pathway and cancer. *Human Molecular Genetics*, **10**, 699–703.

C. Nishiyama, K. Takahashi, M. Nishiyama, K. Okumura, C. Ra, Y. Ohtake and T. Yokota (2000). Splice isoforms of transcription factor Elf-1 affecting its regulatory RT function in transcription-molecular cloning of rat Elf-1. *Bioscience, Biotechnology, and Biochemistry*, **64**, 2601–2607.

C.M. Perou, T. Sorlie, M.B. Eisen, M. van de Rijn, S.S. Jeffrey, C.A. Rees, J.R. Pollack, D.T. Ross, H. Johnsen, L.A. Akslen, O. Fluge, A. Pergamenschikov, C. Williams, S.X. Zhu, P.E. Lnning, A.-L. Brresen-Dale, P.O. Brown, and D. Botstein (2000). Molecular portraits of human breast tumors. *Nature*, **406**, 747–752.

J. Pittman, E. Huang, H. Dressman, C.F. Horng, S.H. Cheng, M.H. Tsou, C.M. Chen, A. Bild, E.S. Iversen, A.T. Huang, J.R. Nevins and M. West (2004). Integrated modeling of clinical and gene expression information for personalized prediction of disease outcomes. *Proceedings of the National Academy of Sciences*, **101**, 8431–8436.

J. Rich, B. Jones, C. Hans, E.S. Iversen, R. McClendon, A. Rasheed, D. Bigner, A. Dobra, H.K. Dressman, J.R. Nevins and M. West (2005). Gene expression profiling and genetic markers in glioblastoma survival *Cancer Research*, **65**, 4051–4058.

K.T. Rogers, P.D.R. Higgins, M.M. Milla and R.S. Phillips (1996). DP-2, a heterodimeric partner of E2F: Identification and characterization of DP-2 proteins expressed *in vivo. Proceedings of the National Academy of Sciences*, **93**, 7594–7599.

D.M. Seo, T. Wang, H. Dressman, E.E. Herderick, E.S. Iversen, C. Dong, K. Vata, C.A. Milano, J.R. Nevins, J. Pittman, M. West and P.J. Goldschmidt-Clermont (2004). Gene expression phenotypes of atherosclerosis. *Arteriosclerosis, Thrombosis and Vascular Biology*, **24**, 1922–1927.

D.M. Seo, P.J. Goldschmidt-Clermont and M. West (2007). Of mice and men: Sparse statistical modeling in cardiovascular genomics. *Annals of Applied Statistics*, **1**, 152–178.

C.Y. Wang, B. Petryniak, C.B. Thompson, W.G. Kaelin and J.M. Leiden (1993). Regulation of the Ets-related transcription factor Elf-1 by binding to the retinoblastoma protein. *Science*, **260**, 1330–1335.

Q. Wang, C. Carvalho, J.E. Lucas and M. West (2007). BFRM: Bayesian factor regression modeling. *Bulletin of the International Society for Bayesian Analysis*, **14**, 4–5.

M. West, C. Blanchette, H. Dressman, E. Huang, S. Ishida, R.Spang, H. Zuzan, J.R. Marks and J.R. Nevins (2001). Predicting the clinical status of human breast cancer utilizing gene expression profiles. *Proceedings of the National Academy of Sciences*, **98**, 11462-11467.

M. West (2003). Bayesian factor regression models in the "large p, small n" paradigm. In *Bayesian Statistics 7*,(J.M. Bernardo, M.J. Bayarri, J.O. Berger, A.P. Dawid, D. Heckerman, A.F.M. Smith and M. West, eds.), Oxford University Press, 723–732.

W. Zhu, P.H. Giangrande and J.R. Nevins (2004). E2Fs link the control of G1/S and G2/M transcription. *EMBO Journal*, **23**, 4615–4626.

Z. Zhu, J. Shendure and G.M. Church (2005). Discovering functional transcription-factor combinations in the human cell cycle. *Genome Research*, **15**, 848–855.

Chapter 13

Proportional Hazards Regression Using Bayesian Kernel Machines

Arnab Maity and Bani K. Mallick

Dept. of Statistics, Texas A&M University, College Station, TX 77843-3143, U.S.A.

13.1 Introduction

One of the main goals of survival analysis is to predict patient survival from the available medical information. In general, suppose T denotes a lifetime or time to certain event of interest and a set of covariates or risk factors denoted by x of dimension $p \times 1$ are available. We develop survival distribution which aids to predict the survival time of a new patient with covariates x_{new}.

Survival distributions are usually characterized by their hazard function which is the conditional density function at time t given survival up to time t. In customary modeling, the hazard of an individual is assumed to be a function of the covariates x as well as t. To accommodate censored survival time we also require the survivor function which is the upper tail probability function of the survival distribution. A well known way to conduct such an analysis is Cox's proportional hazards model (Cox, 1972). In brief, if T is the survival time, subject to possible right censorship, and $X = (X_1, \ldots, X_p)$ is the vector of explanatory variables then the Cox model can be written as $\lambda(t|X) = \lambda_0(t)\exp(\beta.X)$ where t is a realization of T, λ_0 is the unknown baseline hazard function and β is the vector of unknown regression coefficients. The integrated hazard associated with $\lambda(t)$ is $\Lambda(t) = \Lambda_0(t)g(X)$ where $\Lambda_0(t) = \int_0^t \lambda_0(u)du$. The survivor function is $\mathcal{F}(t) = \exp\{-\Lambda_0(t)g(x)\}$ and the density for t takes the form $f(t) = \lambda_0(t)g(\boldsymbol{X})e^{-\Lambda_0(t)g(\boldsymbol{X})}$ where $g(\boldsymbol{X}) = \exp(\beta.\boldsymbol{X})$ is the regression function. The unknown baseline hazard function λ_0 is the nonparametric part and the unknown β is the parametric part of this model, which together create a semiparametric model.

There are two steps in the modeling approach (i) modeling the baseline hazard function λ, (ii) modeling the regression function g. Various semiparametric Bayesian methods have been developed to model λ. These require some prior random process which generates a distribution function, $\Lambda(t)$, or a hazard function, $\lambda(t)$, and then involves modeling either $\Lambda(t)$ or $-\log\mathcal{F}(t)$. Note that if $\mathcal{F}$ is not differentiable then these are not the same (Hjort, 1990). Then $\Lambda(t)$, or $-\log\mathcal{F}(t)$, is viewed as a random sample path from a Levy process, i.e., a nonnegative nondecreasing process on $[0, \infty]$ which starts at 0 and has independent increments. For example, Kalbfleisch (1978) models $-\log\mathcal{F}(t)$ as a draw from Gamma process while Hjort (1990) models $\Lambda(t)$ as a draw from a Beta process. Note that the earliest work, which modeled $1 - \mathcal{F}(t)$ by a Dirichlet process (Ferguson, 1973), implies that $-\log\mathcal{F}(t)$ is a Levy process. Arjas and Gasberra (1994) introduced correlated Gamma process likewise Sinha and Dey (1997), Sinha (1998) developed autocorrelated process to model the base line hazard. A detailed discussion of these procedures has been presented in Ibrahim et al. (2002).

On the contrary, the Bayesian literature on modeling the regression function g is sparse. Standard Bayesian nonlinear regression models based on spline, wavelet or other basis functions (Denison et al., 2002; Smith and Kohn, 1996; Shively et al., 1999) are convenient to use under a conjugate structure of the likelihood and the prior distributions which does not exist in the proportional hazards model setup. Gelfand and Mallick (1995) extended these models by introducing a random finite mixture process prior and modifying the usual proportional hazard form as $\lambda(t|\boldsymbol{X}) = \lambda_0(t)g(\beta.\boldsymbol{X})$ where g is an unknown link function. More general version of this model is $\lambda(t|\boldsymbol{X}) = \lambda_0(t)exp[f(\boldsymbol{x})]$ where f is an unknown function of the covariates. Mallick et al. (1999) utilized multivariate adaptive regression spline (MARS) to model f and to predict the future survival curves.

Machine learning techniques have received considerable attention to construct nonlinear prediction models. Even if they have wide range of applications in the field of classification and regression, their application to survival analysis is limited. Recently, the machine learning community has developed some tools that have been successfully used in the construction of classification models, including medical prognostic models (Lavrac et al., 1997; Lucas et al., 1999). Street (1999) built neural networks for predicting recurrence or survival time. Friedman and Meulman (2003) used additive regression trees to distinguish between different cervical dysplasia. Delen et al. (2005) empirically compared neural networks, decision trees and logistic regression for the task of predicting 60 months breast cancer survival. Zupan et al. (2000) modified Cox proportional hazards model using machine learning approaches.

All these methods including neural nets, decision trees and multivariate adaptive regression spline (MARS) are suitable when the sample size n is much larger than the number of covariates p in the model. The problem becomes much more complex when p, the dimension of x is very large, possibly larger than the sample size n. This is known as the "large p, small n" problem (West, 2003). In this situation, dimension reduction is needed to reduce the high-dimensional covariate space; existing techniques like variable selection, principal component regression or partial least squares can be utilized for this purpose. However, these methods are based on linear relationship between the response and the covariate which may not be very realistic. If actual f is nonlinear, these models may fail to produce reasonable prediction due to lack of flexibility.

To avoid this difficulty, we propose to use the reproducing kernel Hilbert space (RKHS) methodology (Aronszajn, 1950; Parzen, 1970) for these problems. One nice property of RKHS methods is they allow us to project the prediction problem into a space which is of dimension $n \ll p$. That way, dimension reduction is inherent within this method. One of the models which we will consider for this purpose is support vector machine (SVM). The classical support vector machine (SVM) algorithm, despite its success in regression and classification, suffers from a serious disadvantage: it cannot provide any probabilistic outputs. Law and Kwok (2001) introduced a Bayesian formulation of SVM for regression, but they did not carry out a full Bayesian analysis and used instead certain approximations for the posterior. A similar remark applies to Sollich (2001) who considered SVM in the classification context. More recently, Mallick et al. (2005) considered a Markov chain Monte Carlo (MCMC) based full Bayesian analysis for classification where the number of covariates was far greater than the sample size.

As an alternative to SVM, Tipping (2000, 2001) and Bishop and Tipping (2000) introduced relevance vector machines (RVMs). RVMs are suitable for both regression and classification, and are amenable to probabilistic interpretation. However, these authors did not perform a full Bayesian analysis. They obtained type II maximum likelihood (Good, 1965) estimates of the prior parameters, which do not provide predictive distribution of the future observations. Also, their procedure cannot provide measures of precision associated with the estimates.

The present article addresses a full Bayesian analysis of proportional hazards (PH) regression problem when p, the number of covariates far outnumber n, the sample size. The SVM approach is based on reproducing kernel Hilbert spaces (RKHS) as well as Vapnik's (1995) ϵ-insensitive loss function. We also consider the RVM-based analysis in this context, once again by using RKHS. All existing RKHS based approaches are developed for regression and classification problems. Our present approach extends them to cope with censored survival data. This is a significant extension of the existing methods as the survival data distribution is not at all Gaussian so least squares theory (or Gaussian likelihood method) is no more valid here. In the Cox model context, we consider a nonparametric prior for the baseline hazards which make the survival distribution nonstandard and complex. Moreover, the presence of censored observations make this problem more challenging than standard regression or classification problems. In this non-conjugate problem, explicit calculation of the marginal likelihood is not possible so type II MLE method is a daunting task. Instead of relying on the type II maximum likelihood estimates of the prior parameters, we assign distributions to these parameters. In this way, we can capture the uncertainty in estimating these parameters, and consequently get more reliable measures of precision associated with the Bayes estimates. A key feature of our method is to treat the kernel parameter in the model as unknown and infer about it with all other parameters. We obtain a more accurate prediction by the mixing of several RVM (or SVM) models through the kernel parameter. Due to analytical intractability of posteriors, we use the MCMC numerical integration technique for implementation of the Bayes procedure.

As with other regularization methods, there are smoothing or regularization parameter(s) which need to be tuned for efficient classification. One popular approach is to use generalized approximate cross validation (GACV) to tune the smoothing parameters. In this article we take a different approach, by developing a hierarchical model where the unknown smoothing parameter will be interpreted as a shrinkage parameter (Denison *et al.*, 2002). We assign a prior distribution to it and obtain its posterior distribution via the Bayesian paradigm. In this way, we obtain not only the point predictors but also the associated measures of uncertainty. Furthermore, we extend the model to incorporate multiple smoothing parameters, leading to significant improvements in prediction for the examples considered.

We utilize our method to analyze gene expression data. The advent of microarrays makes it possible to provide thousands of gene expression at once (Duggan *et al.*, 1999; Schena *et al.*, 1999). In recent years, several microarray studies with time-to-event outcomes have been collected. The major challenge here is to predict the survival time of a patient observing his gene expression profiles and other phenotypes. The models proposed in this article are applied to the Breast-carcinomas data set (Sørlie *et al.* 2001) and the Diffuse Large B-cell lymphoma (DLBCL) data set (Alizadeh *et al.*, 2000). Both of these data sets contain very large numbers of predictors compared to the number of samples: the Breast-carcinomas data set contains 76 samples and 2300 genes per sample and the DLBCL data set consists of survival times of 40 patients and expression level measurement of 4513 genes for each patient. The relevant survival curves are estimated. Also we compare the two methods to the classical proportional hazards model which assumes linearity of covariates.

The article is organized as follows. A brief overview of the Reproducing Kernel Hilbert Space (RKHS), and SVM and RVM regression is presented in Section 13.2. Section 13.3 introduces proportional hazards model based on kernel regression. The hierarchical relevance vector machine for PH regression is introduced in Section 13.4 and has been extended to support vector machine in Section 13.5. In Section 13.6, the prediction method has been explained.

Section 14.2 contains real data analysis. We present our analysis of the Breast-carcinomas data set and the Diffuse Large B-cell lymphoma (DLBCL) data set and also the comparison of the methods.

13.2 Background and Framework

In this section we provide brief overview of the support vector machine (SVM) and relevance vector machine (RVM), the two main methods we will use in later sections. For the sake of completeness we also present a brief overview of the reproducing kernel Hilbert space (RKHS) setup.

13.2.1 Reproducing Kernel Hilbert Space

A Hilbert space $\mathcal{H}$ is a vector space endowed with an inner product $\langle f, g \rangle$ and the corresponding norm $||f|| = \langle f, f \rangle^{1/2}$. A Hilbert space can be finite dimensional or infinite dimensional. Example of finite dimensional Hilbert space is R^n with inner product $\langle x, y \rangle = x^T y$ and example of infinite dimensional Hilbert space is L_2, the space of all square integrable functions, with inner product $\langle f, g \rangle = \int f(t)g(t)dt$.

However, we will use the so called Reproducing Kernel Hilbert Space (RKHS), which are widely used in machine learning. A formal definition of RKHS is given in Aronszajn (1950), see also Parzen (1970) and Wahba (1990). Let I be any general index set. For example, $I = \{1, 2, \ldots, N\}$ or I could be any d−dimensional Euclidean space. Let $\mathcal{H}$ be a Hilbert space of functions on the set I. A complex valued function $K(x, y)$, $x, y \in I$ is said to be a *reproducing kernel* of $\mathcal{H}$ if (i) for every $x \in I$, $K_x(y)$ as a function of y belongs to $\mathcal{H}$ and (ii) for every $x \in I$ and every $f \in \mathcal{H}$, $f(x) = \langle f, K_x \rangle$. If such a reproducing kernel $K(\cdot, \cdot)$ exists for $\mathcal{H}$ then $\mathcal{H}$ is called a reproducing kernel Hilbert space (RKHS).

A real symmetric function $K(y, x)$ is said to be positive definite on $I \times I$ if for every $n = 1, 2, \ldots$ and every set of real numbers $\{a_1, a_2, \ldots, a_n\}$ and $\{x_1, x_2, \ldots x_n\}$, $x_i \in I$, we have $\sum_{i,j=1}^{n} a_i a_j K(x_i, x_j) \geq 0$. By the Moore-Aronszajn theorem (Aronszajn, 1950), for every positive definite K on $I \times I$, there exists a corresponding unique RKHS $\mathcal{H}_K$ of real-valued functions on I and vice-versa. If K is the reproducing kernel for the Hilbert space $\mathcal{H}$, then the family of functions $\{K(y, x) : y \in I\}$ spans $\mathcal{H}$.

13.2.2 Support Vector Machine

The support vector machine (SVM) is a highly sophisticated tool for classification and regression. As such it is firmly grounded in the framework of statistical learning theory, or *VC theory*, see Vapnik (1995). We will present an overview of SVM in the regression context. Let $(y_1, x_1), \ldots, (y_n, x_n) \in R \times \mathcal{H}$ be n independently and identically distributed observations where y_i's are the responses and x_i's are the predictors, commonly known as input patterns in machine learning theory. In the classical ϵ−SV regression (Vapnik, 1995) the main goal is to find a function $f(x)$ that has at most ϵ deviation from the actually obtained targets y_i for all the observed data but at the same time is as "flat" as possible. For example, in the linear regression case (f is a linear function), the basic SVM regression algorithm seeks to estimate linear functions of the form

$$f(x) = \langle x, \beta \rangle + b,$$

where $\beta, x \in \mathcal{H}$ and $b \in R$. Here, $\mathcal{H}$ is a dot product space in which the covariates x_i reside. "Flatness" in this case means that one seeks small β. In this case, SVM minimizes $||\beta||^2/2$ subject to $|y_i - \langle x_i, \beta \rangle - b| \leq \epsilon$. In a more general formulation, SVM seeks to minimize the regularized risk functional (Schölkopf and Smola, 2002)

$$||\beta||^2 + Cn^{-1} \sum_{i=1}^{n} |y_i - f(x_i)|_\epsilon,$$

where $|y - f(x)|_\epsilon = \max\{0, |y - f(x)| - \epsilon\}$ is called the Vapnik's ϵ−insensitive loss function and C is a constant determining the trade-off with the complexity penalizer.

In general, one is not restricted to the case where f is linear. For general nonlinear functions, SVM makes prediction based on a function of the form

$$f(x) = \beta_0 + \sum_{i=1}^{n} \beta_n K(x, x_n),$$

where $\beta = (\beta_0, \ldots, \beta_n)^{\mathrm{T}}$ is the vector of model weights and $K(\cdot, \cdot)$ is a *kernel function*, see Tipping (2000). The basic idea is that instead of working with the original inputs $\{x_i\}$, one first maps them into a feature space using the kernel K and constructs the target function using these mapped features. In this formulation, one minimizes

$$n^{-1} \sum_{i=1}^{n} \mathcal{L}\{y_i, f(x_i)\} + \lambda ||f||,$$

where $\mathcal{L}\{y, f(x)\}$ is a loss function determining how one will penalize the estimation errors based on the empirical data. From the point of view of the RKHS theory, $f(x)$ is assumed to belong to a RKHS $\mathcal{H}$ with kernel K. In practice, many different kernel functions are used; some wide known examples are (1) polynomial kernel $K(x, x') = (\langle x, x' \rangle + c)^p$, where p is a positive integer, (2) Gaussian kernel $K(x, x') = \exp\{-||x - x'||^2/(2\sigma^2)\}$ and (3) the hyperbolic tangent kernel $K(x, x') = \tanh(\phi \langle x, x' \rangle + c)$.

13.2.3 Relevance Vector Machine

Despite the widespread success of the classical SVM, it exhibits some significant disadvantages. One of the most important disadvantages is that SVM only produces point predictions rather than generating predictive distributions. To remedy this situation, Tipping (2000) proposed the so-called Relevance Vector Machine (RVM), a probabilistic model whose functional form is equivalent to SVM. The main idea of RVM is that a prior distribution is introduced on the model weights $\beta_0, \ldots, \beta_n$ governed by a set of hyperparameters, one for each weight, whose most probable values are iteratively estimated from the data.

Following Tipping (2000) and assuming that y follows Normal$\{f(x), \sigma^2\}$, the function $f(x)$ is represented as $f(x) = \beta_0 + \sum_{i=1}^{n} \beta_n K(x, x_n)$ and the likelihood of the data is written as

$$L = (2\pi\sigma^2)^{-n/2} \exp\{||\mathbf{y} - \mathbf{K}\beta||^2/(2\sigma^2)\},$$

where $\mathbf{y} = (y_1, \ldots, y_n)$, $\beta = (\beta_0, \ldots, \beta_n)$ and $\mathbf{K}$ is an $n \times (n+1)$ design matrix with the (i,j)-th element being $\mathbf{K}_{ij} = K(x_i, x_{j-1}), j = 2, \ldots, n+1$ and $\mathbf{K}_{i1} = 1$. The models weights are then assigned a Gaussian prior

$$p(\beta|\alpha) \sim \prod_{i=0}^{n} N(0, \alpha_i^{-1}),$$

where $\alpha = (\alpha_0, \ldots, \alpha_n)$ are a set of hyper parameters. The resulting posterior distribution of β is given by

$$p(\beta|y, \alpha, \sigma^2) \propto |\Sigma|^{-1/2} \exp\{(\beta - \mu)^{\mathrm{T}}\Sigma^{-1}(\beta - \mu)\},$$

where $\sigma = (\mathbf{K}^{\mathrm{T}}B\mathbf{K} + A)^{-1}$ and $\mu = \Sigma\mathbf{K}^{\mathrm{T}}By$ with $A = diag(\alpha_0, \ldots, \alpha_n)$ and $B = \sigma^{-2}I_n$. It is interesting to note that σ^2 is treated as a hyperparameter as well. To obtain α and σ^2, Tipping (2000) used type II maximum likelihood (Good, 1965) estimates. However, one can also perform a full Bayesian analysis where priors are assigned on α and σ^2 as well, see Chakraborty, Ghosh and Mallick (2007) for example.

13.3 Proportional Hazards Model Based on Kernel Regression

For the survival analysis problem, we have n patients and let the survival time of the ith patient be denoted by T_i with censoring indicators δ_i (0 if data is right censored, 1 otherwise), and $X_i = [X_{i1}, X_{i2}, ..., X_{ip}]^{\mathrm{T}}$ be the p

covariates associated with it. So our combined data is given by

$$D = (n, \boldsymbol{t}, \delta);$$
$$\boldsymbol{X} = [X_1, ..., X_n]^{\mathrm{T}},$$

where $\boldsymbol{t} = (t_1, ..., t_n)$ are conditionally independently distributed survival times and $\delta = (\delta_1, ..., \delta_n)$. One of the objectives is to predict the survival probabilities given a set of new covariates, based on the training data. We utilize the training data $\boldsymbol{t}$ and $\boldsymbol{X}$ to fit a model $p(\boldsymbol{t}|\boldsymbol{X})$ and use it to obtain the survival distribution $P(T_*|\boldsymbol{t}, X_*)$ for a future patient with covariate X_*.

The proportional hazards (PH) model (Cox, 1972) is a frequently used model in survival analysis. Under this model the hazard function is given by

$$h(t_i|X_i) = h_0(t_i) \exp(X_i^{\mathrm{T}}\beta), \tag{13.1}$$

where $h_0(t)$ is baseline hazard function. For added flexibility, we introduce the latent variables $\boldsymbol{W} = (W_1, \ldots, W_n)$ and extend the PH model as

$$h(t_i|x_i) = h_0(t_i) \exp(W_i).$$

In the next stage, the latent variables W_i are modeled as $W_i = f(X_i) + \epsilon_i$, $i = 1, \ldots, n$, where f is not necessarily a linear function, and ϵ_i, the random residual effects, are independent and identically distributed. The use of a residual component is consistent with the belief that there may be unexplained sources of variation in the data. To develop the complete model, we need to specify $p(\boldsymbol{t}|\boldsymbol{W})$, $p(\boldsymbol{W}|f)$ and f.

13.3.1 Modeling of $p(\boldsymbol{t}|\boldsymbol{W})$ and the Baseline Hazard Function

To model $p(\boldsymbol{t}|\boldsymbol{W})$, we have to specify the distribution of h_0 in the PH setup as in (13.1). We followed the nonparametric Bayesian method for PH model which was suggested by Kalbfleisch (1978). In the PH regression model, T_i has a conditional survival function

$$P(T_i > t_i|W_i, \Lambda) = \exp[-\Lambda_0(t_i) \exp(W_i)],$$

where Λ_0 is the baseline cumulative hazard function. A Gamma process prior was suggested by Kalbfleisch (1978)

$$\Lambda_0 \sim GP(a\Lambda^*, a),$$

where Λ^* is the mean process and a is a weight parameter about the mean (Ibrahim *et al.* 2001). Kalbfleisch (1978) showed that if $a \approx 0$ then the likelihood is approximately proportional to the partial likelihood whereas $a \to \infty$, the limit of likelihood is the same as the likelihood when the Gamma process is replaced by Λ^*. Since we have for given t, $\Lambda_0(t) \sim Gamma(a\Lambda^*(t), a)$, the marginal survival function (after integrating out Λ_0) is given by

$$P(T_i > t|W) = \left(\frac{a}{a + \exp(W)} \right)^{a\Lambda^*(t)}.$$

The joint survival function conditional on Λ is

$$P(T_1 > t_1, ..., T_n > t_n|\boldsymbol{W}, \Lambda) = \exp\{-\sum \Lambda(t_i) \exp(W_i)\}.$$

Kalbfleisch (1978) showed that the likelihood with some right censoring is given by

$$L(t|\boldsymbol{W}) = \exp\{-\sum aB_i\Lambda^*(t_i)\} \prod \{a\lambda^*(t_i)B_i\}^{\delta_i},$$

with $A_i = \sum_{k \in R(t_i)} exp(W_k)$, where $R(t_i)$ is the set of individuals at risk at time $t_i - 0$ and $B_i = -\log\{1 - exp(W_i)/(a + A_i)\}$.

13.3.2   Modeling of $p(\boldsymbol{W}|f)$ and f Using RKHS

We can consider this part as a latent regression problem with the training set $\{W_i, X_i\}$, $i = 1, \ldots, n$; where W_i is the latent response variable and X_i is the vector of covariate of size p corresponding to the ith subject. To obtain an estimate of f, we can use a regularization framework with

$$\min_{f \in \mathcal{H}} \left[\sum_{i=1}^n \mathcal{L}(W_i, f(\boldsymbol{x}_i)) + \lambda J(f) \right], \tag{13.2}$$

where $\mathcal{L}(W, f(X))$ is a loss function, $J(f)$ is a penalty functional, $\lambda > 0$ is the smoothing parameter, and $\mathcal{H}$ is a space of functions on which $J(f)$ is defined. In this article, we consider $\mathcal{H}$ to be a reproducing kernel Hilbert space (RKHS) with kernel K, and we denote it by $\mathcal{H}_K$. A formal definition of RKHS in given in Aronszajn (1950), Parzen (1970), and Wahba (1990).

For an $h \in \mathcal{H}_K$, if $f(X) = \beta_0 + h(X)$, we take $J(f) = \| h \|_{\mathcal{H}_K}^2$ and rewrite (13.2) as

$$\min_{f \in \mathcal{H}_K} \left[\sum_{i=1}^n \mathcal{L}(W_i, f(\boldsymbol{x}_i)) + \lambda \| h \|_{\mathcal{H}_K}^2 \right]. \tag{13.3}$$

The estimate of h is obtained as a solution of Equation 13.3. It can be shown that the solution is finite-dimensional (Wahba, 1990) and leads to a representation of f (Kimeldorf and Wahba, 1971; Wahba 2002) as

$$f_\lambda(\boldsymbol{x}) = \beta_0 + \sum_{i=1}^n \beta_i K(\boldsymbol{x}, \boldsymbol{x}_i). \tag{13.4}$$

It is also a property of RKHS that

$$\| h \|_{\mathcal{H}_K}^2 = \sum_{i,j=1}^{n} \beta_i \beta_j K(\boldsymbol{x}_i, \boldsymbol{x}_j).$$

Representation of f in above form is of special interest to us, because cases when the number of covariates p is much larger than the number of data points, we effectively reduce the dimension of covariates from p to n. To obtain the estimate of f_λ we substitute (13.3) and (13.4) in (13.2) and then minimize it with respect to $\boldsymbol{\beta} = (\beta_0, \ldots, \beta_n)$ and the smoothing parameter λ. The other parameters inside the kernel K may be chosen by generalized approximate cross validation (GACV).

One can view the loss function $\mathcal{L}(W, f(X))$ as the negative of the log-likelihood. For example, the squared error loss $\mathcal{L}(W, f(X)) = \{W - f(x)\}^2$ corresponds to the Gaussian negative log-likelihood and the absolute error loss $\mathcal{L}(W, f(X)) = |W - f(x)|$ corresponds to the Laplace negative log-likelihood. Using this formulation, our problem is equivalent to maximization of the penalized log-likelihood

$$-\sum_{i=1}^{n} \mathcal{L}(W_i, f(X_i)) - \lambda \| h \|_{H_k}^2 . \tag{13.5}$$

This duality between "loss" and "likelihood", particularly viewing the loss as the negative of the log-likelihood, is referred in the Bayesian literature as the "logarithmic scoring rule" (Bernardo and Smith, 1994, p. 688).

13.4   Relevance Vector Machine–Based PH Regression

For different choices of $\mathcal{L}$ in (13.3), we can obtain varieties of kernel machine regression. Two most popular choices are relevance vector machine (RVM) and Support vector machine (SVM). In RVM, we use squared error loss as $\mathcal{L}$, so that the corresponding likelihood becomes Gaussian in (13.5). In this framework the distribution of W is given by

$$W_i \sim N(K_i^{\mathrm{T}} \beta, \sigma^2). \tag{13.6}$$

We assign hierarchical prior to unknown parameters β, θ and σ^2. For β a conjugate prior is assigned

$$\beta | \sigma^2 \sim N(0, \sigma^2 S^{-1}),$$

where $S = diag(\lambda_0, \lambda_1, \ldots, \lambda_n)$ is a diagonal matrix with precision parameters. Here λ_0 is fixed at a small value and other λ's are given Gamma priors

$$\lambda_i \sim Gamma(c, d).$$

An inverse gamma prior is assigned to σ^2

$$\sigma^2 \sim IG(\gamma_1, \gamma_2).$$

The parameter θ in the kernel K is assigned a uniform prior

$$\theta \sim Unif(a_L, a_U).$$

The matrix with (i, j)th element $K_{ij} = K(x_i, x_j|\theta)$ is called the kernel matrix. The usual choices are

(i) Gaussian kernel : $K(x_i, x_j|\theta) = exp(\|\frac{x_i - x_j}{\theta}\|)$,

(ii) Polynomial kernel : $K(x_i, x_j|\theta) = (x_i^T x_j + 1)^\theta$.

In the RVM model, the joint posterior is given by

$$\pi(\beta, \lambda, \sigma^2, \theta|W) \propto \frac{1}{\sigma^n} \exp\left(-\sum \frac{(W_i - K_i^T\beta)^2}{2\sigma^2}\right)$$

$$\times \frac{1}{|\sigma^2 S^{-1}|^{1/2}} \exp\left(-\frac{\beta^T S\beta}{2\sigma^2}\right)$$

$$\times \exp(-\gamma_2/\sigma^2)(\sigma^2)^{-\gamma_1-1}$$

$$\times \prod_{i=1}^n exp(-d\lambda_i)\lambda_i^{c-1}.$$

Due to the complexity of the posterior distribution, Bayesian procedure requires MCMC sampling techniques such as Gibbs sampling (Gelfand and Smith, 1990) and Metropolis-Hastings algorithm (Metropolis, 1953).

13.4.1   Conditional Distribution and Posterior Sampling

The full conditional distribution of W is

$$p(W|D, \beta, \sigma^2, \lambda, \theta) \propto \exp\{-\sum aB_i\Lambda^*(t_i)\} \prod\{a\lambda^*(t_i)B_i\}^{\delta_i}$$

$$\times \exp\left(-\sum \frac{(W_i - K_i^T\beta)^2}{2\sigma^2}\right).$$

Also the prior distribution specified for β and σ^2 are conjugate whose posterior distribution is Normal-Inverse Gamma:

$$p(\beta, \sigma^2|W, \lambda, \theta) = N(\tilde{m}, \sigma^2\tilde{V})IG(\tilde{\gamma}_1, \tilde{\gamma}_2),$$

where

$$\tilde{m} = (K_0^T K_0 + S^{-1})^{-1}(K_0^T W);$$
$$\tilde{V} = (K_0^T K_0 + S^{-1})^{-1};$$
$$\tilde{\gamma}_1 = \gamma_1 + n/2;$$
$$\tilde{\gamma}_2 = \gamma_2 + \frac{1}{2}(W'W - \tilde{m}^T\tilde{V}\tilde{m});$$

and $K_0^T = [K_1, \ldots, K_n]$.

The conditional distribution of the precision parameter λ_i given β is Gamma and given by

$$p(\lambda_i|\beta_i) = Gamma(c + \frac{1}{2}, d + \frac{1}{2}\beta_i).$$

Now we use these conditional distributions to construct a Gibbs sampler. The steps are given below

- Update W

- Update β and σ^2

- Update K

- Update λ

Updating of W is done by updating each W_i in turn conditional on the rest. As the conditional distribution does not have any explicit form, Metropolis-Hastings procedure can be used with a symmetric proposal density T.

Updating K is equivalent to updating θ. Here also we need a Metropolis-Hastings procedure. Here the marginal distribution of $\theta|W$ is needed. We can write

$$p(\theta|W) = p(W|\theta)p(\theta).$$

Let θ^* be the proposed change to the parameter. Then we accept this change with acceptance probability

$$\alpha = \min\left(1, \frac{p(W|\theta^*)}{p(W|\theta)}\right).$$

The ratio of marginal likelihoods is given by

$$\frac{p(W|\theta^*)}{p(W|\theta)} = \frac{|\tilde{V}^*|^{1/2}}{|\tilde{V}|^{1/2}}\left(\frac{\tilde{\gamma}_2}{\tilde{\gamma}_2^*}\right)^{\tilde{\gamma}_1},$$

where $\tilde{V}^*$ and $\tilde{\gamma}_2^*$ are similar to $\tilde{V}^*$ and $\tilde{\gamma}_2$ with θ is replaced by θ^*. Updating β, σ^2 and λ is easy as they are coming from standard distributions.

13.5   Support Vector Machine–Based PH Regression

In this section, we develop a support vector machine (SVM) based model for PH regression. The basic setup is similar to that of relevance vector machine but the main difference is in the relation between W_i and $f(x_i)$. In Bayesian relevance vector machine W_i and $f(x_i)$ are connected through the squared error loss function : $L(W_i, f(x_i)) = (W_i - f(x_i))^2$ and hence the likelihood induced by this loss function becomes proportional to a Gaussian density.

In support vector machine (SVM) model, we relate W_i and $f(x_i)$ by a general loss function: Vapnik's ϵ-insensitive loss function,

$$|W_i, f(x_i)|_\epsilon = (|W_i - f(x_i)| - \epsilon)I(|W_i - f(x_i)| > \epsilon).$$

This loss function ignores errors of size less than ϵ but penalizes in a linear fashion when the function deviates more than ϵ amount. This makes fitting less sensitive to outliers.

Now as before we assume that $f(x)$ is generated from a RKHS with reproducing kernel K and hence, for $i = 1, ..., n$, $f(x_i)$ can be written as $K_i^T \beta$ where $K_i^T = [1, K(x_i, x_1|\theta), ..., K(x_i, x_n|\theta)]$.

The likelihood corresponding to Vapnik's loss function is given by

$$p(W_i|\beta, K) \propto \exp\{-\rho|W_i - K_i^T\beta|_\epsilon\}.$$

This likelihood can be written as a mixture of a truncated Laplace distribution and a uniform distribution as follows:

$$
\begin{aligned}
p(W_i|\beta, K) &= \frac{\rho}{2(1 + \epsilon\rho)} \exp\{-\rho|W_i - K_i^T\beta|_\epsilon\} \\
&= \frac{\rho}{2(1 + \epsilon\rho)} \exp\{-\rho[|W_i - K_i^T\beta| - \epsilon]I(|W_i - K_i^T\beta| > \epsilon)\} \\
&= \frac{\rho}{2(1 + \epsilon\rho)} \{I(|W_i - K_i^T\beta| > \epsilon) \exp\{-\rho[|W_i - K_i^T\beta| - \epsilon]\} \\
&\quad + I(|W_i - K_i^T\beta| \le \epsilon)\} \\
&= p_1(\text{Truncated Laplace}(K_i^T\beta, \rho)) \\
&\quad + p_2(\text{Uniform}(K_i^T\beta - \epsilon, K_i^T\beta + \epsilon)),
\end{aligned}
\tag{13.7}
$$

where $p_1 = \frac{1}{1 + \epsilon\rho}$ and $p_2 = \frac{\epsilon\rho}{1 + \epsilon\rho}$.

We assign priors on unknown parameters $\beta, \theta, \rho, \lambda_i$ as follows

$$\beta|S \sim N(0, S^{-1})$$

where $S = \text{diag}(\lambda_0, \lambda_1, \ldots, \lambda_n)$ is a diagonal matrix with λ_0 fixed to a small value and

$$
\begin{aligned}
\lambda_i &\sim \text{Gamma}(c, d); \\
\theta &\sim \text{Uniform}(a_L, a_U); \\
\rho &\sim \text{Uniform}(r_L, r_U).
\end{aligned}
$$

The joint posterior is given by

$$\pi(\beta, \lambda, \theta, \rho | W) \propto \exp\{-\rho \sum_{i=1}^{n} |W_i - K_i^{\mathrm{T}} \beta|_\epsilon\}$$

$$\times \frac{1}{|S^{-1}|^{1/2}} \exp\left(-\beta^T S \beta\right)$$

$$\times \prod_{i=1}^{n} exp(-d\lambda_i) \lambda_i^{c-1}.$$

The full conditional distribution of W is

$$p(W | D, \beta, \sigma^2, \lambda, \theta) \propto \exp\{-\sum_{i=1}^{n} a B_i \Lambda^*(t_i)\} \prod_{i=1}^{n} \{a\lambda^*(t_i) B_i\}^{\delta_i}$$

$$\times \frac{\rho^n}{2^n (1 + \epsilon\rho)^n} \exp\left(-\rho \sum_{i=1}^{n} |W_i - K_i^{\mathrm{T}} \beta|_\epsilon\right).$$

Due to its complex form, we can not use this posterior distribution to generate samples. So Metropolis-Hastings procedure is applied to draw samples from β and λ.

13.6　Prediction

It is often of great interest to predict the survival time when the corresponding gene expressions are given. Due to the non-linearity of our model it is not obvious how to do prediction. In this section a prediction procedure is outlined.

Using results from previous section we see that the conditional survival curve is given by,

$$P(T_i > t | W) = \left(\frac{a}{a + \exp(W)}\right)^{a\Lambda^*(t)}.$$

Hence the conditional density of T is given by

$$f_T(t | W) = -[a \frac{d\Lambda^*(t)}{dt} \log\{\frac{a}{a + \exp(W)}\} \left(\frac{a}{a + \exp(W)}\right)^{a\Lambda^*(t)}].$$

So the algorithm of the prediction procedure can be given as follows:

- Compute estimates of parameters from MCMC samples.

- From the observed gene expressions, x_{obs}, create the vector $M = (M_1, \ldots, M_n)^T$ with

$$M_i = K(x_{obs}, X_i),$$

where $K(\cdot)$ is the kernel function.

- Generate samples of W using (13.6) for RVM or (13.7) for SVM, where posterior estimates are plugged in place of parameters.

- Compute the conditional density $f_T(t|W)$.

- Take the median of the density as the predicted value,

$$T_{pred} = \text{Median}\{f_T(\cdot|W)\}.$$

To form a $(1 - \alpha)\%$ confidence interval for the predicted value, we can use the interval: $(Q_{\alpha/2}, Q_{100-\alpha/2})$, where Q_ζ denotes the ζth quantile of $f_T(\cdot|W)$.

13.7   Application

13.7.1   Breast Carcinomas Data Set

Methods proposed in this article are applied to breast carcinomas data set discussed in Sørlie *et al.* (2001). The purpose of their study was to classify breast carcinomas based on variations in gene expression patterns derived from cDNA microarray. Sørlie *et al.* found some novel sub-classes of breast carcinomas based on gene expression patterns and proved robustness of sub-classes using separate gene sets. The data set consisted of survival times of 76 samples and expression levels of 2300 genes per sample. There are additional subgroup information: Basal-like, ERBB2+, Normal Breast-like, Luminal subtype A, B or C. In this demonstration, the main goal of our analysis is to investigate whether the survival functions for the different classes of cancer (Luminal subtype A, B or C versus rest) are different or not. For this purpose, we consider the binary covariate X_0 as: $X_{i0} = 1$ if the ith sample is Luminal subtype A, B or C and 0 otherwise. We choose 200 genes out of the 2300 genes provided using two-sample t-test. Specifically, we construct t-statistic for each gene using the two classes of samples ($X_0 = 1$ vs. $X_0 = 0$) and select the top 200 genes having largest values of the t-statistic. This selection is done to choose the genes which are most differentially expressed between the two classes considered in the example and to increase the numerical stability of our computations. Lee and Mallick (2004) analyzed this data set in the context of gene selection. However they pre-selected 1000 genes using two-sample t-test.

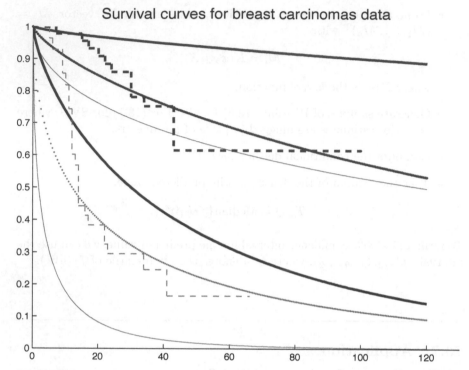

FIGURE 13.1: Estimated survival function for breast carcinomas data using RVM model(thin line = group 1 and thick line = group 2). Plotted are the Kaplan-Meier estimates (dashed line) and our estimate (dotted line) with 95% credible interval (solid line).

To develop the semi-parametric model, the baseline function $\Lambda^*(\cdot)$ is chosen as Weibull distribution for the Gamma process on the cumulative hazard function, that is $\Lambda^*(t) = \eta_0 t^{k_0}$, where η_0 and k_0 are estimated for the data. Following Lee and Mallick (2004), a moderate value of hyperparameter is chosen as $a = 10$. We chose a tight inverse gamma prior for σ^2 choosing $\gamma_1 = 1$ and $\gamma_2 = 10$. For prior of λ we choose $c = 1$ and $d = 1000$. Also, a Gaussian kernel was used for K. The survival function in the Cox proportional hazards model is

$$ S(t|W) = \left(\frac{a}{a + \exp(W)} \right)^{a\Lambda^*}. $$

The algorithm was run for $80,000$ times and the first $30,000$ samples were discarded as burn-in. We used the samples from the posterior distribution to obtain the MCMC estimates of the function. The posterior estimates of the survival curves are given in Figure 13.1 and Figure 13.2 and it is evident that RVM and SVM models are good fit for this data set.

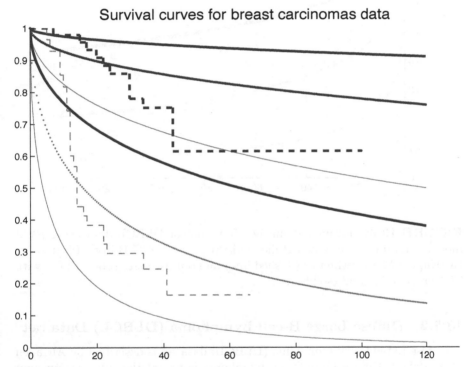

FIGURE 13.2: Estimated survival function for breast carcinomas data using SVM model(thin line = group 1 and thick line = group 2). Plotted are the Kaplan-Meier estimates (dashed line) and our estimate (dotted line) with 95% credible interval (solid line).

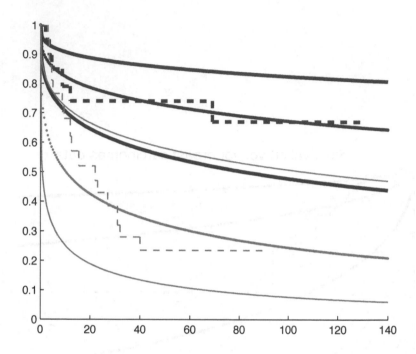

FIGURE 13.3:   Estimated survival function for DLBCL data using RVM model (thin line = activated B-like and thick line = GC B-like). Plotted are the Kaplan-Meier estimates (dashed line) and our estimate (dotted line) with 95% credible interval (solid line).

13.7.2   Diffuse Large B-cell Lymphoma (DLBCL) Data Set

Diffuse Large B-cell lymphoma (DLBCL) data set is described by Alizadeh *et al.* (2000). Diffuse large B-cell lymphoma is one of the subtypes on non-Hodgkin's lymphoma. Using DNA microarray experiment and hierarchical clustering Alizadeh *et al.* (2000) discovered two distinct forms of DLBCL, namely activated B-like and GC-B like DLBCL and showed that these two subgroups of DLBCL had different survival time patterns and they can be distinguished by distinct gene expression of hundreds of different genes.

The data set consists of survival times of 40 patients and expression level measurement of 4513 genes for each patient. To incorporate grouping a indicator variable X_0 is introduced as $X_{i0} = 1$ if the ith sample is Activated B-like and $X_{i0} = 0$ otherwise for $i = 1, \ldots, 40$. Also we have X_{ij} as the normalized log scale measurement of the expression level of the jth gene in the ith sample where $i = 1, \ldots, 40$ and $j = 1, \ldots, 4513$.

Prior and hyper-parameters specifications are as in the earlier example.

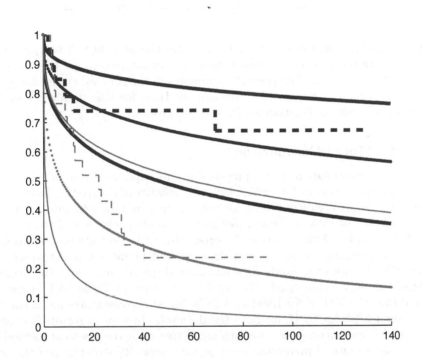

FIGURE 13.4: Estimated survival function for DLBCL data using SVM model (thin line = activated B-like and thick line = GC B-like). Plotted are the Kaplan-Meier estimates (dashed line) and our estimate (dotted line) with 95% credible interval (solid line).

TABLE 13.1: Predictive Mean Square Errors for the Proportional Cox Hazard (CPH) Model, Relevance Vector Machine (RVM) Model, and Support Vector Machine (SVM) Model for the Breast Carcinomas Data and DLBCL Data

Model	Breast carcinomas data	DLBCL data
CPH	7.24	7.03
RVM	1.49	1.97
SVM	1.67	2.11

The algorithm was run for $80,000$ times and the first $30,000$ samples were discarded as burn-in. We exploited the posterior samples to get the MC estimates of the function. The posterior estimates of the survival curves are given in Figure 13.3 and Figure 13.4. We can see that, for this data set too, our methods provide good estimates for the two groups.

13.7.3 Model Comparison

Finally, we compare methods proposed in this paper to the classical proportional hazards model (CPH) which assumes linearity of covariates. To compare the methods, we divide our data, for each of the two data sets, into a training set consisting of 75% of the samples and a test set consisting of the remaining 25% samples. After we train the procedure with the training set, survival times are predicted using the test data set and prediction mean square errors (PMSE) are computed. This has been done 20 times and the average PMSE values are computed. The results are shown in Table 13.1. It can be noted that the PMSE for RVM and SVM based regression are much smaller than the PMSE for CPH, for both the data sets. In view of results it is clear that non-linear methods are performing better than the classical method in the sense of reduced prediction mean square error. RVM performed the best among these three procedures.

13.8 Discussion

In this article, we have introduced kernel based proportional hazards regression models for relating covariates to censored survival data. The method is particularly useful when the number of covariates exceeds the sample size. The methods developed here generalize the idea of kernel regression for binary or multi-category data to censored survival data. Specifically, we propose relevance vector machine based fully Bayesian framework for proportional hazard regression with censored data. The main idea is to start from the likelihood arising from standard proportional hazard regression but model the joint effect of the covariates using a unknown nonparametric function $f(\cdot)$. Next, $f(\cdot)$ is assumed to belong to a RKHS with reproducing kernel $K(\cdot, \cdot)$ and $f(\cdot)$ is expressed using the kernel representation. This representation automatically reduces the dimension of the design matrix to n from $p >> n$. Using this formulation, we then build a fully Bayesian model and draw inference using the posterior sample.

There are various advantages of this method. One important advantage is that it enables the user to create predictive distributions for future observations. In addition, there is no computational or methodological limitation in terms of number of covariates.

One important possible future research is to examine how different number of covariates used in model affects the results of prediction, extending this method for multivariate survival data is another.

References

Alizadeh, A., Eisen, Michael B., Davis, R. E., et al. 2000. Distinct types of diffuse large B-cell lymphoma identified by gene expression profiling, *Nature*, **403**, 503-511.

Arjas, E., and Gasbarra, D. 1994. Nonparametric Bayesian inference from right-censored survival data, using the gibbs sampler. *Statistica Sinica* 4, 505-524.

Aronszajn, N. 1950. Theory of reproducing kernels. *Transactions of the American Mathematical Society* **68**, 337–404.

Bernardo, J. M., and Smith, A. F. M. 1994. *Bayesian theory.* London:Wiley.

Bishop, C. and Tipping, M. 2000. Variational relevance vector machines. In *Proceedings of the 16th conference in uncertainty and artificial intelligence*, ed. C. Boutilier and M. Goldszmidt, 46–53. Morgan Kauffman.

Chakraborty, S., Ghosh, M. and Mallick, B. K. 2007. Bayesian Non Linear Regression for Large p Small n Problems. Technical report.

Cox, D.R. 1972. Regression models and life tables. *J. R. Statist. Soc. B* **34**, 187-220.

Delen, D., Walker, G.,and Kadam, A. 2005. Predicting breast cancer survivability: a comparison of three data mining methods. *Artificial Intelligence in Medicine* **34**, 113-127.

Denison, D., Holmes, C., Mallick, B. K. and Smith, A. F. M. 2002. *Bayesian methods for nonlinear classification and regression.* London: Wiley.

Duggan, D. J., Bittner, M., Chen, Y., Meltzer, P., and Trent., J. M. 1999. Expression profiling using cDNA microarrays. *Nature Genetics* **21**, 10-14.

Ferguson, T. 1973. A Bayesian analysis of some nonparametric problems *The Annals of Statistics* **1**, 209–230.

Friedman, J. H. and Meulman, J. J. 2003. Multiple additive regression trees with application in epidemiology. *Statistics in Medicine* **22**, 1365–1381.

Gelfand, A. and Smith, A. F. M. 1990. Sampling-based approaches to calculating marginal densities. *Journal of the American Statistical Association* **85**,398–409.

Gelfand, A.E. and Mallick, B.K. 1995. Bayesian analysis of proportional hazards models built from monotone functions. *Biometrics* **51**, 843–852.

Good, I. J. 1965. *The estimation of probabilities: An essay on modern Bayesian methods.* Cambridge: Mass.

Hjort, N.L. 1990. Nonparametric Bayes estimation based on beta processes in models for life history data. *The Annals of Statistics* **18**, 1259–1294.

Ibrahim, J.G., Chen, M.H. and Sinha, D. 2001. *Bayesian survival analysis.* Springer.

Kalbfleisch, J.D. 1978. Nonparametric Bayesian analysis of survival time data. *Journal of the Royal Statistical Society*, Series B **40**, 214–221.

Kimeldorf, G. and Wahba, G. 1971. Some results on tchebycheffian spline functions. *Journal of Mathematical Analysis and Applications* **33**, 82–95.

Lavrac, N., Keravnou, E., Zupan, B. 1997. *Intelligent data analysis in Medicine and pharmacology.* Boston:Kluwer.

Law, M. H, and Kwok, J. T. 2001. Bayesian support vector regression. In *Proceedings of the Eighth International Workshop on Artificial Intelligence and Statistics (AISTATS)* 239–244. Key West, Florida, USA.

Lee, K.E and Mallick, B. 2004. Bayesian methods for gene selection in the survival model with application to DNA microarray data. *Sankhya* **66**, 756-778.

Lucas, P. and Abu-Hanna, A. 1999. Prognostic methods in medicine. *Artificial Intelligence in Medicine* **15**, 105-19.

Mallick, B. K., Denison, D. G. T., and Smith, A. F. M. 1999. Bayesian survival analysis using a mars model. *Biometrics* **55**, 1071-1077.

Mallick, B. K., Ghosh, D., and Ghosh, M. 2005. Bayesian classification of tumors using gene expression data. *Journal of the Royal Statistical Society, Series B* **67**, 219–232.

Metropolis, N., Rosenbluth, A. W., Rosenbluth, M. N., Teller, A. H., and Teller, E. 1953. Equations of state calculations by fast computing machines. *Journal of Chemical Physics* **21**, 1087–1092.

Parzen, E. 1970. Statistical inferences on time series by RKHS methods. In *Proceedings of the 12th Biennial Seminar*, 1–37. Canadian Mathematical Congress, Montreal, Canada.

Schölkopf, B. and Smola, A. 2002. *Learning with kernels.* Cambridge, MA: MIT Press.

Schena, M., Shalon, D., Davis, R. W. and Brown, P. O. 1995. Quantitative monitoring of gene expression patterns with a complementary DNA microarray. *Science* **270**, 567570.

Shively, T., Kohn, R. and Wood, S. 1999. Variable selection and function estimation in additive nonparametric regression using a data-based prior (with discussion). *Journal of American Statistical Association* **94**, 777-794.

Sinha, D. and Dey, D. 1997. Semiparametric Bayesian analysis of survival data. *Journal of American Statistical Association* **92**, 1195-1212.

Sinha, D. 1998. Posterior likelihood methods for multivariate survival data. *Journal of American Statistical Association* **54**, 1463-1474.

Smith and Kohn 1996. Nonparametric regression using Bayesian variable selection. *Journal of Econometrics* **54**, 1463-1474.

Sollich, P. 2001. Bayesian methods for support vector machines: Evidence and predictive class probabilities. *Machine Learning* **46**, 21–52.

Sørlie, T., Perou, C. M., Tibshirani, R., et al. 2001. Gene expression patterns of breast carcinomas distinguish tumor subclasses with clinical implications. *Proceedings of the National Academy of Sciences* **98**, 10869-10874.

Street, W.N. 1998. A neural network model for prognostic prediction. *Proc. 15th International Conf. on Machine Learning*, 540 - 546.

Tipping, M. 2000. The relevance vector machine. In *Neural Information Processing Systems – NIPS* Vol 12, ed. S. Solla, T. Leen and K. Muller, 652–658. Cambridge, MA: MIT Press.

———, 2001. Sparse Bayesian learning and the relevance vector machine. *Journal of Machine Learning Research* **1**, 211–244.

Vapnik, V.N. 1995. *The nature of statistical learning theory*, 2nd edn. New York: Springer.

Wahba, G. 1990. *Spline models for observational data.* Philadelphia: SIAM.

———, 1999. Support vector machines, reproducing kernel Hilbert spaces and the randomized GACV. In *Advances in Kernel Methods*, ed. B. Schölkopf, C. Burges and A. Smola , 69–88. Cambridge: MIT Press.

Wahba, G., Lin, Y., Lee, Y., Zhang, H. 2002. Optimal properties and adaptive tuning of standard and nonstandard support vector machines. In *Nonlinear Estimation and Classification*, ed Denison, D. D., Hansen, M. H., Holmes, C., Mallick, B., B. Yu, 125-143.

Wahba, G., Lin, Y. and Zhang, H. 2000. Generalized approximate cross validation for

West, M. 2003. Bayesian factor regression models in the "large p, small n" paradigm. Technical Report, Duke University.

Zupan, B., Demsar, J., Kattan, M., Beck, R. and Bratko, I. 2000. Machine learning for survival analysis: A case study on recurrence of prostate cancer. *Artificial Intelligence in Medicine* **20**, 59-75.

Chapter 14

A Bayesian Mixture Model for Protein Biomarker Discovery

Peter Müller[1], Keith Baggerly[1], Kim Anh Do[1] and Raj Bandyopadhyay[2]

[1] *U.T. M.D. Anderson Cancer Center and* [2] *Rice University*

Abstract

Early detection is critical in disease control and prevention. Molecular biomarkers provide valuable information about the status of a cell at any given time point. Biomarker research has benefited from recent advances in technologies such as gene expression microarrays, and more recently, proteomics. Motivated by specific problems involving proteomic profiles generated using Matrix-Assisted Laser Desorption and Ionization (MALDI-TOF) mass spectrometry, we propose model-based inference with mixtures of beta distributions for real-time discrimination in the context of protein biomarker discovery. Most biomarker discovery projects aim at identifying features in the biological proteomic profiles that distinguish cancers from normals, between different stages of disease development, or between experimental conditions (such as different treatment arms). The key to our approach is the use of a fully model-based approach, with coherent joint inference across most steps of the analysis. The end product of the proposed approach is a probability model over a list of protein masses corresponding to peaks in the observed spectra, and a probability model on indicators of differential expression for these proteins. The probability model provides a single coherent summary of the uncertainties in multiple steps of the data analysis, including baseline subtraction, smoothing and peak identification. Some ad-hoc choices remain, including some pre-processing and the solution of the label switching problem when summarizing the simulation output.

KEY WORDS: Density Estimation; Mass Spectrometry; Mixture Models; Nonparametric Bayes; Proteomics; Spectra.

14.1   Introduction

We propose a model-based approach to analyze data from Matrix-Assisted Laser Desorption and Ionization – Time of Flight (MALDI-TOF) experiments. We construct a mixture of Beta model to represent protein peaks in the spectra. An important feature of the proposed model is a hierarchical prior with indicators for differential expression for each protein peak. The posterior distribution on the number of peaks, the locations of the peaks and the indicators for differential expression summarizes (almost) all relevant uncertainty related to the experiment. This is made possible by using one coherent model to implement joint inference for multiple steps in the inference, including baseline subtraction, noise removal, peak detection and comparison across biologic conditions.

Molecular biologists and geneticists have been guided by the central dogma that DNA produces RNA, which makes proteins, the actual agents that perform the cellular biologic functions (Alberts et al., 1994). Researchers are interested in seeking both genetic and protein biomarkers of disease. Advances in genomics have given scientists the ability to assess the simultaneous expression of thousands of genes commonly collected from cDNA microarrays and oligonucleotide gene chips. More recent advances in mass spectrometry have generated new data analytic challenges in proteomics, similar to those created by gene expression array technologies. Proteomics is valuable in the discovery of biomarkers because the proteome reflects both the intrinsic genetic program of the cell and the impact of its immediate environment (Srinivas et al., 2001). Specifically, in the context of medical and cancer research, the clinician's ultimate aim of finding new targets for diagnosis and therapeutic strategies can benefit from a better understanding of the molecular circuitry. Valuable information can be obtained by investigating protein profiles over a wide range of molecular weights in small biological specimens collected from different pathological groups, such as different disease stages, different treatment arms, or normal versus cancer subjects. See, for example, Baggerly et al. (2006) for a review of the experimental setup. In this paper, motivated by specific cancer research challenges involving Matrix-Assisted Laser Desorption and Ionization (MALDI) proteomic spectra, we suggest techniques based on density estimation for purposes of discrimination, classification, and prediction.

14.1.1   Background

Proteomics was originally defined to represent the analysis of the entire protein component of a cell or tissue. Proteomics now encompasses the study of expressed proteins; specifically, this refers to a group of technologies that attempt to separate, identify and characterize a global set of proteins, as discussed in Arthur (2003). Proteomic methods may be used to simultaneously quantify the expression levels of many proteins, thus providing information on biological cell activity such as the structure-function relationship under healthy conditions and disease conditions, for example cancer. Proteomic technologies can be employed to identify markers for cancer diagnosis, to monitor disease progression, and to identify therapeutic targets. These methods may be applied to easily obtained samples from the body (blood serum, saliva, urine, breast nipple aspirate fluid) in order to measure the distribution of proteins in that sample. Among the most important proteomic methods are 2D gel electrophoresis, and mass spectrometry. Mass spectrometry (MS) measures the relative amounts of individual molecules in a sample, converted to ions. The quantity that is actually measured is the mass-to-charge, or the m/z ratio. Masses of atoms and molecules are usually measured in a unit called the *Dalton*, which is defined as $1/12$ the mass of ^{12}C. The unit of charge z is that present on an electron or a proton. The m/z ratio therefore refers to the number of Daltons per unit charge. In the case of singly charged ions, such as (most of) those generated in MALDI, this ratio is numerically equal to the ionic mass of the ions (in Daltons), plus one, due to the added mass of the unbalanced proton. In general, MS works by converting molecules to ions and measuring the proportions of the resulting ions in the mixture, after sorting them by the m/z ratio. This results in a histogram of m/z values, usually termed as a *mass spectrum*. Two common MS technologies are the Surface Enhanced Laser Desorption and Ionization (SELDI) technique and the MALDI method. MS, especially using the MALDI technique, has been very successfully applied to determine the molecular weight of both peptides and proteins, as well as to obtain structural information. This procedure has the advantages of rapid setup, high sensitivity, and tolerance for heterogeneous samples. Details of the experimental setup are described, for example, in Baggerly et al. (2003) or Baggerly et al. (2006). Briefly, the biological sample for which we wish to determine protein abundance is fixed in a matrix. A laser beam is used to break free and ionize individual protein molecules. The experiment is arranged such that ionized proteins are exposed to an electric field that accelerates molecules along a flight tube. On the other end of the flight tube molecules hit a detector that records a histogram of the number of molecules that hit over time. Assuming that all ionized molecules carry a unit charge, the time of flight is deterministically related to the molecule mass. The histogram of detector events over time can therefore be changed to a histogram of detector events over protein masses. Allowing for multiple charges, the mass scale is replaced by a scale of mass/charge ratios. The histogram

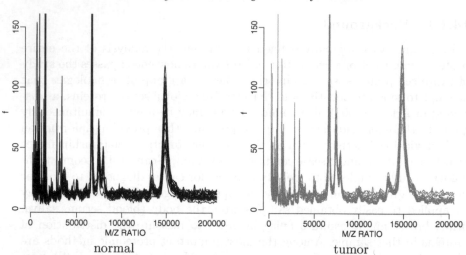

normal   tumor

FIGURE 14.1: Raw data. The left panel shows the recorded spectra for normal samples. The right panel shows the spectra for tumor samples. The X axis are the m/z values.

of detector events is known as mass/charge spectrum. More technical details are available, for example, in Siuzdak (2003) or de Hoffman and Stroobant (2002). Our motivating data set was generated using the MALDI technique.

Figure 14.1 shows typical mass spectra, plotted against mass/charge ratios. Ideally, one would expect a mass spectrum to be mostly flat, with spikes at the masses of proteins in the mixture. However, that is not the case. Firstly, each charged particle, on being energized by the laser, has a random initial velocity, before being accelerated by the electromagnetic field. These phenomena lead to the existence of a fairly smooth peak around the mass of the protein concerned. For lower m/z molecules, the peak is very narrow, and it gets broader with increasing m/z. Moreover, the ions generated may have slightly different masses, due to addition or removal of charged particles. In order to improve the resolution of the mass spectrometer, the sample is usually fractionated. This is achieved, for example, by running the sample through a gel, which separates out different molecules according to their pH values. The gel is then divided into pieces or fractions. Ideally, each fraction should contain a definite and distinct subset of proteins in the sample. The portion of the sample from each fraction is isolated and analyzed by the mass spectrometer. In practice, however, we find that the fractions do not separate as cleanly as expected. In some cases, the same protein may be found in multiple fractions, and the fractions may not be consistent across samples. This is the case, for example, for the motivating data set described later. Various other sources of noise exist in mass spectrometry. In MALDI, noise may arise from ionized molecules from the underlying matrix. This usually occurs at the lower end of the m/z

spectrum. Also, electrical noise may interfere, particularly if the sensitivity is low or extremely small proportions of samples are used. This noise may give rise to spurious peaks. At very low m/z values, large numbers of particles can saturate the detector and introduce additional artifacts.

14.1.2 Statistical Methods for Classification of Mass Spectrometry Data

Recent literature in the cancer classification arena that used MS generated data focused mainly on identifying biomarkers in serum to discriminate cancer from normal samples. Often the process involves a split of the data into a training set and a validation set. The training set is used to identify a subset of interesting biomarkers; the validation set is used to assess the selected biomarkers either individually or simultaneously by their ability to classify samples accurately in the separate test set. A number of statistical methods have been discussed for biomarker selection, including t–statistics by Chen et al. (2002), tree-based classification methods by Adam et al. (2002), genetic algorithms and self-organizing maps by Petricoin et al. (2002) and Baggerly et al. (2003), and artificial neural networks by Ball et al. (2002). The common classification methods used include the classical approaches such as linear discriminant analysis, quadratic discriminant analysis and k-nearest neighbors. More recent publications have discussed the use of bagging and boosting to the construction of a classifier, see Yasui et al. (2003) and Wu et al. (2003). Moreover, Wu et al. (2003) also compared the performance of random forest to other methods of classification and variable selection.

14.2 The Data

We consider the data set used as part of the First Annual Conference on Proteomics and Data Mining at Duke University. The ultimate goal is to identify protein biomarkers that distinguish between lung cancer and normal subjects. Assuming that the up- (or down-) regulation of certain proteins is the consequence of a transformed cancerous cell and its clonal expansion, an early detection research project may focus on the identification of such early molecular signs of lung cancer via the assessment of protein profiles from specific biological specimens. Researchers can thus analyze the collected protein profiles and identify signature *fingerprints* for the classification between lung cancer and normal samples. Based on the identified signature profiles, researchers can ultimately study the biological significance of those specific proteins or peptides. Such advances can potentially lead to clinical detection tools. Different research groups have attempted to develop techniques to classify or cluster this data set. Our analysis initially employs the preprocessing steps described in Baggerly et al. (2003). This motivating data set consists of MALDI-MS spectra of serum for 24 individuals with lung cancer and 17 normal individuals (without cancer). For each subject (sample), the raw data contained recordings of 20 fractions. Each such spectrum had readings for 60831 m/z values.

Traditional inference for mass/charge spectra proceeds in a step-wise fashion, often ignoring uncertainties involved in each step of the process. For example, Baggerly et al. (2003) analyzed this data set using the following steps.

First, a baseline, computed using a windowed local minimum technique, is subtracted from the data. This baseline correction has to be performed separately for each spectrum in each sample. It is a crucial step in the preprocessing, as we cannot combine the spectra otherwise in a meaningful manner. Next, using a Fourier transform, periodic noise most likely associated with electrical activity is removed. Next, the spectra are scaled by dividing by the total current over all the readings. Next, exploratory data analysis showed that several peaks were scattered across fractions, appearing in different fractions for different samples. Therefore, the normalized fractions were combined to generate one spectrum per sample. In a final pre-processing step, the dimensionality of the data was drastically reduced by carrying out a windowed peak identification. Taking the maximum intensity in each window (of 200 readings) and taking windows in which at least 8 of the samples contained a peak, using an ad-hoc definition of peaks, the dimensionality was reduced from 60381 to 506. The combined result of these pre-processing steps is a 506×41 peak matrix.

Baggerly et al. (2003) combined a genetic algorithm and an exhaustive search to extract a small subset of peaks which were good discriminators between cancers and normals. Perfect classification was achieved with a set of 5 peaks with intensities at the following M/Z values: 3077, 12886, 74263, 2476, and 6966 Dalton.

14.3   Likelihood Based Inference for Proteomic Spectra

A common feature of currently used methods is the use of some form of smoothing to separate the observed mass/charge spectrum into noise and signal, by directly smoothing the raw spectrum, by considering principal components, or by using reasonable exploratory data analysis tools like the windowing of the raw data described before. Also, most methods involve multiple steps, including separate steps related to noise removal, baseline subtraction, peak identification, and finally identification of differentially expressed peaks.

Such methods are highly appropriate when the focus of inference is the search for peaks corresponding to specific proteins and the identification of peaks that are correlated with the biologic condition of interest. A critical limitation, however, is the lack of a joint probability model that characterizes the combined uncertainty across all steps, and can be the basis for probability statements related to the desired inference. As an alternative we propose a likelihood-based approach that allows us to implement joint inference across all steps. The estimated peaks differ little from what is obtained with other methods. The main difference is in the full probabilistic description of all relevant uncertainties. Instead of a point estimate we report a posterior distribution on the unknown true spectrum of mass/charge ratios. All uncertainties related to denoising and baseline substraction are appropriately propagated and accounted for. The proposed approach proceeds as follows. We treat the spectrum as a histogram of observed detector events. We assume that the original time-of-flight scale has been transformed to counts on a mass/charge scale. Let $p_k(m)$ denote the frequency of mass/charge ratios m in the k-th sample. We treat the recorded data as an i.i.d. sample of draws from p_k. This naturally turns the inference problem into a density estimation problem of inference on the unknown distribution p_k, conditional on a random sample summarized by the observed spectrum. The problem differs from traditional density estimation problems due to the nature of the prior information. The spectrum p_k is known to be multimodal with sharp peaks corresponding to different proteins, plus a smooth baseline. We are thus lead to consider models for random distributions p_k on a compact interval, allowing for relatively narrow peaks corresponding to different proteins. Without loss of generality we assume that the range of mass/charge ratios is rescaled to the interval $[0, 1]$. A convenient model for random distributions on the unit interval that allows for the desired features is a mixture of Beta distributions. Restricting the Beta distributions in the mixture to integer parameters over a certain range leads to Bernstein priors discussed in Petrone (1999a). We follow Robert and Rousseau (2003) who argue for the use of Beta mixtures with unconstrained parameters to achieve more parsimonious models. We introduce an additional level of mixture to deconvolve the random distribution into one part, f_k, corresponding to the peaks arising from specific proteins and one part, B_k, corresponding to a non-zero baseline arising from background noise in the detector, the matrix used to fix the probe on the sample plate and other unspecified sources unrelated to the biologic effects of interest. A hierarchical prior distribution completes the model. In words, the hierarchical prior probability model is described as follows. We start with a distribution for a random number J of peaks, continue with a prior for the location and scale of the J peaks, and weights for each peak in each of the biologic samples. Samples collected under different biologic conditions, for example, tumor and normal, might require different weights, corresponding to different abundance of the respective protein in the samples. In addition to the peaks related to specific proteins in the probes the mixture also includes terms to represent a smooth baseline. Inference about the baseline is usually not of interest in itself. It is a nuisance parameter. We use J_k to denote the number of Beta terms that constitute the baseline, allowing for a different size mixture for each sample k. Details of the prior are described below. In the context of this hierarchical Beta mixture the desired inference about relative abundance of proteins in the probes reduces to inference about the weights in the Beta mixtures. We implement inference with a reasonably straightforward Markov chain Monte Carlo (MCMC) posterior simulation. The random number J and J_k of terms in the mixtures complicates inference by introducing a variable dimension parameter space. We use a reversible jump MCMC implementation to achieve the desired posterior simulation.

14.4 A Hierarchical Beta Mixture Model

We assume that the mass/charge spectrum is recorded on a grid m_i, $i = 1, \ldots, I$. For convenience we rescale to $m_i \in [0, 1]$. Let $y_k(m_i)$, $k = 1, \ldots, K$ denote the observed count in sample k for mass/charge ratio m_i. Due to the nonlinear nature of the transformation from time-of-flight to mass/charge, the grid on the mass/charge scale is not equally spaced. Sample k is observed under biologic condition x_k. For example, $x_k \in \{0, 1\}$ for a comparison of normal versus tumor tissue samples. We write $y_k = (y_k(m_1), \ldots, y_k(m_I))$ for the data from the k-th sample, $y = (y_1, \ldots, y_K)$ for the entire data set, and θ to generically indicate all unknown parameters in the model. We use an unconventional parametrization of Beta distributions, letting $\mathrm{Be}(m, s)$ denote a Beta distribution with mean m and standard deviation s (with an appropriate constraint on the variance). This notation simplifies the description of reversible jump moves and other technical details below. Finally, we generically use $p(a)$ and $p(a \mid b)$ to denote the distribution of a random variable a and the conditional distribution of a given b. We use notation like $\mathrm{Be}(x; m, s)$ to denote a Beta distribution with parameters (m, s) for the random variable x. We assume the following sampling model:

$$p(y_k \mid \theta) = \prod_{i=1}^{I} p_k(m_i)^{y_k(m_i)}, \quad p_k(m) = p_{0k}\, B_k(m) + (1 - p_{0k})\, f_k(m). \quad (14.1)$$

In words, we assume i.i.d. sampling from an unknown distribution p_k. The distribution p_k is assumed to arise as a convolution of a baseline $B_k(m)$ and a spectrum $f_k(m)$. We refer to the data y_k as the *empirical spectrum*, f_k as the unknown true *spectrum*, B_k as the baseline, and p_k as baseline plus spectrum, or the unknown distribution of mass/charge ratios. Both baseline and spectrum, are represented as mixtures of beta distributions. The means of the Beta kernels in the mixture for f_k are the mass/charge ratios of the detected proteins. Specifically we define

$$f_k(m) = \sum_{j=1}^{J} w_{xj} \, \text{Be}(m; \, \epsilon_j, \alpha_j), \qquad (14.2)$$

where $x = x_k$ denotes the biologic condition of sample k. We assume that $\theta_0 = (J, \epsilon_j, \alpha_j, \, j = 1, \ldots, J)$, the number and location and scale of the Beta terms, is common across all samples. The weights w_{xj} are specific to each biologic condition. This is consistent with the interpretation of the peaks in the spectrum as arising from proteins in the sample. The weights are the abundance of protein j in the k-th sample. For the baseline we use a similar mixture of Beta model:

$$B_k(m) = \sum_{j=1}^{J_k} \upsilon_{kj} \, \text{Be}(m; \, \eta_{kj}, \beta_{kj}). \qquad (14.3)$$

Here J_k is the size of the mixture, υ_{kj} are the weights, and (η_{kj}, β_{kj}) are the parameters of the j-th term in the mixture for the baseline of sample k. The parameters $\theta_k = (J_k, \upsilon_{kj}, \eta_{kj}, \beta_{kj}, \, j = 1, \ldots, J_k)$ describe the Beta mixture model for the baseline in the k-th sample. The choice of the mixture of Beta representation for the baseline is for convenience of the implementation. Any alternative non-linear regression model, such as regression splines, could be used with little change in the remaining discussion.

The likelihood (14.1) describes the assumed sampling model, conditional on the unknown parameters $\theta = (\theta_0, \theta_k, w_k, \, k = 1, \ldots, K)$, with (14.2) and (14.3) defining how the parameters determine the sampling model. The model is completed with a hierarchical prior. Let $\text{Poi}^+(\lambda)$ denote a Poisson distribution with parameter λ, constrained to non-zero values, let $\text{Ga}(a, b)$ denote a Gamma distribution with expectation a/b, and let $U(a, b)$ denote a uniform distribution on $[a, b]$. For the baseline mixture we assume

$$J_k \sim \text{Poi}^+(R_0), \eta_{kj} \sim U(0, 1), \; \beta_{kj} \sim U(\underline{\beta}, \overline{\beta}), \text{ and } \upsilon_k \sim Dir(1, \ldots, 1). \qquad (14.4)$$

Here $R_0, \underline{\beta}$ and $\overline{\beta}$ are fixed hyperparameters. For the peaks in the spectra we assume

$$J \sim \text{Poi}^+(R_1), \epsilon_j \sim U(0, 1), \; \alpha_j \sim U(\underline{\alpha}, \overline{\alpha}), \qquad (14.5)$$

with fixed hyperparameters $R_1, \underline{\alpha}$ and $\overline{\alpha}$. A constraint $\overline{\alpha} < \underline{\beta}$ ensures identifiability by uniquely identifying any given Beta kernel as a term in either the baseline mixture (14.4) or a peak in (14.5).

Finally, for the weights w_{xj} we assume common values for all samples under the same biologic condition. Thus the weights are indexed by biologic condition x and peak j. The prior model includes positive prior probability for ties $w_{0j} = w_{1j}$. Let $\lambda_j = I(w_{0j} = w_{1j})$ be an indicator for a tie and let $\Gamma = \{j : \lambda_j = 1\}$ and $L = \sum \lambda_j$ denote the set of indices and the number of peaks with $\lambda_j = 1$ and $W_1 = \sum_{j \in \Gamma} w_{0j}$. Let $w^{\star 1}$ and $w_x^{\star 0}$, $x = 0, 1$, denote the (standardized) subvectors of the weights w_{xj} defined by λ_j as follows:

$$w^{\star 1} = (w_1^{\star 1}, \ldots, w_L^{\star 1}) \equiv \frac{1}{W_1}(w_{0j}; \ j \in \Gamma) \text{ and}$$

$$w_x^{\star 0} = (w_{x1}^{\star 0}, \ldots, w_{x,J-L}^{\star 0}) \equiv \frac{1}{1 - W_1}(w_{xj}; \ j \notin \Gamma).$$

We assume

$$Pr(\lambda_j = 1) = \pi, \ j = 1, \ldots, J$$

$$w^{\star 1} \sim Dir(C_w, \ldots, C_w), \text{ and } w_x^{\star 0} \sim Dir(C_w, \ldots, C_w), \ x = 0, 1, \quad (14.6)$$

and $p(W_1 \mid \lambda) = Be(L\,C_w, (J - L)\,C_w)$. In words, we assume that the weights $w_0 = (w_{01}, \ldots, w_{0J})$ and $w_1 = (w_{11}, \ldots, w_{1J})$ are generated as product of independent rescaled Dirichlet distributions on the subsets of differentially and non-differentially expressed peaks, with positive prior probability π of any of the peaks $j = 1, \ldots, J$ being identical across $x = 0, 1$. The model is completed with a hyper prior $\pi \sim Be(A_\pi, B_\pi)$ (using the conventional parametrization of a Beta distribution).

We recognize that the model specification includes several simplifying assumptions. For example, we assume equal weights w_{xj} across all samples under the same biologic condition. A more realistic prior would require a hierarchical extension with sample specific weights, centered at distinct means under each biologic condition. Instead, we chose to use the simplified model and subsume the model misspecification error in the multinomial sampling model. Another simplification is the uniform prior on the peak locations ϵ_j and peak widths α_j. A more informative prior might formalize the fact that peaks at higher masses tend to be wider, for reasons related to physics of the experimental arrangement. Such dependent priors could easily be substituted for (14.5).

14.5   Posterior Inference

Inference in the proposed mixture of Beta model is implemented by Markov chain Monte Carlo (MCMC) simulation. Most details of the implementation are straightforward applications of standard MCMC algorithms. See, for example Tierney (1994). We describe the outline of the algorithm by indicating for each step the random variables that are updated, and the random quantities that are conditioned upon their currently imputed values. We use notation like $[x \mid y, z]$ to indicate that x is being updated, conditional on the known or currently imputed values of y and z. We generically use θ^- to indicate all parameters, except the parameter on the left side of the conditioning bar. Each iteration of the MCMC simulation includes the following steps: $[v_k \mid \theta^-, y^*]$, $[\lambda_j \mid \theta^-, y^*]$, $[W_1 \mid \theta^-, y^*]$, $[w^{*1} \mid \theta^-, y^*]$, $[w_x^{*0} \mid \theta^-, y^*]$, $[\pi \mid \theta^-, y^*]$, $[p_{0k} \mid y^*]$, $[J_k \mid \theta^-, y]$, $[J \mid \theta^-, y]$, $[\beta_{kj} \mid \theta^-, y]$, $[\eta_{kj} \mid \theta^-, y]$, $[\alpha_j \mid \theta^-, y]$, and $[\epsilon_j \mid \theta^-, y]$.

All steps except for the steps that update J and J_k are carried out with Metropolis-Hastings transition probabilities. We considered the use of imputed latent variables to replace the mixtures with conditionally conjugate hierarchical models, but found this to be computationally inferior. The transition probabilities used to update J and J_k require more attention. Changing the size J and J_k, respectively, of the mixtures implies a change in dimension of the parameter vector. We implement this by an application of reversible jump MCMC (RJ) transitions (Green, 1995).

The reversible jump moves for changing J and J_k use split/merge and birth/death proposals. The construction of the moves follows Richardson and Green (1997) who define RJ for mixture of normal models. However, the nature of the spectra with multipe highly peaked local modes requires a careful implementation. Below we explain the moves to update J. The moves for J_k are similar. The nature of B_k as relatively smooth baseline makes the transition probabilities to change J_k computationally easier to carry out.

We introduce a matching pair of *split* and *merge* moves to propose increments and decrements in J. The split move implements a proposal to replace a currently imputed peak j in f_k by two daughter peaks j_1 and j_2, maintaining the first two moments, and restricting the move such that the two new peaks are each others' nearest neighbors. The latter restriction simplifies the matching merge move. To select the peak to be split we pick with equal probabilty any of the current peaks. Without loss of generality assume $j = J$, $j_1 = J$ and $j_2 = J + 1$, and we assume $\lambda_J = 0$, i.e., the selected peak is imputed to be differentially expressed across biologic conditions $x = 0, 1$. The modifications for $\lambda_J = 1$ are straightforward. Also, in the following description we mark parameter values for the proposal with tilde, as in $\tilde{\theta}$. To propose new weights we generate two auxiliary variable $u_{1x} \sim Be(2, 2)$, $x = 0, 1$ and define $\tilde{w}_{xJ} = u_{1x} w_{xJ}$ and $\tilde{w}_{x,J+1} = (1-u_{1x}) w_{xJ}$. To propose location and scale for the two new daughter peaks we generate two auxiliary variables $u_2 \sim Be(2, 2)$ and $u_3 \sim Ga(5, 5)$ and define $\tilde{\epsilon}_J = \epsilon_J + u_2 \sqrt{\tilde{w}_{x,J+1}/\tilde{w}_{xJ}}$ and $\tilde{\epsilon}_{J+1} = \epsilon_J - u_2 \sqrt{\tilde{w}_{xJ}/\tilde{w}_{x,J+1}}$ for the locations and $\tilde{\alpha}_J = \sqrt{(1 - u_3)(1 - u_2^2)\alpha_J^2 w_{xJ}/\tilde{w}_{xJ}}$ and $\tilde{\alpha}_{J+1} = \sqrt{u_3(1 - u_2^2)\alpha_J^2 w_{xJ}/\tilde{w}_{xJ}}$. With the appropriate RJMCMC acceptance probability we accept the proposal as the new state of the MCMC simulation. Otherwise we discard the proposal. The matching *merge* proposal starts by selecting a pair of adjacent peaks, (j_1, j_2) and uses the inverse transformation to propose a merge.

Another pair of transition probabilities to change J are *birth* and *death* moves. To prepare a *birth* proposal it is important to slightly modify traditional *birth/death* proposals as used, for example, in Richardson and Green

(1997). Because of the extremely narrow nature of the peaks in the spectrum, a randomly proposed new peak would have practically zero chance of of being accepted when evaluating the ratio of likelihood values in the RJ acceptance probability. Instead we exploit the availability of reasonable ad-hoc solutions. Let $A = \{\alpha_h^o, \epsilon_h^o, w_{xh}^o, x = 0, 1, \pi_h^o; \; h = 1, \ldots, H\}$ denote a list of peaks identified by a preliminary data analysis step, using for example the approach outlined in Section 14.2. We refer to A as the reference solution. In words, the *birth* move proceeds by proposing a new peak as a slightly jittered copy of one of the reference peaks. The new peak is restricted to be, among all currently imputed peaks, the closest neighbor to the identified reference peak. The additional randomness and the restriction to nearest neighbors is important to facilitate the matching *death* move. Specifically, we start by randomly selecting an index $h \in \{1, \ldots, H\}$ to identify a peak in the reference solution. Then evaluate $\Delta = \min\left\{|\epsilon_h^o - \epsilon_j|, j = 1 \ldots J, \frac{1}{2}|\epsilon_h^o - \epsilon_g^o|, g \neq h\right\}$ and $\sigma = \min\{\Delta, \alpha_h^o\}$ and generate auxiliary variables u, s, v_0, v_1, r

$$u \sim N(0, \sigma^2)\, I(|u| \leq \Delta), \;\; s \sim \mathrm{Ga}(c_b, c_b), \;\; Pr(r = 1) = \pi_h^o, v_x \sim \mathrm{Be}(c_b, c_b).$$

Again we use $\tilde{\theta}$, etc., to mark proposed parameter values. We propose $\tilde{\epsilon}_{J+1} = \epsilon_h^o + u$, $\tilde{\alpha}_{J+1} = \alpha_h^o s$ and $\tilde{\lambda}_{J+1} = r$. We define weights for the newly proposed peak as $\tilde{w}_{x,J+1} = w_{xh}^o v_x$, with $x = 0, 1$ if $\tilde{\lambda}_{J+1} = 0$ and $x = 0$ if $\tilde{\lambda}_{J+1} = 1$. Finally, we re-standardize the weights $\tilde{w}_{xj}$ to sum to W_1 over $\{j : \lambda_j = 1\}$ and to $(1 - W_1)$ over $\{j : \lambda_j = 0\}$.

The matching *death* move proceeds by identifying one of the reference peaks, again by $h \sim \mathrm{U}\{1, \ldots, H\}$, and finding the currently imputed peak j that is closest to ϵ_h^o. When evaluating the appropriate RJ acceptance probability we keep in mind that ϵ_j might be nearest neighbor to more than one reference peak.

14.6 Results

We implemented the proposed algorithm to analyze the lung cancer data set described in Section 14.2. We exclude the very low part of the m/z scale to avoid saturation artifacts. Figures 14.2 through 14.8 summarize the inference. We use Y to generically indicate the observed data. Figure 14.2 shows the estimated size of the beta mixtures. The baseline is adequately modeled with mixtures of between 1 and 4 beta kernels. Although the baseline is not of interest by itself, appropriate modeling of the baseline is critical for meaningful inference on the spectrum f_k. The number of distinct proteins seen in the spectrum is *a posteriori* estimated between 15 and 29 (Figure 14.2). The raw spectra include many more local peaks. But only a subset of these are a posteriori identified as protein peaks. Others are attributed to noise, or

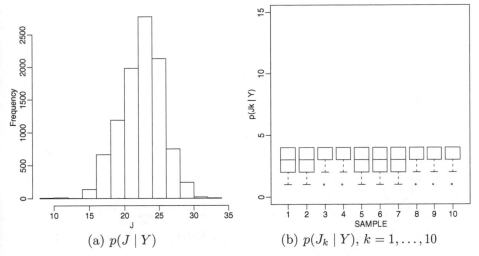

(a) $p(J \mid Y)$ (b) $p(J_k \mid Y)$, $k = 1, \ldots, 10$

FIGURE 14.2: $p(J \mid Y)$ and $p(J_k \mid Y)$. The histogram in the left panel shows the posterior probabilities for the number of distinct peaks. We find the posterior mode around 18 peaks. The boxplots (right panel) summarize the posterior distributions $p(J_k \mid Y)$. The number of terms in the baseline mixtures was constrained at $J_k \leq 5$.

included in the estimated baseline. Figure 14.3 shows the estimated mean spectra. The figure plots $E[f_k(m) \mid Y, x_k = x]$ for a future, $k = (K + 1)$-st sample of normal ($x = 0$) and cancer ($x = 1$) tissue, respectively. The posterior mean is evaluated as

$$E[f_k(m) \mid Y, x_k] = E[\sum_{j=1}^{J} w_{xj} Be(m; \ \epsilon_j, \alpha_j) \mid Y].$$

The posterior expectation is approximated as the ergodic average over the iterations of the MCMC simulation. The model includes the unknown distributions f_k as random quantities, parametrized by θ_0. Thus, in addition to point estimates, posterior inference provides a full probabilistic description of uncertainties. Figure 14.4 illustrates the posterior distribution $p(f_k \mid Y)$ by showing 10 random draws for the random distributions.

Posterior probabilities for any event of interest related to f_k can be reported. In particular, we find posterior probabilities for differential expression across the two biologic conditions. This is shown in Figure 14.5 which summarizes posterior inference on the indicators for differential expression, $1 - \lambda_j$. (Recall that λ_j is defined as indicator for $w_{0j} = w_{1j}$, i.e., non-differential expression). The figure shows estimated marginal probabilities of differential expression for all distinct peaks. A minor complication arises in reporting and summarizing posterior inference about distinct proteins. The mixture f_k only includes exchangeable indices j, leading to the complication that the Beta

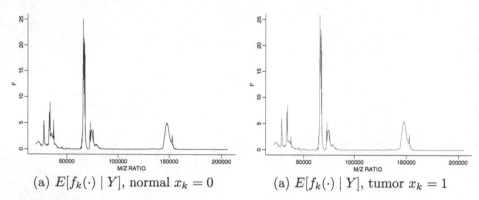

(a) $E[f_k(\cdot) \mid Y]$, normal $x_k = 0$ (a) $E[f_k(\cdot) \mid Y]$, tumor $x_k = 1$

FIGURE 14.3: $E[f_k(m) \mid Y, x_k = x]$. Estimated spectrum for normal and tumor samples.

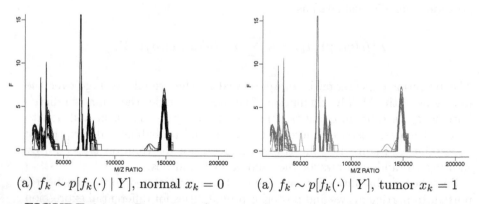

(a) $f_k \sim p[f_k(\cdot) \mid Y]$, normal $x_k = 0$ (a) $f_k \sim p[f_k(\cdot) \mid Y]$, tumor $x_k = 1$

FIGURE 14.4: Posterior inference defines a probability model on the unknown spectra. The two panels show ten draws from $p(f_k \mid Y)$ for $x_k = 0$ (left) and $x_k = 1$ (right).

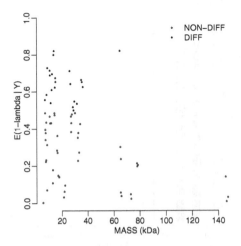

FIGURE 14.5: $E(1 - \lambda_j \mid Y)$. Posterior expected probability of differential expression for the reported peaks.

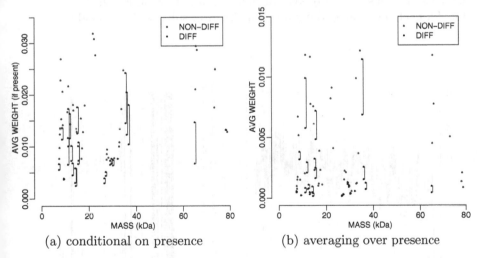

     (a) conditional on presence   (b) averaging over presence

FIGURE 14.6: Detected proteins. The figure plots the posterior estimated weights $E(w_{xj} \mid Y)$ against mass/charge ratios $E(\epsilon_j \mid Y)$. Panel (a) plots the average weight $E(w_{xj} \mid Y)$, averaging over all iterations that reported a protein peak at ϵ_j. Panel (b) scales the average weight by the probability of a peak being detected at ϵ_j (i.e., no detection is counted as $w = 0$). For genes with $E(\lambda_j \mid Y) > 0.5$ we plot the posterior expectation of w_{0j}. For differentially expressed genes we plot posterior expectations for w_{xj} for both conditions, $x = 0, 1$, linked with a line segment. Only proteins with posterior expected weight greater or equal to 0.05 are shown.

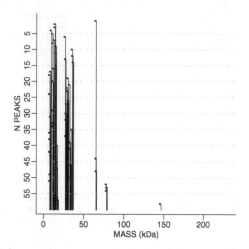

FIGURE 14.7: Proteins included in a panel of biomarkers to classify samples according to biologic condition x. Let $E_R = \{\epsilon_1, \ldots, \epsilon_R\}$ denote the locations of peaks, i.e., proteins, included in a subset of size R. The figure plots for each location ϵ_r the minimum size R of subsets that include ϵ_r.

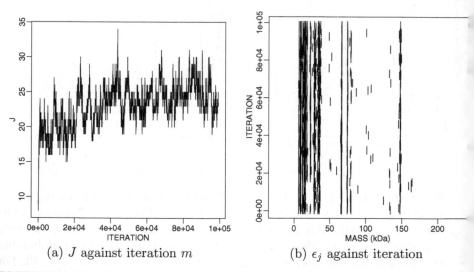

(a) J against iteration m (b) ϵ_j against iteration

FIGURE 14.8: Some aspects of the MCMC simulation. The left panel plots the imputed number of proteins J against iteration. The right panel plots the locations ϵ_j of the imputed beta kernels (x-axis) over iterations (y-axis).

kernel corresponding to a given protein might have different indices at different iterations of the posterior MCMC simulation. In other words, the protein identity is not part of the probability model. To report posterior inference on specific proteins requires additional post-processing to match Beta kernels that correspond to the same protein across iterations. We use an ad-hoc rule. Assume two peaks, j and h are recorded in the MCMC simulations and assume that the j-th peak occurred first in the MCMC output. The two peaks j and h are counted as arising from the same protein if the difference in masses is below a certain threshold. Specifically, we use $|\epsilon_j - \epsilon_h| < 0.5\alpha_j$ to match peaks. Alternatively to this ad-hoc choice one could use a threshold related to the nominal mass accuracy of the instrument. The problem of reporting inference related to the terms in a mixture is known as the label switching problem. Figure 14.5 shows unique peaks using this rule. Also, only proteins that occur in at least 5% of the MCMC simulations are reported. Different sets of peaks appear in different iterations of the MCMC, i.e., the number of peaks shown in Figure 14.5 does not match J. All proteins with $Pr(\lambda_j = 0 \mid Y) > 50\%$ are considered differentially expressed and are are marked as solid dots. Figure 14.6 shows the estimated relative abundance $E(w_{xj} \mid Y)$ for all detected proteins. For peaks corresponding to differentially expressed proteins we plot the estimated abundance for $x = 0$ and $x = 1$, connected by a short line segment. Table 14.1 gives a brief description of known proteins with mass close to the identified peaks.

Inference on λ allows evaluation of posterior expected false discovery rates (FDR). The notion of FDRs is a useful generalization of frequentist type-I error rates to multiple hypothesis testing, introduced in Benjamini and Hochberg (1995). Let δ_j denote an indicator for rejecting the j-th comparison in a multiple comparison problem, i.e., deciding $\lambda_j = 0$ (no tie), in our example. FDR is defined as FDR $= \sum \lambda_j \delta_j / \sum \delta_j$, the fraction of false rejections, relative to the total number of rejections. Applications of FDR to high throughput gene expression data are discussed, among others, by Storey and Tibshirani (2003). Posterior expected FDR is easily evaluated as $\overline{\text{FDR}} = E(\text{FDR} \mid Y) = [\sum E(\lambda_j \mid Y) \delta_j] / \sum \delta_j$. Let $\overline{\lambda}_j = E(\lambda_j \mid Y)$ denote the marginal posterior probability of non-differential expression for peak j. Consider now decision rules that classify a protein as differentially expressed if $1 - \overline{\lambda}_j > \gamma^*$. In analogy to classical hypothesis testing, we fix γ^* as the minimum value that achieves a certain pre-set false discovery rate, $\overline{\text{FDR}} \leq \alpha$. It can be shown (Müller et al., 2004) that under several loss functions that combine false negative and false discovery counts and/or rates the optimal decision rule is of this form. Newton et al. (2004) comment on the dual role of $\overline{\lambda}_j$ in decision rules like $\delta_j = I(1 - \overline{\lambda}_j > \gamma^*)$. It determines the decision, and at the same time already reports the probability of a false discovery as $\overline{\lambda}_j$ for $\delta_j = 1$ and the probability of a false negative as $1 - \overline{\lambda}_j$ for $\delta_j = 0$.

An important feature of the proposed model is the multilevel nature of the hierarchical probability model defined in (14.1) through (14.6). The posterior on the indicators λ_j contains all relevant information about differential levels of protein expression. Thus any inference about patterns of protein expression across different biologic conditions can be derived from $p(\lambda \mid Y)$ only. The discussed inference for the multiple comparison is one example. Another important decision problem related to the λ_j indicators is the identification of a minimal set of of proteins to classify samples according to biologic condition. In contrast to the multiple comparison decision the goal now is to select a small number of differentially expressed proteins. A convenient formalization of this question is to find locations m_i with maximum posterior probability $P(\lambda_j = 1 \mid Y)$ for peaks located at m_i. Figure 14.7 shows the optimal sets of peak locations as a function of the desired size of the set.

Finally, Figure 14.8 shows some aspects of the posterior MCMC, plotting the trajectory of imputed values for J, and for the unique peak locations ϵ_j against iterations.

14.7 Discussion

We have proposed a likelihood-based approach to inference about differential expression of proteins in mass/charge spectra from SELDI or MALDI experiments. We argued that the appropriate likelihood is based on random sampling. The usual approach of smoothing the raw spectrum is reasonable and leads to almost identical point estimates.

An important advantage of the proposed model is the easy generalization to more complicated experimental setups. Conditional on the λ indicators the rest of the model is independent of the biologic conditions for each sample. Consider, for example, an experiment with more than two biologic conditions (more than two tumor types, etc.). For more than two conditions it is convenient to describe ties by configuration indicators $s_{jx} \in \{1, \ldots, S_j\}$. For example, for four conditions a configuration of $s_j = (1, 1, 1, 2)$ would indicate that the first three conditions share the same peak, whereas expression is different under the fourth condition.

Our approach allows added flexibility in addressing a number of statistical challenges presented by MALDI-TOF mass spectra, including the interpretation of multiple tests, modelling and overfitting, and inadequate covariance in estimation, as well as substantial autocorrelation within a spectrum. In particular, a typical characteristic of MS data is that variance (and higher moments) appear to be related to mean intensity, resulting in measured intensities at the highest protein "peaks" exhibiting greater variability than do less abundant species. This property makes the magnitude of high intensity peaks less reliable for ensuing inference.

TABLE 14.1: Detected Proteins

Mass	Name	Description
11662	S10AE	Expressed at moderate level in lung
13016	MAGA5	Expressed in many tumors of several types, such as melanoma, head and neck squamous cell carcinoma, lung carcinoma and breast carcinoma, but not in normal tissues except for testes
13018	MAGB5	Expressed in testis. Not expressed in other normal tissues, but is expressed in tumors of different histological origins
15126	HBA	Involved in oxygen transport from the lung; Defects cause thalassemia
15864	ERG28	Ubiquitous; strongly expressed in testis and some cancer cell lines
15867	HBB	Involved in oxygen transport from the lung; Defects cause sickle-cell anemia
15989	PA2GE	Present in lung
29383	FA57A	Not detected in normal lung
29937	LAPM5	High levels in lymphoid and myeloid tissues. Highly expressed in peripheral blood leukocytes, thymus, spleen and lung
35010	GPR3	Expressed in lung at low level
35049	PLS1	Expressed in lung
35055	MAGA2	Expressed in many tumors of several types, such as melanoma, head and neck squamous cell carcinoma, lung carcinoma and breast carcinoma, but not in normal tissues except for testes
35844	SPON2	Expressed in normal lung tissues but not in lung carcinoma cell lines
65369	SEPT9	Chromosomal aberration involving SEPT9/MSF is found in therapy-related acute myeloid leukemia
65418	IL1AP	Detected in lung

Note: The table reports all human proteins within 0.1% of the reported masses ϵ_j *and* with Swissprot entries reporting terms "tumor, cancer, lung," or "carcinoma."

References

Adam, B., Qu, Y., Davis, J. W., Ward, M. D., Clements, M. A., Cazares, L. H., Semmes, O. J., Schellhammer, P. F., Yaui, Y., Feng, Z., and Wright Jr., G. L. (2002), "Serum protein fingerprinting coupled with a pattern-matching algorithm distinguishes prostate cacner from benign prostate hyperplasia and healthy men," *Cancer Research*, 62, 3609–3614.

Alberts, B., Bray, D., Lewis, J., Ra, M., Roberts, K., and Watson, J. D. (1994), *Molecular biology of the cell (3rd ed.)*, New York, NY: Garland.

Arthur, J. M. (2003), "Proteomics," *Current opinion in nephrology and hypertension*, 12, 423–430.

Baggerly, K. A., Coombes, K. R., and Morris, J. S. (2006), "Bayesian Inference for Gene Expression and Proteomics," Cambridge University Press, chap. An Introduction to High-Throughput Bioinformatics Data, pp. 1–39.

Baggerly, K. A., Morris, J. S., Wang, J., Gold, D., Xiao, L. C., and Coombes, K. R. (2003), "A comprehensive approach to analysis of MALDI-TOF proteomics spectra from serum samples," *Proteomics*, 9, 1667–1672.

Ball, G. S., Mian, F., Holding, F., Allibone, R. O., Lowe, J., Ali, S., G., L., McCardle, S., Ellis, I. O., Creaser, C., and Rees, R. C. (2002), "An integrated approach utilizing artificial neural networks and SELDI mass spectrometry for the classification of human tumors and rapid identification of potential biomarkers," *Bioinformatics*, 18, 395–404.

Benjamini, Y. and Hochberg, Y. (1995), "Controlling the false discovery rate: A practical and powerful approach to multiple testing," *Journal of the Royal Statistical Society B*, 57, 289–300.

Chen, G., Gharib, T. G., Huang, C.-C., Thomas, D. G., Shedden, K. A., Taylor, J. M. G., Kardia, S. L. R., Misek, D. E., Giordano, T. J., Iannettoni, M. D., Orringer, M. B., Hanash, S. M., and Beer, D. G. (2002), "Proteomic analysis of lung adenocarcinoma: identification of a highly expressed set of proteins in tumors," *Clinical Cancer Research*, 8, 2298–2305.

de Hoffman, E. and Stroobant, V. (2002), *Mass Spectrometry: Principles and Applications*, John Wiley.

Green, P. J. (1995), "Reversible jump Markov chain Monte Carlo computation and Bayesian model determination," *Biometrika*, 82, 711–732.

Müller, P., Parmigiani, G., Robert, C., and Rousseau, J. (2004), "Optimal Sample Size for Multiple Testing: the Case of Gene Expression Microarrays," *Journal of the American Statistical Association*, 99, 990-1001.

Newton, M., Noueriry, A., Sarkar, D., and Ahlquist, P. (2004), "Detecting differential gene expression with a semiparametric heirarchical mixture model," *Biostatistics*, 5, 155-176.

Petricoin, E. F., Ardekani, A. M., Hitt, B. A., Levine, P. J., Fusaro, V. A., Steinberg, S. M., Mill, G. B., Simone, C., Fishman, D. A., Kohn, E. C., and Liotta, L. A. (2002), "Use of proteomic patterns in serum to identify ovarian cancer," *The Lancet*, 359, 572-577.

Petrone, S. (1999a), "Bayesian density estimation using Bernstein polynomials," *Canadian Journal of Statistics*, 27, 105-126.

Richardson, S. and Green, P. J. (1997), "On Bayesian Analysis of Mixtures with an Unknown Number of Components," *Journal of the Royal Statistical Society B*, 59, 731-792.

Robert, C. and Rousseau, J. (2003), "A mixture approach to Bayesian goodness of fit," Tech. rep., CREST/INSEE, Paris.

Siuzdak, G. (2003), *The Expanding Role of Mass Spectrometry in Biotechnology*, MCC Press.

Srinivas, P. R., Srivastava, S., Hanash, S., and Wright, Jr., G. L. (2001), "Proteomics in early detection of cancer," *Clinical Chemistry*, 47, 1901-1911.

Storey, J. S. and Tibshirani, R. (2003), "SAM Thresholding and False Discovery Rates for Detecting Differential Gene Expression in DNA Microarrays," in *The analysis of gene expression data: methods and software*, eds. Parmigiani, G., Garrett, E. S., Irizarry, R. A., and Zeger, S. L., New York: Springer.

Tierney, L. (1994), "Markov chains for exploring posterior distributions," *The Annals of Statistics*, 22, 1701-1762.

Wu, B., Abbott, T., Fishman, D., McMurray, W., More, G., Stone, K., Ward, D., Williams, K., and Zhao, H. (2003), "Comparison of statistical methods for classification of ovarian cancer using mass spectrometry data," *Bioinformatics*, 19, 1636-1643.

Yasui, Y., Pepe, M., Thompson, M. L., Adam, B. L., Wright Jr., G. L., Qu, Y., Potter, J. D., Winget, M., Thornquist, M., and FEng, Z. (2003), "A data-analytic strategy for protein biomarker discovery: profiling of high-dimensional proteomic data for cancer detection," *Biostatistics*, 4, 449-463.

Chapter 15

Bayesian Methods for Detecting Differentially Expressed Genes

Fang Yu, Ming-Hui Chen, and Lynn Kuo
University of Nebraska Medical Center and University of Connecticut

15.1 Introduction

An important and common task in microarray analysis is to detect differentially expressed genes by comparing the expression levels of thousands of genes in samples collected at two different conditions. The microarray data usually contain thousands of genes, but only a very limited number of samples on each gene. Thus, the microarray experiments present statisticians a big challenge with its "large p small n" paradigm (West *et al.*, 2000). To solve this issue, many researches have been conducted from both the frequentist and the Bayesian framework, to propose powerful models for analyzing the microarray data, and efficient model selection algorithms for identifying the best subset of genes to be differentially expressed (DE). The detailed descriptions of the existing methods are available in several review articles and books, including Pan (2002), Kuo *et al.* (2008), Parmigiani *et al.* (2003), Speed (2003), Wit and McClure (2004), and Lee (2004).

In the frequentist framework, Schena *et al.* (1996), DeRisi *et al.* (1997), Chen *et al.* (1997), and Draghici (2002) proposed to use simple fold change methods for identifying DE genes. An earlier method for detecting DE genes was purely based on the size of the mean fold change ratio between the treatment and control. Specifically, when the mean fold change ratio is at least 2 or at most 1/2, then the gene is called to be a DE gene. This method is not adopted anymore, because it ignores the variability of the expression levels over replicates, and is biased toward genes with low expression levels. Dudoit *et al.* (2002) applied t statistics in assessing DE genes. However, it is common to observe a number of large t statistics driven by very small standard deviations given in the denominator. Even though the numerators in the t statistics measure the differences in the gene expressions between two biological conditions, the sizes of the differences may also be small. To overcome these shortcomings, a variety of approaches have been proposed to adjust standard t statistics with more reliable variance estimates. For example, Rocke and Durbin (2001), Huber *et al.* (2002), and Jain *et al.* (2003) proposed models to fit the variance of the expression of a gene as a function of the mean expression of the gene. Tusher *et al.* (2001) and Storey and Tibshirani (2003) proposed SAM (significant analysis of microarray) to add a variance stabilizing fudge factor to the denominator for each t statistics. Microarray ANOVA methods were also proposed by Kerr *et al.* (2000) under multiple biological conditions.

In the Bayesian framework, Baldi and Long (2001) proposed Cyber-T, which is a Bayesian version of t test that uses the posterior variance. Other approaches based on t statistics via either Bayesian or empirical Bayesian were proposed by Wright and Simon (2003), Newton *et al.* (2001), and Lönnstedt and Speed (2002). Smyth (2004) extended the method of Lönnstedt and Speed (2002) by considering the hierarchical Bayesian model in the context of general linear models, and used the moderated t or F statistics for identifying DE genes. There are several other Bayesian approaches built in the context of ANOVA or linear models, including Kerr *et al.* (2000), Wolfinger *et al.* (2001), Smyth(2004), Gottardo *et al.* (2006a, 2006b), and McLachlan *et al.* (2006). Considering that the same gene may be either expressed or not expressed even under the same biological condition, Ibrahim *et al.* (2002) and Chen *et al.* (2008) proposed mixture distributions to model the expression data at gene expression levels. We note that the mixture models are more commonly used to model gene-specific summary statistics or parameters in the models for gene expression levels. In the latter case, the components in the mixture model represent the status of the genes (equally expressed (EE) or DE, sometimes split into up-regulated (UP) and down-regulated (DN)). The literature on this development is quite rich, including Efron *et al.* (2001), Lee *et al.* (2000), Pan *et al.* (2002), Allison *et al.* (2002), Ghosh (2004), Kendziorski *et al.* (2003), Newton *et al.* (2004), Liu *et al.* (2004), and Yu *et al.* (2008).

The multiple comparison issue arises when the tests need to be conducted simultaneously for tens of thousands of genes. It is important to find good gene selection algorithms to determine which genes are differentially expressed and to control the overall error rates at the same time. Westfall and Young (1993) controlled FWER (family-wise error rate) via step-down maxT permutation adjusted p-values. Benjamini and Hochberg (BH)(1995, 2000) provided a well known approach to control the FDR (false discovery rate), which equals the false positive rate among the rejected hypotheses. Efron and Tibshirani (2002) provided an empirical Bayes justification for the BH procedure. Efron (2004, 2005, 2007a, 2007b) provided a gene-specific measure called local FDR to bound the global FDR and to estimate the false negative rate. Newton *et al.* (2004) proposed a Bayesian false discovery rate. Ibrahim *et al.* (2002) and Chen *et al.* (2008) proposed Bayesian criteria to declare a gene to be DE based on the size of posterior probability that the change in means of raw intensities across two conditions exceeds a certain threshold. Chen *et al.* (2008) empirically observed that the Bayesian criterion based algorithm yields good false positive and negative rates. Yu *et al.* (2008) proposed a calibrated Bayes factor approach, which weights two types of prior predictive error probabilities differently for each gene and at the same time controls the overall error rate for all genes. Dudoit *et al.* (2003) provided guidance over the multiple comparison methods through simulation. More discussions on multiple comparison issues can be found in Dudoit *et al.* (2004) and van der Laan *et al.* (2004a, 2004b).

The rest of the chapter is organized as follows. In Section 15.2, we present several models used for gene expression data. Section 15.3 addresses the prior elicitation for the models discussed in Section 15.2. Several recently developed gene selection algorithms are given in Section 15.4. A simulation study is presented in Section 15.5 to evaluate and compare the performances of the seven methods, including SAM (Tusher *et al.*, 2001), linear models for microarray data (LIMMA) (Smyth, 2004), a semiparametric hierarchical Bayes (SHB) approach based on mixtures of gamma models (Newton *et al.*, 2004), parametric empirical Bayes (EBarrays) approach for microarray data (Kendziorski *et al.*, 2003), the calibrated Bayes factor approach (Yu *et al.*, 2008), the ordered Bayes factor approach (Liu *et al.*, 2004), and the two-criterion (Chen *et al.*, 2008). Section 15.6 provides an analysis of a real data set from an affymetrix chip experiment. We conclude the chapter with a brief discussion in Section 15.7.

15.2   Models for Microarray Gene Expression Data

Log-normal, linear, and gamma models are frequently used to fit gene expression data. In this section, we discuss each of these three types of models in detail.

We assume that the data have already been properly normalized and converted to gene expression matrix. Let X_{1gj} and X_{2gj} denote the expression intensities of the gth gene in the jth sample under the first and second conditions, respectively. We also assume that there are a total of G genes and there are n_{1g} and n_{2g} replicates under conditions 1 and 2 for the gth gene. Thus, the data on gene g can be summarized in two vectors: $\mathbf{X}_{1g} = (x_{1g1}, ..., x_{1gn_{1g}})$ and $\mathbf{X}_{2g} = (x_{2g1}, ..., x_{2gn_{2g}})$ under the two conditions, respectively.

To detect whether the gth gene is DE or not, we consider two hypotheses for each gene: H_{0g}: gene g is EE and H_{1g}: gene g is DE. The problem then becomes to determine which hypothesis is more supported by the data.

15.2.1 Normal Model

Normal distributions have been widely used for modeling the large scale gene expression data. Based on whether the variances are the same or not for the expression level across different biological conditions, equal variance and unequal variance normal distributions can be constructed.

Equal Variance. The simplest form of the normal model assumes that the log-scale intensities from different biological conditions share the same variance on each gene. For example, in Liu *et al.* (2004) and Yu *et al.* (2008), they assumed equal replication numbers, say n_g, from both conditions. Then they further assumed that the observed expression measurements x_{1gj} and x_{2gj} come from independent normal distributions with

$$x_{1gj}|\mu_g, \sigma_g^2, \delta_g \overset{iid}{\sim} \mathcal{N}(\mu_g - \frac{1}{2}\delta_g, \sigma_g^2), x_{2gj}|\mu_g, \sigma_g^2, \delta_g \overset{iid}{\sim} \mathcal{N}(\mu_g + \frac{1}{2}\delta_g, \sigma_g^2), \forall j.$$
(15.1)

Under Model 15.1, gene g is EE if $\delta_g = 0$ and DE if $\delta_g \neq 0$.

A similar model is assumed in EBarrays (Kendziorski *et al.*, 2003). However, in EBarrays, all genes are assumed to share the same variance with:

$$x_{1gj}|\mu_{1g}, \sigma^2 \overset{iid}{\sim} \mathcal{N}(\mu_{1g}, \sigma^2), x_{2gj}|\mu_{2g}, \sigma^2 \overset{iid}{\sim} \mathcal{N}(\mu_{2g}, \sigma^2).$$
(15.2)

EBarrays conducts a test on comparison between $H_{0g}: \mu_{1g} = \mu_{2g}$ and $H_{1g}: \mu_{1g} \neq \mu_{2g}$.

Unequal Variances. A more complex normal model allows the intensities under two conditions to have different variances. In Yu *et al.* (2008), the intensities are assumed to have

$$H_{0g}: x_{1gj}, x_{2gj}|\mu_g, \sigma_g^2 \overset{iid}{\sim} \mathcal{N}(\mu_g, \sigma_g^2),$$
$$H_{1g}: x_{1gj}|\mu_{1g}, \sigma_{1g}^2 \overset{iid}{\sim} \mathcal{N}(\mu_{1g}, \sigma_{1g}^2), x_{2gj}|\mu_{2g}, \sigma_{2g}^2 \overset{iid}{\sim} \mathcal{N}(\mu_{2g}, \sigma_{2g}^2).$$
(15.3)

Yu *et al.* (2008) used the Bayes factor approach to compare the distributions of intensities under H_{0g} and H_{1g}.

15.2.2 Linear Model

In LIMMA (Smyth, 2004), a hierarchical linear model is assumed for log-scale intensities from multiple conditions. In this chapter, we consider its special case with only two conditions, i.e. control and treatment.

Let the column vector $\mathbf{X}_g$ contain the log intensities from $\mathbf{X}_{1g}$ and $\mathbf{X}_{2g}$ in order. Assume that $\mathbf{X}_g$ is modeled via a normal linear regression, and the corresponding expectation $E(\mathbf{X}_g)$ and variance $\text{Var}(\mathbf{X}_g)$ are given by

$$E(\mathbf{X}_g) = \mathbf{Z_g}\boldsymbol{\alpha}_g, \text{Var}(\mathbf{X}_g) = \mathbf{W}_g\sigma_g^2.$$
(15.4)

The matrix $\mathbf{Z_g}$ in Equation 15.4 is a design matrix of full rank with

$$\mathbf{Z_g}' = \begin{pmatrix} 1 \cdots 1 \, 0 \cdots 0 \\ 0 \cdots 0 \, 1 \cdots 1 \end{pmatrix}_{2*(n_{1g}+n_{2g})}.$$

The vector $\alpha_g' = (\alpha_{1g}, \alpha_{2g})$. Its elements represent the effects of the control and treatment groups, respectively, for gene g. Also, in Equation 15.4, $\mathbf{W}_g$ is a known non-negative definite weight matrix. Under this linear model, we have

$$\mathbf{X}_g \sim \mathcal{N}(\mathbf{Z}_g \alpha_g, \mathbf{W}_g \sigma_g^2).$$

LIMMA conducts a test on hypotheses whether the difference $\alpha_{1g} - \alpha_{2g}$ is zero or not. Let $\mathbf{C}' = (-1, 1)$ and $\beta_g = \mathbf{C}' \alpha_g$, then LIMMA tests hypotheses $H_{0g}: \beta_g = 0$ vs. $H_{1g}: \beta_g \neq 0$.

15.2.3 Gamma Model

In SHB (Newton *et al.*, 2004), the intensities on the raw scale are assumed to follow gamma distributions for both conditions. Specifically, we have

$$\exp(x_{kgj})|\mu_{kg} \overset{iid}{\sim} \mathcal{G}(a_k, \frac{a_k}{\mu_{kg}}), k = 1, 2. \tag{15.5}$$

Note that μ_{kg} is the mean of raw intensity $\exp(x_{kgj})$, and the shape parameter a_k measures the variations within the replicated experiments under the kth condition across all genes. SHB tests on whether the means of raw intensities are the same, i.e., $H_{0g}: \mu_{1g} = \mu_{2g}$. The same form of the gamma model with different priors is also implemented in EBarrays (Kendziorski *et al.*, 2003).

15.3 Prior Elicitation

Prior specification is a very important aspect of Bayesian analysis. Based on the setting of the gene-level models, in this section, we discuss three interesting types of priors, namely, the traditional conjugate prior, the power prior and the mixture of non-parametric priors.

15.3.1 Traditional Conjugate Prior

To ease computation, many methods adopt traditional conjugate priors. For example, EBarrays (Kendziorski *et al.*, 2003) assumes that in Equation 15.2, σ^2 is an unknown but fixed parameter and as apriori, μ_{1g}, μ_{2g}, and μ_g are assumed to follow from the prior distribution $\mathcal{N}(\mu_0, \tau_0^2)$, independently. Under the gamma models given by Equation 15.5, EBarrays assumes that a_k is a fixed constant, and $\frac{a_k}{\mu_{kg}} \sim \mathcal{G}(\alpha_0, \nu)$ for $k = 1, 2$.

In EBarrays, empirical Bayesian is used to obtain the estimates, denoted by $\hat{\mu}_0$, $\hat{\sigma}^2$ and $\hat{\tau}_0^2$, for hyperparameters μ_0, σ^2 and τ_0^2. Then the value of posterior odds in favor of a gene being DE on each gene is calculated using the formula:

$$odds_g = \sqrt{\frac{|\Sigma_{n_{1g}+n_{2g}}|}{|\Sigma_*|}} \exp\left\{ -\frac{1}{2}\xi_g'(\Sigma_*^{-1} - \Sigma_{n_{1g}+n_{2g}}^{-1})\xi_g \right\}, \qquad (15.6)$$

where $\xi_g = (\mathbf{X}_{1g}, \mathbf{X}_{2g})' - \hat{\mu}_0$, $\Sigma_{n_{1g}+n_{2g}} = \hat{\sigma}^2 I_{n_{1g}+n_{2g}} + \hat{\tau}_0^2 J_{n_{1g}+n_{2g}}$, $I_{n_{1g}+n_{2g}}$ is the $n_{1g}+n_{2g}$ identity matrix, and $J_{n_{1g}+n_{2g}}$ is the square matrix of size $n_{1g}+n_{2g}$ with all ones. Σ_* is a block-diagonal square matrix of dimension $n_{1g} + n_{2g}$, its upper left block is $\Sigma_{n_{1g}}$ and its lower right block is $\Sigma_{n_{2g}}$. EBarrays declare a gene with a large posterior odds value to be DE.

LIMMA (Smyth, 2004) is another method using the traditional conjugate prior. First it fits the linear model on each gene to obtain the estimates for the parameters defined in Equation 15.4. The estimates are denoted as $\hat{\alpha}_g$ for the vector of coefficients, s_g^2 for variances σ_g^2, and $\widehat{\text{Var}}(\hat{\alpha}_g) = s_g^2 \mathbf{V}_g$ for the covariance matrix. $\mathbf{V}_g$ is a positive definite matrix independent of s_g^2. So the estimator for the contrast is given by $\hat{\beta}_g = \mathbf{C}^T\hat{\alpha}_g = \hat{\alpha}_{2g} - \hat{\alpha}_{1g}$ and its estimated variance $\text{Var}(\hat{\beta}_g) = v_g s_g^2$ with $v_g = \mathbf{C}'\mathbf{V}_g\mathbf{C}$. The distributions of $\hat{\beta}_g$ and the random element in its variance have:

$$\hat{\beta}_g|\beta_g, \sigma_g^2 \sim \mathcal{N}(\beta_g, v_g\sigma_g^2) \text{ and } s_g^2|\sigma_g^2 \sim \frac{\sigma_g^2}{d_g}\chi_{d_g}^2,$$

where d_g is the residual degrees of freedom for the linear model on gene g. LIMMA specifies prior distributions on the parameters β_g and σ_g^2 as follows:

$$P(\beta_g \neq 0) = p \ , \ \beta_g|\sigma_g^2, \beta_g \neq 0 \sim \mathcal{N}(0, v_0\sigma_g^2), \text{ and } \frac{1}{\sigma_g^2} \sim \frac{1}{d_0 s_0^2}\chi_{d_0}^2.$$

Under this hierarchical model, the posterior mean of σ_g^2 given s_g^2 is

$$\tilde{s}_g^2 = \frac{d_0 s_0^2 + d_g s_g^2}{d_0 + d_g}.$$

LIMMA assesses the evidence of being DE based on a large moderated t statistics

$$t_g = \frac{\hat{\beta}_g}{\tilde{s}_g\sqrt{v_g}}.$$

15.3.2 Power Prior

The power prior proposed by Chen and Ibrahim (2003) is constructed based on prior predictive data. For the models discussed in Section 15.2.1, the power priors are conjugate. Compared to the traditional conjugate prior, it has a more general setting by allowing the incorporation of information from historical data or expert's knowledge. The power prior was firstly considered in microarray data analysis in Yu *et al.* (2008).

We discuss how to construct power priors for parameters $(\mu_{1g}, \sigma_{1g}^2, \mu_{2g}, \sigma_{2g}^2, \mu_g, \sigma_g^2)$ in the models defined by Equation 15.3. For models defined under H_{1g}, we take

$$\pi(\mu_{1g}, \mu_{2g}, \sigma_{1g}^2, \sigma_{2g}^2) \propto \left(\sigma_{1g}^2\right)^{-\frac{a_{01g}}{2}} \exp\left\{-\frac{a_{01g}}{2\sigma_{1g}^2}[(\mu_{1g} - \overline{x}_{01g})^2 + s_{01g}^2]\right\}$$

$$\times \left(\sigma_{2g}^2\right)^{-\frac{a_{02g}}{2}} \exp\left\{-\frac{a_{02g}}{2\sigma_{2g}^2}[(\mu_{2g} - \overline{x}_{02g})^2 + s_{02g}^2]\right\} \pi_0(\mu_{1g}, \mu_{2g}, \sigma_{1g}^2, \sigma_{2g}^2).$$

$$(15.7)$$

For models defined under H_{0g}, we specify

$$\pi(\mu_g, \sigma_g^2) \propto (\sigma_g^2)^{-\frac{a_{0g}}{2}} \exp(-\frac{a_{0g}}{2\sigma_g^2}[(\mu_g - \overline{x}_{01g})^2 + s_{01g}^2])$$

$$\times \exp(-\frac{a_{0g}}{2\sigma_g^2}[(\mu_g - \overline{x}_{02g})^2 + s_{02g}^2])\pi_0(\mu_g, \sigma_g^2). \qquad (15.8)$$

Here, $\overline{x}_{01g}$ and $\overline{x}_{02g}$ are the prior predictive means, s_{01g}^2 and s_{02g}^2 are the prior predictive variances, $\pi_0(\mu_{1g}, \mu_{2g}, \sigma_{1g}^2, \sigma_{2g}^2)$ and $\pi_0(\mu_g, \sigma_g^2)$ are initial priors. The positive scalar parameters a_{01g}, a_{02g} and a_{0g} quantify one's belief on the prior predictors denoted by $(\overline{x}_{01g}, \overline{x}_{02g}, s_{01g}^2, s_{02g}^2)$ under each hypothesis. When $a_{01g} \to 0$, $a_{02g} \to 0$, and $a_{0g} \to 0$, the prior defined in Equations 15.7 and 15.8 reduce to the initial priors $\pi_0(\mu_{1g}, \mu_{2g}, \sigma_{1g}^2, \sigma_{2g}^2)$ and $\pi_0(\mu_g, \sigma_g^2)$. Note that when $\pi_0 \propto 1$, both Equations 15.7 and 15.8 reduce to a traditional normal-inverse gamma conjugate prior. In the power prior, the hyperparameters $(\overline{x}_{01g}, \overline{x}_{02g}, s_{01g}^2, s_{02g}^2)$ are simply summary statistics of the historical data. Also the initial prior is added in case there is no prior predictive information available.

For simplicity, Yu *et al.* (2008) specified the initial priors separately as $\pi_0(\mu_{1g}, \mu_{2g}, \sigma_{1g}^2, \sigma_{2g}^2) \propto \pi_0(\sigma_{1g}^2, \sigma_{2g}^2) \propto \mathcal{IG}(\alpha_{1g}, \beta_{1g}) \times \mathcal{IG}(\alpha_{2g}, \beta_{2g})$, and $\pi_0(\mu_g, \sigma_g^2) \propto \pi_0(\sigma_g^2) \propto \mathcal{IG}(\alpha_g, \beta_g)$. The hyperparameters $\alpha_{1g}, \alpha_{2g}, \alpha_g, \beta_{1g}, \beta_{2g}$ and β_g are chosen to ensure the existence of the prior moments for σ_g^2, σ_{1g}^2 and σ_{2g}^2. They claimed that $(\beta_g, \beta_{1g}, \beta_{2g})$ are the variance-stabilized parameters and play a role similar to the "fudge" factor in SAM. In Yu *et al.* (2008), the Bayes factor values are used to assess the evidence of DE on each gene.

Denoting the prior predictive distributions by $m_{0g}(\mathbf{X}_{1g}, \mathbf{X}_{2g})$ under H_{0g} and $m_{1g}(\mathbf{X}_{1g}, \mathbf{X}_{2g})$ under H_{1g}, the Bayes factor for H_{0g} against H_{1g} is calculated as

$$BF_{01}(g) = \frac{m_{0g}(\mathbf{X}_{1g}, \mathbf{X}_{2g})}{m_{1g}(\mathbf{X}_{1g}, \mathbf{X}_{2g})}. \qquad (15.9)$$

Note that the more evidence for the gene to be DE, the smaller the $BF_{01}(g)$ is. Define $c_{1g} = \frac{1}{2}(a_{01g} + n_{1g} + 2\alpha_{1g} - 1)$, $c_{2g} = \frac{1}{2}(a_{02g} + n_{2g} + 2\alpha_{2g} - 1)$ and $c_{0g} = \frac{1}{2}(a_{0g} + n_{1g} + n_{2g}) + \alpha_g - 1$. Under the given model, the Bayes factor is given by

$$BF_{01}(g) \propto \left(\sum_j x_{1gj}^2 + a_{01g}(s_{01g}^2 + \overline{x}_{01g}^2) + \frac{\beta_{1g}}{2} - \frac{(n_{1g}\overline{x}_{1g} + a_{01g}\overline{x}_{01g})^2}{a_{01g} + n_{1g}} \right)^{c_{1g}}$$

$$\times \left(\sum_j x_{2gj}^2 + a_{02g}(s_{02g}^2 + \overline{x}_{02g}^2) + \frac{\beta_{2g}}{2} - \frac{(n_{2g}\overline{x}_{2g} + a_{02g}\overline{x}_{02g})^2}{a_{02g} + n_{2g}} \right)^{c_{2g}}$$

$$\times \left(\sum_j x_{1gj}^2 + \sum_j x_{2gj}^2 + \frac{n_{1g}a_{0g}}{n_{1g} + n_{2g}}(s_{01g}^2 + \overline{x}_{01g}^2) + \frac{n_{2g}a_{0g}}{n_{1g} + n_{2g}}(s_{02g}^2 + \overline{x}_{02g}^2) \right.$$

$$\left. + \frac{\beta_g}{2} - \frac{(\sum_j x_{1gj} + \sum_j x_{2gj} + \frac{a_{0g}n_{1g}}{n_{1g}+n_{2g}}\overline{x}_{01g} + \frac{a_{0g}n_{1g}}{n_{1g}+n_{2g}}\overline{x}_{02g})^2}{a_{0g} + n_{g1} + n_{g2}} \right)^{-c_{0g}}. \quad (15.10)$$

15.3.3   Mixture Non-Parametric Prior

Consider that the gene expression can be classified into three states: EE, DN (under-expressed in the first condition) and UP (over-expressed in the first condition), different from all other methods, SHB (Newton et al., 2004) conducts a test on three hypotheses: H_{0g}: $\mu_{1g} = \mu_{2g}$, H_{1g}: $\mu_{1g} < \mu_{2g}$, and H_{2g}: $\mu_{1g} > \mu_{2g}$. Accordingly, SHB defines a mixture prior for parameters μ_{1g}, μ_{2g} of Equation 15.5 in a form of:

$$f(\mu_{1g}, \mu_{2g}) = p_0 f_0(\mu_{1g}, \mu_{2g}) + p_1 f_1(\mu_{1g}, \mu_{2g}) + p_2 f_2(\mu_{1g}, \mu_{2g}), \quad (15.11)$$

where f_0, f_1, and f_2 are the joint densities evaluating the fluctuations of the means μ_{1g}, μ_{2g} under each hypothesis. Values p_0, p_1 and p_2 measure the marginal proportions of genes supporting each of the three hypotheses. To make sure all components are estimable, SHB relates the joint densities f_0, f_1, and f_2 with a univariate density π, which represents a probability vector on the support of equally spaced finite grids of log mean intensities. So the densities are rewritten as:

$$f_0(\mu_{1g}, \mu_{2g}) = \pi(\mu_{1g})1[\mu_{1g} = \mu_{2g}],$$
$$f_1(\mu_{1g}, \mu_{2g}) = 2\pi(\mu_{1g})\pi(\mu_{2g})1[\mu_{1g} < \mu_{2g}],$$
$$f_2(\mu_{1g}, \mu_{2g}) = 2\pi(\mu_{1g})\pi(\mu_{2g})1[\mu_{1g} > \mu_{2g}], \quad (15.12)$$

here 1 is an indicator function, for example, $1[\mu_{1g} = \mu_{2g}] = 1$ if $\mu_{1g} = \mu_{2g}$. Otherwise, it equals 0.

In SHB, the marginal proportions of the hypotheses (p_0, p_1, p_2) and the density π are estimated using a non-parametric EM algorithm to obtain the densities f_0, f_1, and f_2 given by Equation 15.12. Let $p(\mathbf{x}_{kg}|\mu_{kg})$, $k = 1, 2$ be the density of observing intensities $\mathbf{x}_{kg}$ under condition k on gene g, as defined in Equation 15.5. Then the posterior probabilities $P(H_{gl}|\mathbf{x}_{1g}, \mathbf{x}_{2g})$ with $l = 0, 1, 2$ are calculated using

$$Pr(H_{lg}|\mathbf{x}_{1g}, \mathbf{x}_{2g}) = p_l p(\mathbf{x}_{1g}, \mathbf{x}_{2g}|H_{gl})/p(\mathbf{x}_{1g}, \mathbf{x}_{2g}), \quad (15.13)$$

where

$$p(\mathbf{x}_{1g}, \mathbf{x}_{2g}|H_{lg}) = \int \int p(\mathbf{x}_{1g}|\mu_{1g})p(\mathbf{x}_{2g}|\mu_{2g})f_l(\mu_{1g}, \mu_{2g})d\mu_{1g}d\mu_{2g},$$

and the marginal density of the data $p(\mathbf{x}_{1g}, \mathbf{x}_{2g}) = \sum_l p_l p(\mathbf{x}_{1g}, \mathbf{x}_{2g}|H_{lg})$.

Different from the other methods, SHB evaluates the evidence of DE on each gene through the posterior probabilities given in Equation 15.13. More specifically, the smaller the posterior probability $P(H_{0g}|\mathbf{x}_{1g}, \mathbf{x}_{2g})$, the stronger evidence for the gene being DE.

15.4 Gene Selection Algorithms

It is important but difficult to develop an efficient gene selection algorithm to identify DE genes with a good control of overall error rate. Towards this goal, several gene selection algorithms have been developed in the Bayesian framework. In this section, we discuss six of those algorithms.

15.4.1 Ordered Bayes Factor

Assume that the Bayes factor values $BF_{01}(g)$ have been calculated on all genes $g = 1, \cdots, G$ based on Equation 15.9. Denote these calculated Bayes factor values by $BF^*_{01|X}(g)$. The smaller the $BF^*_{01|X}(g)$ is, the stronger evidence for gene g being DE. A simple approach proposed by Liu *et al.* (2004) is to order the calculated Bayes factors in a non-decreasing order and to declare top genes with small Bayes factors as DE genes. We call this approach as the ordered Bayes factor approach.

15.4.2 Calibrated Bayes Factor

Note that the ordered Bayes factor approach requires that the distribution of Bayes factors is exchangeable, which means that the distributions of Bayes factor are not gene dependent. We observe that the exchangeability usually is not satisfied. For example, in Figure 15.1, the data is assumed to follow the equal variance normal model defined by Equation 15.1, $\sigma_g^2 = \sigma^2$, $\delta_g \sim N(\mu_\delta, \lambda^2)$, and $\mu_g \sim N(0, \tau^2)$. We set $n_g = 20$, $\lambda^2 = 10$, $\mu_\delta = 0$, and $p = 0.05$, then 10000 Bayes factor values were sampled from its mixture distribution for each σ^2 value. Figure 15.1 shows that the median of the Bayes factors monotonically decreases as a function of σ^2 and the inter quartile range (IQR) increases as a function of σ^2. Note that as the variance of intensities varies greatly from gene to gene in the real data, the exchangeability does not hold. It is very crucial to find a good gene selection algorithm that does not require the assumption of exchangeability.

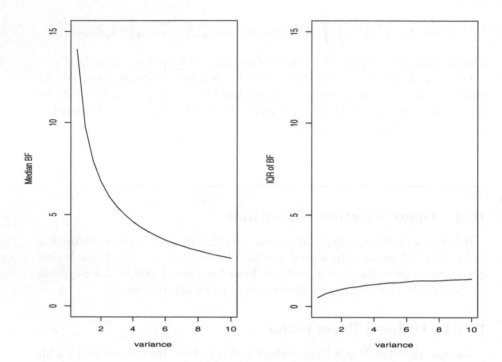

FIGURE 15.1:   Plots of medians and inter quartile ranges (IQR) of 10,000 Bayes factors with the variance σ^2 of each gene ranging from 0.5 to 10.

The Bayes factor based calibrated method (Yu *et al.*, 2008) is a new gene selection algorithm that protects against the violation of exchangeability on the distribution of Bayes factors across different genes. First, the Bayes factor based calibrated method assumes that all genes have the same probability of being DE, more specifically, it defines $H_g = 1$ if the gth gene is DE, $H_g = 0$ if the gth gene is EE, and $H_g \sim$ Bernoulli (p) for $g = 1, \ldots, G$ (Storey, 2002). With an expectation to declare a p proportion of genes to be DE, it chooses the pth percentile of the Bayes factor on each gene, denoted as c_g, to be the threshold. The calibrated method claims the gene g to be DE if $BF_{01}(g) \leq c_g$, and EE otherwise. The threshold c_g is also named as the calibration value. Note that c_g has a few attractive characteristics: (1) c_g defines different cut-off values on different genes and always uniquely exists. (2) When all genes are DE $(p = 1)$, c_g will be so large that $BF^*_{01|X}(g) \leq c_g$. All genes will be declared as DE genes. When all genes are EE $(p = 0)$, c_g will be very small so that $BF^*_{01|X}(g) > c_g$ and all genes are not declared as DE genes.

Since the calibration value c_g ensures

$$Pr(BF_{01}(g) > c_g | H_g = 1) \times p = Pr(BF_{01}(g) \leq c_g | H_g = 0) \times (1 - p),$$

where $Pr(BF_{01}(g) \leq c_g | H_g = 0)$ and $Pr(BF_{01}(g) > c_g | H_g = 1)$ are the prior predictive probabilities of making wrong decisions under $H_g = 0$ and $H_g = 1$, respectively. So the calibration value weighs two types of prior predictive error probabilities differently for each gene and at the same time controls the overall error rate for all genes.

Note that it is very computationally expensive to compute c_g, especially when the distributions of Bayes factors are not analytically tractable. To lessen the burden due to the calculation of c_g on each gene, Yu *et al.* (2008) proposed an alternative but equivalent gene selection algorithm based on p_g^*, which is calculated via

$$
p_g^* = \frac{Pr(BF_{01}(g) \leq BF_{01|\mathbf{X}}^*(g) | H_g = 0)}{Pr(BF_{01}(g) \leq BF_{01|\mathbf{X}}^*(g) | H_g = 0) + Pr(BF_{01}(g) > BF_{01|\mathbf{X}}^*(g) | H_g = 1)}.
$$
(15.14)

Then, a gene is claimed to be DE if and only if the calculated $p_g^* \leq p$. Observe that p_g^* behaves like the traditional p-value, and it measures the *relative probability* of making a mistake in declaring gene g to be DE against the mistake in declaring gene g to be EE. Yu *et al.* (2008) also proposed two methods to determine a guide value for p to control the overall error rate.

Method 1. Control the type I error at some pre-specified level α.
First c_g is calculated using $Pr(BF_{01}(g) < c_g | H_g = 0) = \alpha$. Then p_g is calculated via Equation 15.14 by replacing $BF_{01|\mathbf{X}}^*(g)$ with c_g. Finally we find the first quartile of the calculated p_y values across all genes as the guide value for p. Denote the quartile as p^*, then gene g is declared to be DE if and only if $p_g^* \leq p^*$.

Method 2. Use SAM to determine the number of genes selected to be DE. First use SAM to select genes by controlling the FDR rate at some pre-specified level γ. Denote the number of DE genes by SAM as k_γ. Then the guide value equals the k_γ^{th} smallest ordered value of the p_g^*, $g = 1, \cdots, G$, calculated based on Equation 15.14.

15.4.3   Two-Criterion

To identify genes with differential gene expressions between two biological conditions, Ibrahim *et al.* (2002) applied a gene selection algorithm, named as one-criterion, to compare the averages of raw scale intensities from different biological conditions given the observed data. Under the one-criterion, a gene is declared to be DE if the posterior probability that the gene has at least one fold change in means of raw intensities is larger than a pre-specified threshold value. To obtain a better control over the false positive and false negative rates, Chen *et al.* (2008) extended the one-criterion method to the two-criterion. Under the two-criterion, we calculate the posterior probability of having at least two fold changes in means of raw intensities across two biological conditions on each gene. This posterior probability measures the evidence of a gene being DE. The false negative rate is the proportion of wrongly detected DE genes. The false positive rate is the proportion of wrongly declared EE genes. We use the normal model with unequal variances as an example to illustrate this criterion based gene selection algorithm. The model is described by Equation 15.3 and the priors are given by Equations 15.7 and 15.8. The posterior densities for the parameters μ_{1g}, μ_{2g}, σ_{1g}^2 and σ_{2g}^2 are:

$$\mu_{1g}|\mathbf{X}_{1g},\sigma_{1g}^2 \sim \mathcal{N}(\frac{a_{01g}\bar{x}_{01g} + n_{1g}\bar{x}_{1g}}{a_{01g} + n_{1g}}, \frac{\sigma_{1g}^2}{a_{01g} + n_{1g}}),$$

$$\mu_{2g}|\mathbf{X}_{2g},\sigma_{2g}^2 \sim \mathcal{N}(\frac{a_{02g}\bar{x}_{02g} + n_{2g}\bar{x}_{2g}}{a_{02g} + n_{2g}}, \frac{\sigma_{2g}^2}{a_{02g} + n_{2g}}),$$

$$\sigma_{1g}^2|\mathbf{X}_{1g} \sim \mathcal{G}(\frac{a_{01g} + n_{1g} - 1}{2}, \frac{n_{1g}}{2}s_{1g}^2 + \frac{a_{01g}}{2}s_{01g}^2 + \frac{a_{01g}n_{1g}}{2(a_{01g} + n_{1g})}(\bar{x}_{1g} - \bar{x}_{01g})^2),$$

$$\sigma_{2g}^2|\mathbf{X}_{2g} \sim \mathcal{G}(\frac{a_{02g} + n_{2g} - 1}{2}, \frac{n_{2g}}{2}s_{2g}^2 + \frac{a_{02g}}{2}s_{02g}^2 + \frac{a_{02g}n_{2g}}{2(a_{02g} + n_{2g})}(\bar{x}_{2g} - \bar{x}_{02g})^2).$$

$$(15.15)$$

Denote $\tilde{\mu}_{1g}$, $\tilde{\mu}_{2g}$, $\tilde{\sigma}_{1g}^2$ and $\tilde{\sigma}_{2g}^2$ as sampled values of parameters μ_{1g}, μ_{2g}, σ_{1g}^2 and σ_{2g}^2 from their posterior distributions, respectively. Then the posterior sampled mean of the raw intensities $\psi_{kg} = E_x[\exp(x_{kgj})]$ for $k = 1, 2$ can be written as: $\psi_{1g} = \exp(\tilde{\mu}_{1g} + \frac{1}{2}\tilde{\sigma}_{1g}^2)$, and $\psi_{2g} = \exp(\tilde{\mu}_{2g} + \frac{1}{2}\tilde{\sigma}_{2g}^2)$. The two-criterion focuses on ψ_{kg} to select DE genes since the means of raw intensities ψ_{kg} provide combined information on both the location and scale parameters in the normal distribution model for intensities $\mathbf{x}_{kg}$. So the two-criterion is more informative than gene selection criteria that focus on only the comparison between the mean parameters μ_{kg} of the log-intensities.

To compare the means of raw gene expression levels from two biological conditions, according to Ibrahim *et al.* (2002) and Chen *et al.* (2008), a log-ratio ζ_g is defined with

$$\zeta_g = \log(\psi_{2g}) - \log(\psi_{1g}) = \mu_{1g} - \mu_{2g} + \frac{1}{2}(\sigma_{1g}^2 - \sigma_{2g}^2), \quad g = 1, 2, \ldots, G. \quad (15.16)$$

Let γ_g be the posterior probability of having at least two fold changes in means of raw intensities across two biological conditions on each gene. Mathematically, γ_g is of the form

$$\gamma_g = P(|\zeta_g| > \log2 \,|D), \forall\, g. \qquad (15.17)$$

Under the two-criterion, the larger the γ_g, the stronger evidence for gene g being DE. Specifically, the two-criterion claims a gene to be DE if $\gamma_g \geq \gamma_0$, where γ_0 is a pre-specified cut-off value for γ_g obtained from all genes. Common choices of γ_0 are 0.7, 0.8, and 0.9.

15.4.4 BH Method

BH procedure (Benjamini and Hochberg, 1995) controls the FDR on the average to be no more than some pre-specified α based on a sequence of p values. First, a list of p-values from gene-wise tests have been obtained to present the evidence of each gene being DE. Then the p values are sorted in an increasing order. Denote p_{r_i} as the ith gene in the sorted gene list. Let k be the largest integer i for which $p_{r_i} \leq \frac{i}{G}\alpha$ for all i. Then we declare all the genes with labels $r_1, \ldots, r_k$ to be DE. We call the value $p_{r_i}\frac{G}{i}$ for each i as the BH adjusted p-value. Observe that the BH method controls FDR at a quite low level by a factor of G_0/G, where G_0 and G are number of truly EE genes and total number of genes respectively, an adaptive BH procedure by Benjamini and Hochberg (2000) estimates G_0 and use $\alpha^* = \alpha(\frac{\hat{G_0}}{G})$ instead as the target FDR level to gain more power. Compared to the 1995 BH procedure, the adaptive procedure performs better as the prevalence of DE increases. Both BH methods can be applied to the methods producing test statistics and p-values, as done in LIMMA (Smyth, 2004).

15.4.5 Bayesian FDR

Bayesian FDR proposed in Newton *et al.* (2004) captures wide attention and great popularity since it provides a method of evaluating a Bayesian version of FDR. Assume that the posterior probabilities defined by Equation 15.13 have been calculated. Recall that the smaller $P(H_{0g}|\mathbf{X}_{1g}, \mathbf{X}_{2g})$, the stronger evidence the gene being DE. Therefore, genes are selected as DE genes if they are located on the top of the sorted gene list sorted in an increasing order of $P(H_{0g}|\mathbf{X}_{1g}\mathbf{X}_{2g})$. Note that when gene g gets selected as a DE gene, $P(H_{0g}|\mathbf{X}_{1g}, \mathbf{X}_{2g})$ can be interpreted as the conditional probability of wrongly placing gene g on the DE gene list. The expected FDR given the data can be estimated by the average of conditional probabilities $P(H_{0g}|\mathbf{X}_{1g}, \mathbf{X}_{2g})$ in the selected DE gene list. We can decide the number of DE genes by including as many top genes in the sorted gene list as possible with a control of the expected FDR to be bounded at the target level of α.

15.4.6 Posterior Odds

The posterior odds for differential expressions is used in EBarrays (Kendziorski *et al.*, 2003) to detect DE genes. The values of the posterior odds can be calculated via Equation 15.6. The larger the $odds_g$, the stronger evidence the gene g being DE. To reduce the type I error, Kendziorski *et al.* (2003) suggested that a gene is declared to be DE if the posterior odds is larger than 1.

15.5   A Simulation Study

We use a simulation study to evaluate the performance of seven methods mentioned above for detecting differential gene expressions: Calibrated Bayes factor approach (CBF), ordered Bayes factor approach (BF), two-criterion, SAM, SHB, LIMMA, and EBarrays. Except for the ordered Bayes factor approach, each of the remaining six methods has certain control of overall error rate. For the ordered Bayes factor approach, the number of DE genes to be declared must be pre-specified. For the two-criterion method, the genes are declared to be DE if its posterior probability defined by Equation 15.17 is no more than some pre-specified value γ_0. SAM, SHB, and LIMMA allow us to calculate the realized false discovery rate associated with different numbers of selected DE genes. However, the false discovery rates appear in different forms. For example, SAM provides the median FDR, SHB provides Bayesian FDR, and LIMMA provides BH adjusted FDR. Under EBarrays, we calculate the values of posterior odds and choose the genes with values of posterior odds larger than some chosen value (say 1) to be DE. Those chosen values are denoted as "Cut-Off" values.

The performances of the methods are compared via four estimated different error rates: false negative rate, false positive rate, false discovery rate (FDR), and false non-discovery rate (FNDR). FNDR is the realized rate of false non-detections in the non-detected genes. We expect a good method will have smaller error rates than the others.

We used two different simulation settings. In each setting, 50 simulations were generated with 250 DE genes set among 5000 genes in total. Ten replications were generated under each condition.

Simulation I. We borrowed the simulation setting from Yu *et al.* (2008). For $j = 1, \ldots, 10$, we independently generated $x_{1gj} \sim \mathcal{N}(\mu_g - 0.5, 0.3^2)$ and $x_{2gj} \sim \mathcal{N}(\mu_g + 0.5, 1.2^2)$ for $g = 1, \cdots, 125$; $x_{1gj} \sim \mathcal{N}(\mu_g + 0.5, 0.3^2)$ and $x_{2gj} \sim \mathcal{N}(\mu_g - 0.5, 1.2^2)$ for $g = 126, \cdots, 250$; $x_{1gj}, x_{2gj} \sim \mathcal{N}(\mu_g, 0.7^2)$ for $g = 251, \cdots, 5000$; and $\mu_g \sim \mathcal{U}(5, 11)$ for all g. The normal setting with unequal variance power prior was used for the calibrated Bayes factor approach, the ordered Bayes factor approach, and the two-criterion approach. The parameters involved in the power prior were set to be: $\bar{x}_{01g} = \bar{x}_{02g} = 0$, and $s_{01g} = s_{02g} = 0.5$ or 6 for every 50 genes alternatively. $a_{01g} = a_{02g} = a_{0g}n_{0g}$, where $n_{0g} = n_g$ and $a_{0g} = 0.05$ when the gene index g is a multiple of 8, and 0.005 otherwise. Since there are in total 250 true DE genes, we allowed each method to select 250 DE genes in the first analysis to compare the ranking of DE genes among all methods. Then each method was allowed to have two choices of their own for the cut-off value in the second analysis, whose purpose is to compare the performance of each method as well as the relevant gene selection algorithm.

TABLE 15.1: Method Comparison on Selecting 250 DE Genes under Simulation I

Method	CCDE	CCEE	False Neg.	False Pos.	FDR	FNDR
CBF	209.0	4709.0	0.164	0.009	0.164	0.009
BF	174.9	4674.9	0.300	0.016	0.300	0.016
SAM	156.9	4657.1	0.372	0.020	0.372	0.020
SHB	138.9	4638.9	0.444	0.023	0.444	0.024
LIMMA	179.7	4679.7	0.281	0.015	0.281	0.015
EBarrays	178.9	4678.9	0.285	0.015	0.285	0.015

Table 15.1 summarizes the analysis result of each method from the first analysis, including the number of correctly claimed DE genes (CCDE), the number of correctly claimed EE genes (CCEE), and all four error rates. Among all six methods, the calibrated Bayes factor approach selects the largest number of true DE genes and the smallest values for all error rates. So the calibrated Bayes factor performs the best in this analysis. LIMMA, EBarrays and the ordered Bayes factor approach follow immediately after the calibrated Bayes factor approach, and performs similarly well. SHB performs the worst. The possible reason is that SHB requires the assumption of constant coefficient of variation (CV) across genes within the same condition.

TABLE 15.2: Method Comparison with Error Control under Simulation I

Method	Cut-Off	CDE	CCDE	CCEE	False Neg.	False Pos.	FDR	FNDR
CBF	0.01	80.4	79.9	4749.4	0.681	0.000	0.007	0.035
	0.05	169.8	162.1	4742.3	0.352	0.002	0.045	0.018
Two-Crit.	0.9	81.8	81.3	4749.5	0.675	0.000	0.006	0.034
	0.8	188.8	116.9	4678.0	0.532	0.015	0.380	0.028
SAM	0.05	65.1	61.7	4746.6	0.753	0.001	0.052	0.040
	0.1	95.1	85.1	4739.9	0.660	0.002	0.104	0.035
SHB	0.05	97.2	95.1	4747.9	0.620	0.000	0.021	0.033
	0.1	110.8	105.0	4744.2	0.580	0.001	0.051	0.031
LIMMA	0.05	115.1	111.0	4745.9	0.556	0.001	0.035	0.029
	0.1	145.5	133.7	4738.2	0.465	0.002	0.081	0.025
EBarrays	1	108.2	105.3	4747.1	0.579	0.001	0.027	0.030
	0.5	128.8	122.3	4743.6	0.511	0.001	0.050	0.027

In the second analysis, different gene selection algorithms are applied for each method, therefore different number of DE genes are selected accordingly. So when we summarized the results using Table 15.2, we also include the number of genes claimed to be DE (CDE). All methods except for the two-

TABLE 15.3: Method Comparison on Selecting 250 DE
Genes under Simulation II

Method	CCDE	CCEE	False Neg.	False Pos.	FDR	FNDR
CBF	213.9	4713.9	0.144	0.008	0.144	0.008
BF	184.7	4684.7	0.261	0.014	0.261	0.014
SAM	181.0	4681.3	0.276	0.014	0.275	0.015
SHB	222.4	4722.4	0.110	0.006	0.110	0.006
LIMMA	168.8	4668.8	0.325	0.017	0.325	0.017
EBarrays	177.9	4677.9	0.289	0.015	0.289	0.015

criterion with γ_0 set to be 0.8 provide a small FDR, that is, when a gene is selected to be DE, it has a very large chance to be true DE. We also observe that all methods are more conservative compared to themselves in the first analysis by selecting less numbers of genes to be DE. Among all methods, the calibrated Bayes factor approach with its Type I error controlled to be no more than 0.05 performs the best by providing the smallest false positive, false negative, FNDR rates, and simultaneously controlling FDR rate to be less than 0.05. SAM selects the smallest number of genes to be DE. When the type I error are controlled to be less than 0.01, the calibrated Bayes factor performs similar to SAM with the FDR controlled at 0.05.

Simulation II. In this setting, we set both means and variances of the log-scale intensities to be larger under the first condition compared to the second condition on each gene. Particularly, we independently generated $x_{1gj} \sim N(\mu_g - 0.5, 0.25)$ and $x_{2gj} \sim N(\mu_g + 0.5, 4)$ for $g = 1, \cdots, 250$; and $x_{1gj}, x_{2gj} \sim N(\mu_g, 1)$ for $g = 251, \cdots, 5000$. μ_g still were generated via $\mu_g \sim Unif(5, 11)$ for all g. All methods were applied to the simulated data in the same way as Simulation I. Two analysis were carried out the same as before by either letting all methods select 250 DE genes, or by setting the cut-off values the same as Simulation I.

Simulation results for the first analysis are summarized by Table 15.3. SHB performs better in this setting compared to Simulation I. Both SHB and the calibrated Bayes factor approach perform better than all other four methods.

Table 15.4 summarizes the results from the second analysis. Again, all methods have good controls of the FDR rate except for the two-criterion method with γ_0 set to be 0.8. SHB performs the best with the FDR controlled at either 0.05 or 0.1 among all methods. The calibrated Bayes factor approach with the type I error controlled at 0.05 takes the third place. When the type I error is controlled to be no more than 0.01, the calibrated Bayes factor approach selects a similar number of genes as SAM with FDR controlled at 0.05. Therefore, under this scenario, the two methods providing guide value for p discussed in Section 15.4.2 for the calibrated Bayes factor approach agree well with each other.

TABLE 15.4: Method Comparison with Error Control under Simulation II

Method	Cut-Off	CDE	CCDE	CCEE	False Neg.	False Pos.	FDR	FNDR
CBF	0.01	91.5	91.0	4749.5	0.636	0.000	0.005	0.032
	0.05	181.9	174.6	4742.7	0.302	0.002	0.040	0.016
Two-Crit.	0.9	160.0	159.4	4749.4	0.362	0.000	0.004	0.019
	0.8	287.4	215.1	4677.7	0.139	0.015	0.251	0.007
SAM	0.05	106.2	101.4	4745.2	0.594	0.001	0.044	0.031
	0.1	143.4	130.0	4736.5	0.480	0.003	0.093	0.025
SHB	0.05	204.9	199.3	4744.4	0.203	0.001	0.027	0.011
	0.1	230.3	214.7	4734.5	0.141	0.003	0.067	0.007
LIMMA	0.05	98.8	94.5	4745.7	0.622	0.001	0.042	0.033
	0.1	129.9	117.7	4737.8	0.529	0.003	0.092	0.028
EBarrays	1	107.3	104.2	4746.9	0.583	0.001	0.028	0.031
	0.5	128.2	121.0	4742.9	0.484	0.001	0.055	0.027

15.6 Real Data Example

We consider a data set from Kalajzic *et al.* (2005) based on all methods mentioned in the simulation study. The dataset includes samples from mouse calvarial cultures at day 17 with Affymetrix microarray experiment to understand the gene expression profile of osteoblast lineage at defined stages of differentiation. colα1 promoter-GFP transgenic mouse lines are used to demonstrate the feasibilities of generating more homogeneous populations of cells at distinct stages of osteoprogenitor maturation and obtain valid microarray interpretations. As a result, the expression of genes are compared between cells cultured with 2.3GFP[pos] and cells cultured with 2.3GFP[neg].

Before the application of each method, we observed the characteristics of the data through the summary statistics for the ratios of the sample standard deviations across two conditions for all the genes. We observed that the ratios range from 1 to 1.289e+5. We also observed that the distribution of the ratio is strongly right-skewed, with the median ratio equal to 2.844, and the mean ratio equal to 42.4. Therefore, we adopted the normal model with unequal variances of intensities between the two conditions and used the calibrated Bayes factor approach, the two-criterion and the order Bayes factor approach to select DE genes. The values of a_{01g} and a_{02g} were chosen to be $a_0 n_g$, with $a_0 = 0.5$ and $n_g = 3$. To use a moderate informative prior in the analysis, we set $\alpha_{1g} = \alpha_{2g} = 1.0$ and $\beta_{1g} = \beta_{2g} = 1.0$ in the initial prior $\pi_0(\sigma_{1g}^2, \sigma_{2g}^2)$. Both guidance methods discussed by Section 15.4.2 were used for p^* for the calibrated Bayes factor approach. Accordingly, two analysis were conducted under each method by either letting each method select the same number of DE genes as SAM with a control of FDR at 0.05; or allowing each method

TABLE 15.5:   Number of Genes Shared between 253 DE
Genes Selected by Two Methods

	CBF	BF	SAM	SHB	LIMMA	EBarrays
CBF	253	229	223	194	180	178
BF	229	253	226	185	180	162
SAM	223	226	253	196	173	177
SHB	194	185	196	253	171	214
LIMMA	180	180	173	171	253	156
EBarrays	178	162	177	214	156	253

have their own cut-off values.

Since SAM selects 253 DE genes with the FDR controlled at 0.05, we first
let each method detect 253 DE genes. A summary of the results are shown in
Table 15.5, which uses the diagonal elements to record the numbers of genes
claimed to be DE by each method, and the off-diagonal element to record
the number of commonly claimed DE genes by both methods recorded by
the relevant row and the relevant column. Since all methods select 253 DE
genes, the diagonal entries all share a value of 253. Note that the calibrated
Bayes factor approach, the ordered Bayes factor approach, and SAM make
very similar decisions. Any two methods of these three methods select at least
223 genes to be DE in common, and all three methods detect 212 DE genes
in common. We also note that SHB and EBarrays detect 214 genes to be DE
in common.

In the second analysis, we detected DE genes by using different cut-off value
for different methods. Specifically, we controlled the type I error at 0.01 for
the calibrated method, γ_0 at 0.8 for the two-criterion, false discovery rate to
be no more than 0.05 for SAM, SHB and LIMMA, and posterior odds to be
larger than 1 for EBarrays. The summary of the analysis results is presented
by Table 15.6 in the same form as Table 15.5. Table 15.6 shows that the
calibrated Bayes factor approach selects 292 genes, LIMMA selects 295 genes,
and EBarrays selects 294 genes to be DE, while SHB selects 247 genes and
SAM selects 253 genes, to be DE. The same 237 genes are selected to be DE

TABLE 15.6:   Number of DE Genes Selected by Two Methods
with Error Control

	CBF	Two-Crit.	SAM	SHB	LIMMA	EBarrays
CBF	292	233	243	204	207	216
Two-Crit.	233	262	209	167	198	169
SAM	243	209	253	191	189	198
SHB	204	167	191	247	182	237
LIMMA	207	198	189	182	295	188
EBarrays	216	169	198	237	188	294

by both SHB and EBarrays. The calibrated Bayes factor approach, SAM and the two-criterion make similar decisions and they select 205 genes as DE genes in common. Note that the gene list identified by the calibrated Bayes factor approach overlaps more with another paired method compared to all other methods. Therefore the genes selected by the calibrated Bayes factor approach have larger chances to be picked by other methods, and are expected to be true DE genes with larger probabilities.

15.7   Discussion

In this chapter, we considered seven methods for detecting DE genes across two biological conditions. SAM is the only approach proposed in the frequentist framework. The calibrated Bayes factor approach is very attractive and flexible in evaluating evidence for a gene to be DE as it allows us to compare not only two population means but also two population distributions. When the means and variances change in different directions across two conditions, the calibrated Bayes factor approach performs better than all other considered methods based on the simulation I. SHB shows great performance when the means and standard deviations of intensities change in the same direction across conditions. Compared the results obtained from two simulation settings in Table 15.1 and Table 15.3, 24 more genes are correctly declared as DE by SAM under Simulation II compared to that under Simulation I. So there is a larger difference in the numbers of correctly claimed DE genes by SAM under these two simulation settings compared to all other methods except SHB. Note that almost the same number of genes are correctly claimed by EBarrays under the two simulation settings.

Sometimes gene expression data can have very complex characteristics. We discuss two such types of data and the related literature.

15.7.1   Bimodal Data

Chen *et al.* (2003) published a typical dataset, in which cDNA samples were collected to study gastric cancer. More specifically, sample mRNA and reference mRNA were first labeled with Cy5-dUTP and Cy3-dUTP (Amersham, Piscataway, NJ), respectively. Then the two labeled cDNA samples were filtered, mixed and hybridized to microarray. The gene expression is measured by the logarithm of the red to green channel $log(R/G)$. The dataset contains 90 tumor samples and 22 normal samples on 6688 genes. The objective of the project is to determine which genes are DE between the tumor samples and the normal samples. Figure 15.2 includes two sub-figures, which draw the non-parametric density estimates of $log(R/G)$ for two selected genes respectively

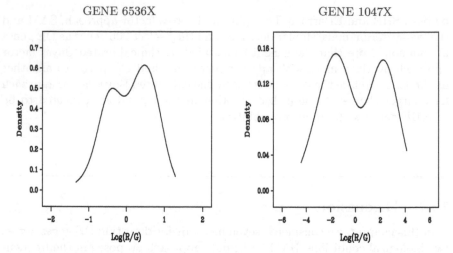

FIGURE 15.2:   Densities of gene expressions for selected genes.

in the tumor sample group. The horizontal axis in Figure 15.2 measures the value of $\log(R/G)$. The vertical axis measures the density value. All 90 tumors sampled are used for the estimator. The distributions of the gene expression drawn in Figure 15.2 are shown to have bimodality.

Chen *et al.* (2008) considered such data and proposed a bimodal normal model for the log-scale intensities for each group on each gene. As discussed in Chen *et al.* (2008), the bimodality may be due to the fact that certain genes are expressed for certain subjects and not expressed for other subjects. Since the gene can be either expressed or not expressed under the same biological condition, a mixture distribution with two components is fit on the log-scale gene intensities under each biological condition. Let x_{kgj} denote the log-scale gene intensity for the jth subject and g^{th} gene under condition k. Mathematically, the distribution of the intensity x_{kgj} are written as:

$$p(x_{kgj}|\alpha_{kg}, \tau_{kg}^2, \mu_{kg}, \sigma_{kg}^2) = p_{kg}p_1(x_{kgj}|\alpha_{kg}, \tau_{kg}^2) + (1 - p_{kg})p_2(x_{kgj}|\mu_{kg}, \sigma_{kg}^2),$$
$$(15.18)$$

where p_{kg} represents the probability that the gth gene is not expressed for condition k, $p_1(x_{kgj}|\alpha_{kg}, \tau_{kg}^2)$ and $p_2(x_{kgj}|\mu_{kg}, \sigma_{kg}^2)$ are the probability density functions of x_{kgj} for the unexpressed and expressed genes, respectively. Assume that $h(x_{kgj})$ is a known differentiable transformation of x_{kgj} to achieve normality, the probability density functions $p_1(x_{kgj}|\alpha_{kg}, \tau_{kg}^2)$ and $p_2(x_{kgj}|\mu_{kg}, \sigma_{kg}^2)$ are written, respectively, as $p_1(x_{kgj}|\alpha_{kg}, \tau_{kg}^2) = (2\pi)^{-1/2}|h'(x_{kgj})| \tau_{kg}^{-1}$ $\exp\left\{-\frac{1}{2\tau_{kg}^2}(h(x_{kgj}) - \alpha_{kg})^2\right\}$ and $p_2(x_{kgj}|\mu_{kg}, \sigma_{kg}^2) = (2\pi)^{-1/2}|h'(x_{kgj})|\sigma_{kg}^{-1}$ $\exp\left\{-\frac{1}{2\sigma_{kg}^2}(h(x_{kgj}) - \mu_{kg})^2\right\}$. In Chen *et al.* (2008), prior elicitation is carried out using empirical Bayes methodology due to the complexity of the mixture model.

15.7.2   Time Course Data

Time course microarray data is another type of frequently studied microarray data. Different from the above dataset, the gene expression levels in the time course data are observed at different time points on each gene. The goal of time course data analysis is to identify DE genes over time among different biological conditions. The time course data adds complexity to the study due to its time dependency of the gene expression measurements over time and its unbalanced design, which is caused by missing of follow-up at some time points. There is an extensive literature on time course data analysis. In the frequentist framework, Park *et al.* (2003) proposed to fit ANOVA model with both time and biological conditions as factors on the log-scale intensities. Then they applied a permutation test on the residuals of the ANOVA model with only time effects to identify DE genes among different biological conditions.

In the Bayesian framework, Storey *et al.* (2005) proposed a Bayesian hierarchical model based on basis in the linear space. Denote x_{klgj} as the expression level of gene g obtained on individual j under condition k, at time point l observed at time t_{klj}, for $g = 1, \cdots, G$, $k = 1, 2, l = 1, \cdots, L$, and $j = 1, \cdots, n_{kg}$. Storey *et al.* (2005) modeled the expression values by $x_{klgj} = \mu_g(t_{klj}) + \gamma_{kgj} + \epsilon_{klgj}$. Here $\mu_g(t_{klj})$ measures the population average time curve, with individual j deviated by γ_{kgj}. The γ_{kgj} and ϵ_{klgj} are assumed to have gene dependent normal distributions with zero mean and their own variances. The population time curve $\mu_g(t_{klj})$ is written in a linear form with $\mu_g(t_{klj}) = \alpha_g + \boldsymbol{\beta}'_g \times \mathbf{s}(t_{klj})$. The $\mathbf{s}(t_{klj})$ is a pre-specified vector acting like the basis in the linear space. Tests are conducted on whether every element of $\boldsymbol{\beta}'_g$ is zero for each gene. The evidence of the gene being DE is evaluated using an analogue of F test statistics based on the sum square of the residuals from models with or without the time effect. Chi *et al.* (2007) proposed another Bayesian hierarchical model for time course data to characterize gene expression patterns with time and experimental factors, and simultaneously account for the correlation among the gene expressions within the same subject. Denote $\mathbf{x}_{kgj}$ as a vector including all m_{kgj} log-scale expression levels of gene g obtained on individual j under condition k for $g = 1, \cdots, G$, $k = 1, 2$, and $j = 1, \cdots, n_k$. The vector size m_{kgj} is the total number of replications observed over time on individual j and gene g under condition k. Let $\mathbf{z}_{kgj}$ be the design matrix with dimension $m_{kgj} \times p$ including both baseline and time-varying covariates. Chi *et al.* (2007) fit the vector of gene expressions $\mathbf{x}_{kgj}$ via $\mathbf{x}_{kgj} = \mathbf{z}_{kgj}\boldsymbol{\beta}_{kg} + \gamma_{kj}\mathbf{1}_{m_{kgj}} + \boldsymbol{\epsilon}_{kgj}$. The $\mathbf{1}_{m_{kgj}}$ is a vector of m_{kgj} ones. The coefficient γ_{kj} represents correlation between gene expression levels within the same subject j in the kth biological condition, and $\gamma_{kj} \overset{iid}{\sim} \mathcal{N}(0, \tau^2)$. The vector $\boldsymbol{\epsilon}_{kgj} \overset{ind}{\sim} \mathcal{N}_{m_{kgj}}(\mathbf{0}, \sigma_g^2 \mathbf{I}_{m_{kgj}})$, here $\mathbf{I}_{m_{kgj}}$ is an $m_{kgj} \times m_{kgj}$ identity matrix. Denote $\hat{\beta}_{kgl}$ as the estimator for the lth covariate defined by β_{kg}. Chi *et al.* (2007) calculated and used $\max(\frac{\hat{\beta}_{kgl}}{\hat{\beta}_{k^*gl}}; k \neq k^* = 1, \cdots, p)$ to evaluate the differences in gene expression levels on gene g corresponding to the lth covariate between the biological conditions k and k^*. The two-criterion is applied in Chi *et al.* (2007) to identify DE genes.

Note that the methods discussed in this chapter can be used or be extended to the analysis of the bimodal data or time-course data. SAM, SHB, LIMMA, and EBarrays have been implemented in R and they are available for download from Web sites http://www-stat.stanford.edu/~tibs/SAM, http://www.stat.wisc.edu/~newton/research/arrays.html and http://www.bio conductor.org/packages/devel/Software.html, respectively. Currently, the R package for the calibrated Bayes factor approach is under development and it will be available to public in the near future.

References

Allison, D. B., Gadbury, G. L., Heo, M., Fernández, J. R., Lee, C.-K., Prolla, T. A., and Weindruch, R. (2002), A mixture model approach for the analysis of microarray gene expression data. *Computational Statistics & Data Analysis*, **39**, 1-20.

Baldi, P. and Long, A. D. (2001), A Bayesian framework for the analysis of microarray expression data: regularized t-test and statistical inferences of gene changes. *Bioinformatics*, **17(6)**, 509-519.

Benjamini, Y. and Hochberg, Y. (1995), Controlling the false discovery rate: a practical and powerful approach to multiple testing. *Journal of the Royal Statistical Society B*, **57**, 289-300.

Benjamini, Y. and Hochberg, Y. (2000), On the adaptive control of the false discovery rate in multiple testing with independent statistics. *Journal of Educational and Behavioral Statistics*, **25**, 60-83.

Chen, M.-H. and Ibrahim, J. G. (2003), Conjugate priors for generalized linear models. *Statistica Sinica*, **13(2)**, 461-476.

Chen, M.-H., Ibrahim, J. G., and Chi, Y.-Y. (2008), A new class of mixture models for differential expression in DNA microarray data. *Journal of Statistical Planning and Inference*, **138**, 387-404.

Chen, M.-H., Shao, Q.-M., and Ibrahim, J. G. (2000), *Monte Carlo Methods in Bayesian Computation*, New York: Springer-Verlag.

Chen, X., Leung, S., Yuen, S. T., Chu, K.-M., Ji, J., Li, R., Chan, S. Y., Law, S., Troyanskaya, O. G., Wong, J., Samuel, S., Botstein, D., and Brown, P. O. (2003), Variation in gene expression patterns in human gastric cancers. *Molecular Biology of the Cell*, **14**, 3208-3215.

Chen, Y., Dougherty, E. R., and Bittner, M. L. (1997), Ratio-based decisions and the quantitative analysis of cDNA micaroarray images. *Journal of Biomedical Optics*, **2**, 364-374.

Chi, Y.-Y., Ibrahim, J. G., Bissahoyo, A., and Threadgill, D. W. (2007), Bayesian hierarchical modeling for time course microarray experiments. *Biometrics*, **63**, 496-504.

DeRisi, J. L., Iyer, V. R., and Brown, P. O. (1997), Exploring the metabolic and genetic control of gene expression on a genomic scale. *Science*, **278**, 680-686.

Draghici, S. (2002), Statistical intelligence: effective analysis of high-density microarray data. *Drug Discovery Today*, **7**, S55-S63.

Dudoit, S., Shaffer, J., and Boldrick, J. C. (2003), Multiple hypothesis testing in microarray experiments. *Statistical Science*, **18**, 71-103.

Dudoit, S., van der Laan, M., and Pollard, K. (2004), Multiple testing. part I. Single-step procedures for control of general type I error rate. *Statistical Applications in Genetics and Molecular Biology*, **3**, No.1, Article 13.

Dudoit, S., Yang, Y. H., Callow, M. J., and Speed, T. P. (2002), Statistical methods for identifying differentially expressed genes in replicated cDNA microarray experiments. *Statistica Sinica*, **12**, 111-139.

Efron, B. (2004), Large-scale simultaneous hypothesis testing: the choice of a null hypothesis. *Journal of the American Statistical Association*, **99**, 96-104.

Efron, B. (2005), Local false discovery rates. *Technical Report*, Stanford, CA: Department of Statistics, Stanford University.

Efron, B. (2007a), Correlation and large-scale simultaneous significance testing. *Journal of the American Statistical Association*, **103**, 93-103.

Efron, B. (2007b), Size, power, and false discovery rates. *Annal of Statistics*, **35(4)**, 1351-1377.

Efron, B. and Tibshirani, R. (2002), Empirical bayes methods and false discovery rates for microarrays. *Genetic Epidemiology*, **23**, 70-86.

Efron, B., Tibshirani, R., Storey, J., and Tusher, V. (2001), Empirical Bayes analysis of a microarray experiment. *Journal of the American Statistical Association*, **96**, 1151-1160.

Ghosh, D. (2004), Mixture models for assessing differential expression in complex tissues using microarray data. *Bioinformatics*, **20(11)**, 1663-1669.

Gottardo, R., Raftery, A. E., Yeung, K. Y., and Bumgarner, R. E. (2006a), Robust estimation of cDNA microarray intensities with replicates. *Journal of the American Statistical Association*, **101**, 30-40.

Gottardo, R., Raftery, A. E., Yeung, K. Y., and Bumgarner, R. E. (2006b), Bayesian robust inference for differential gene expression in cDNA microarrays with multiple samples. *Biometrics*, **62**, 10-18.

Huber, W., von Heydebreck, A., Sültmann, H., Poustka, A., and Vingron, M. (2002), Variance stabilization applied to microarray data calibration and to the quantification of differential expression. *Bioinformatics*, **1**, 1-9.

Ibrahim, J. G., Chen, M.-H., and Gray, R. J., (2002), Bayesian models for gene expression with DNA microarray data. *Journal of the American Statistical Association*, **97**, 88-99.

Jain, N., Thatte, J., Braciale, T., Ley, K., O'Connell, M., and Lee, J. K. (2003), Local-poolederror test for identifying differentially expressed genes with a small number of replicated microarrays. *Bioinformatics*, **19**, 1945-1951.

Kalajzic, I., Staale, A., Yang, W.-P., Wu, Y., Johnson, S. E., Feyen, J. H. M., Krueger, W., Maye, P., Yu., F., Zhao, Y., Kuo, L., Gupta, R. R., Achenie, L. E. K., Wang, H.-W. Shin, D.-G., and Rowe, D. W. (2005), Expression profile of osteoblast lineage at defined stages of differentiation. *Journal of Biological Chemistry*, **280 (26)**, 24618-24626.

Kendziorski, C. M., Newton, M. A., Lan, H., and Gould, M. N. (2003), On parametric empirical Bayes methods for comparing multiple groups using replicated gene expression profiles. *Statistics in Medicine*, **22**, 3899-3914.

Kerr, M. K., Martin, M., and Churchill, G. A. (2000), Analysis of variance for gene expression microarray data. *Journal of Computational Biology*, **7**, 819-837.

Kuo, L., Yu, F., and Zhao, Y. (2008), Statistical methods for identifying differentially expressed genes in replicated microarray experiments: A review in *Statistical Advances in the Biomedical Sciences* (Biswas, A. Data, S., Fine, J. and Segal, M. eds.), Wiley, 341-363.

Lee, M. T. (2004), *Analysis of Microarray Gene Expression Data*, Kluwer Academic, Boston.

Lee, M. T., Kuo, F. C., Whitmore, G. A., and Sklar, J. (2000), Importance of replication in microarray gene expression studies: statistical methods and evidence from repetitive cDNA hybridizations. *Proceedings of the National Academy of Sciences USA*, **97**, 9834-9839.

Liu, D.-M., Parmigiani, G., and Caffo, B. (2004), Screening for differentially expressed genes: are multilevel models helpful? *Johns Hopkins University Tech Report*.

Lönnstedt, I. and Speed, T. (2002), Replicated microarray data. *Statistica Sinica*, **12**, 31-46.

McLachlan, G. J., Bean, R. W., and Jones, L. B. (2006), A simple implementation of a normal mixture approach to differential gene expression in multiclass microarrays. *Bioinformatics*, **22**, 1608-1615.

Newton, M. A., Kendziorski, C. M., Richmond, C. S., Blattner, F. R., and Tsui, K. W. (2001), On differential variability of expression ratios: improving statistical inference about gene expression changes from microarray data. *Journal of Computational Biology*, **8**, 37-52.

Newton, M. A., Noueiry, A., Sarkar, D., and Ahlquist, P. (2004), Detecting differential gene expression with a semiparametric hierarchical mixture method. *Biostatistics*, **5**, 155-176.

Pan, W. (2002), A comparative review of statistical methods for discovering differentially expressed genes in replicated microarray experiments. *Bioinformatics*, **18**, 546-554.

Pan, W., Lin, J., and Le, C. (2002), How many replicates of arrays are required to detect gene expression changes in microarray experiments? A mixture model approach. *Genome Biology*, **3(5)**, 0022.1-0022.10.

Parmigiani, G., Garrett, E. S., Irizarry, R. A., and Zeger, S. L. (2003), *The Analysis of Gene Expression Data: An Overview of Methods and Software* (Ed.), New York: Springer-Verlag.

Park, T., Yi, S.-G., Lee, S., Lee, S.-Y., Yoo, D.-H., Ahn, J.-I., and Lee, Y.-S. (2003), Statistical tests for identifying differentially expressed genes in time-course microarray experiments. *Bioinformatics*, **19(6)**, 694-703.

Rocke, D. M. and Durbin, B. (2001), A model for measurement error for gene expression arrays. *Journal of Computational Biology*, **8**, 557-570.

Schena, M., Shalon, D., Heller, R., Chai, A., Brown, P. O., and Davis, R. W. (1996), Parallel human genome analysis: microarray-based expression monitoring of 1000 genes. *Proceedings of the National Academy of Sciences USA*, **93**, 10614-9.

Smyth, G. K. (2004), Linear models and empirical Bayes methods for assessing differential expression in microarray experiments. *Statistical Applications in Genetics and Molecular Biology*, **3**, No. 1, Article 3.

Speed, T. (2003), *Statistical Analysis of Gene Expression Microarray Data* (Ed.), Chapman & Hall/CRC, Boca Raton.

Storey, J. D. (2002), A direct approach to false discovery rates. *Journal of the Royal Statistical Society, Series B*, **64**, 479-498.

Storey, J. D. and Tibshirani, R. (2003), SAM thresholding and false discovery rates for detecting differential gene expression in DNA microarrays. In *The Analysis of Gene Expression Data: Methods and Software* (Parmigani, G., Garrett, E. S., Irizarry, R. A. and Zeger, S. L. eds.), Springer, New York.

Storey, J., Xiao, W., Leek, J. T., Dai, J. Y., Tompkins, R. G., and Davis, R. W. (2005), Significance analysis of time course microarray experiments. *Proceedings of the National Academy of Sciences USA*, **102**, 12837-12842.

Tusher, V. G., Tibshirani, R., and Chu, G. (2001), Significance analysis of microarrays applied to the ionizing radiation response. *Proceedings of the National Academy of Sciences USA*, **98(9)**, 5116-5121.

van der Laan, M., Dudoit, S., and Pollard, K. (2004a), Multiple testing part II. step-down procedures for control of the family-wise error rate. *Statistical Applications in Genetics and Molecular Biology*, **3**, No.1, Article 14.

van der Laan, M., Dudoit, S., and Pollard, K. (2004b), Augmentation procedures for control of the generalized family-wise error rate and tail probabilities for the proportion of false positives. *Statistical Applications in Genetics and Molecular Biology*, **3**, No.1, Article 15.

West, M., Nevins, J. R., Marks, J. R., Spang, R., Blanchette, C., and Zuzan, H. (2000), DNA microarray data analysis and regression modeling for genetic expression profiling. *Institute Statistics and Decision Sciences Working Paper, #15*.

Westfall, P. H. and Young, S. S. (1993), *Resampling-Based Multiple Testing: Examples and Methods for P-Value Adjustment*, Wiley, New York.

Wit, E. and McClure, J. (2004), *Statistics for Microarrays : Design, Analysis, and Inference.* John Wiley & Sons, The Atrium, England.

Wolfinger, R. D., Gibson, G., Wolfinger, E. D., Bennett, L., Hamadeh, H., Bushel, P., Afshari, C., and Paules, R. S. (2001), Assessing gene significance from cDNA microarray expression data via mixed models. *Journal of Computational Biology*, **8(6)**, 625-637.

Wright, G. W. and Simon, R. M. (2003), A random variance model for detection of differential gene expression in small microarray experiments. *Bioinformatics*, **19**, 2448-2455.

Yu, F., Chen, M.-H., and Kuo, L. (2008), Detecting differentially expressed genes using calibrated Bayes factors. *Statistica Sinica*, **18**, 783-802.

Chapter 16

Bayes and Empirical Bayes Methods for Spotted Microarray Data Analysis

Dabao Zhang

Department of Statistics, Purdue University

16.1 Introduction

With a two-color competitive hybridization process, spotted microarrays provide a genome-wide measure of differential gene expression levels in two or more samples as well as controlling for undesirable effects (Schena *et al.*;[24] Brown and Botstein[3]). While the microarray experiment is powerful in profiling thousands or even millions of genes simultaneously, it holds challenges in technology as presenting reliable reading for each gene, and in statistics as extracting useful information and essentially making sensible conclusions from the enormous amount of data.

The microarray technology has been rapidly evolving and advancing, with the endeavor of a better platform to study gene expression, and therefore understand functions of genes in biological activities. Spotted microarrays, privileged with easily customized designs, allow researchers to spot project-specific probes (e.g., oligonucleotides, cDNA, and small fragments of PCR products corresponding to mRNAs) on the array surface. The most popular spotted microarrays are designed with two channels where cDNAs from the two samples are labeled with two different fluorophores, i.e., Cy3 and Cy5, respectively. In two-channel spotted microarray experiments, the cDNA samples labeled with two dyes are mixed together and then hybridized to an identical microarray to compare the gene expressions under two different conditions.

In this chapter, we focus on statistical analysis of data from two-channel spotted microarray experiments. Typically, statistical issues arise in experimental design (Yang and Speed;[31] Kerr and Churchill;[14] Kerr and Churchill,[15]) image analysis (Yang *et al.*,[28]) normalization and inference for differential expression on a probe-by-probe basis (Chen *et al.*;[4] Kepler *et al.*;[13] Kerr *et al.*;[16] Finkelstein *et al.*;[10] Newton *et al.*;[19] Tseng *et al.*;[26] Wolfinger *et al.*;[27] Yang *et al.*;[30] Shadt *et al.*;[23] Yang *et al.*;[29] Dudoit *et al.*,[7]) and finally, the information on regulatory network of genes (Friedman *et al.*[11]; Pe'er *et al.*[21]). While each issue has been addressed by statisticians, difficulties in background correction and normalization remain the major barriers for analysts to set up routine statistical analyses. The lack of reliable statistical protocol may endanger the validity of subsequent analyses. We suggest taking the multiplicative background correction (MBC) approach proposed by Zhang *et al.*[34], and then normalize the microarray data via an empirical Bayesian inference with a semi-parametric measurement-error model (Zhang *et al.*[32]).

The MBC is a simple but effective solution to a perplexing problem in spotted microarray data analysis. As a conventional approach, the additive background correction (ABC) directly subtracts the background intensities from the foreground ones, which may provide unreliable estimates of the differential gene expression levels, especially for those probes with overwhelming background noises. As foreground intensities of many probes marginally dominate the background intensities, the ABC approach presents $M - A$ plots with "fishtails" or "fans". As a result, it is preferred to ignore the background noise, which will increase the number of false positives. On the basis of the more realistic multiplicative assumption, the MBC minimizes the skewness of the intensity by logarithmically transforming the intensity readings before background correction. This approach not only precludes the "fishtails" and "fans" in the $M - A$ plots, but also provides highly reproducible background-corrected intensities for both strongly and weakly expressed genes.

Intensity-dependent normalization is commonly employed to remove systematic measurement errors occurred in the differential expression ratios. To remove the patterns in M-A plots, conventional normalization takes account of measurement errors in the differential expression ratio but ignores measurement errors in the total intensity. We propose a measurement-error model for intensity-dependent normalization that accounts for the measurement errors in both total intensities and differential expression ratios. A Bayesian inference for the measurement-error model will provide intensity-dependent normalization and identification of differentially expressed genes in the case of single-array experiment.

If multiple microarrays are available to identify differentially expressed genes, the Bayesian normalization provides the posterior means to estimate the differential gene expression levels within each array. To identify differentially expressed genes with the multiple microarrays, we need to analyze high dimensional data and simultaneously make inference on many parameters. Shrinkage and thresholding methods have been quite successful in improving various component-wise estimators. Zhang *et al.*[35] developed a class of generalized thresholding estimators, which are adaptive to both sparsity and asymmetry of the parameter space. The proposed estimators have bounded shrinkage property under a slightly broader condition than the one proposed by Johnstone and Silverman.[12] An empirical Bayes construction of the generalized thresholding estimators is suggested to estimate the differential gene expression levels and identify differentially expressed genes.

Chromatin immunoprecipitation coupled with DNA microarray (ChIP-chip) is a powerful technique to scan the whole genome for binding sites of specific proteins of interest. However, the high-dimensionality of ChIP-chip data challenges the statistical modeling and the downstream identification of binding sites. Reducing ChIP-chip data is equivalent to estimating nonnegative location parameters in high-dimensional yet sparse spaces. Generalized thresholding estimators are developed via setting the lower thresholds at $-\infty$ and optimizing upper thresholds on the basis of empirical Bayes approaches. The priors are chosen to account for the special properties of ChIP-chip data, i.e., nonnegativity of all location parameters and sparsity of these parameters. Unlike naive extensions of classical thresholding estimators which need pre-specifying certain parameters, the empirical Bayes-based estimators provide data-driven hyperparameters, a characteristic appealing to reducing ChIP-chip data. These estimators' adaptability to sparsity and asymmetry of parameter spaces provides effective reduction of ChIP-chip data.

16.2   Multiplicative Background Correction

Spotted microarrays are quantified into image files, with each pixel segmented into either the spotted or unspotted regions. As shown in Figure 16.1, for each probe on the array surface, pixel intensities of the spotted region are summarized into the median feature intensities for the two channels (traditionally red and green), say R_f and G_f, while pixel intensities of the unspotted region are summarized into the median background intensities, say R_b and G_b. The feature intensities R_f and G_f measure the fluorescence intensities caused by specific hybridization of the mRNA samples to the spotted probe, the background intensities R_b and G_b measure the fluorescence intensities of the background noise. The goal of background correction is to correct the foreground intensities for the background noise within the spotted region. With R_b and G_b being estimates of the background noise within R_f and G_f, it is appropriate to correct R_f and G_f with R_b and G_b, respectively.

While empirical observations reveal a nonadditive relationship between the foreground and background intensities (Brown *et al.*[2]), the principles of fluorescence spectroscopy indicate that the background noise affects the feature intensities multiplicatively. Therefore, instead of subtracting the background in-

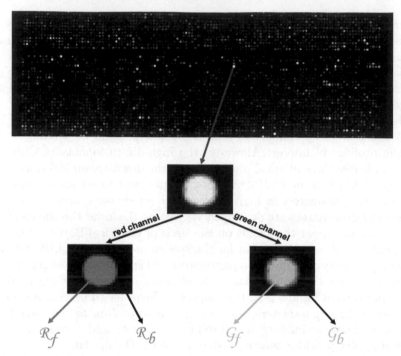

FIGURE 16.1: Image analysis of microarray experiment. Each probe will provide at least four readings: a pair of statistics R_f and R_b indicating the feature and background intensities in the red channel, and a pair of summary statistics G_f and G_b indicating the feature and background intensities in the green channel, respectively.

tensities directly from the feature intensities as ABC, we propose MBC to subtract the estimates of the logarithmic background intensities from the logarithmic feature intensities. First, because the median is invariant to any monotonic transformation, we suggest summarizing the feature (or background) intensities of a probe with the median of the corresponding spotted (or unspotted) region for each channel, i.e., R_f (or R_b) for the traditional red channel and G_f (or G_b) for the traditional green channel. Second, we assume that the logarithmic intensity medians $r_f = \log_2(R_f)$ and $g_f = \log_2(G_f)$ have additive noise effects $\tilde{r}_b = \log_2(\tilde{R}_b)$ and $\tilde{g}_b = \log_2(\tilde{G}_b)$, respectively, where $\tilde{R}_b$ and $\tilde{G}_b$ are the unobservable median background intensities of the spotted region for the two distinct channels. Third, it is reasonable to estimate $\tilde{r}_b$ by $r_b = \log_2(R_b)$ and $\tilde{g}_b$ by $g_b = \log_2(G_b)$. Consequently, the background-corrected logarithmic spot intensities are calculated as $r = r_f - r_b$ and $g = g_f - g_b$. Therefore, the *log-ratio* M (the logarithm of the ratio between the background-corrected spot intensities) and *log-intensity* A (the average of the logarithmic background-corrected spot intensities) are calculated as follows,

$$M = r - g = log_2 \left(\frac{R_f/R_b}{G_f/G_b} \right),$$

$$A = \frac{r+g}{2} = log_2 \sqrt{\frac{R_f}{R_b} \times \frac{G_f}{G_b}}.$$

The immediate advantage of MBC is that the background-corrected log-ratio and log-intensity for each spot are well defined. In addition, the logarithmic transformation roughly symmetrizes the background intensities, since background intensities are skewed to the right (Kim *et al.*[17]) and are approximately distributed lognormally (Zhang *et al.*[34]). Obviously, for genes expressed weakly or not at all, we have background-corrected logarithmic spot intensities fluctuating around zero due to the facts that the paired r_f and r_b, g_f and g_b have similar values and that both r and g will stay close to zero even before any normalization. Negative A implies that the corresponding gene is expressed weakly or not at all in both mRNA samples.

Assuming the feature intensities are affected additively by the background noise, the conventional additive background correction (ABC) directly subtract R_b and G_b from R_f and G_f, respectively. For genes expressed very weakly (or not at all) in either channel, the feature intensities of the corresponding probes are overwhelmed by the background intensities. Henceforth, missing log-ratios and missing log-intensities will be reported for such spots in all downstream analyses. The differential expression information of these genes is essentially lost by discarding non-positive background-corrected spot intensities, even though these genes may be highly differentially expressed or their expression patterns may change significantly across time. Furthermore, those probes with small positive background-corrected intensities usually have unreliable estimates of M and A, which manifest as "fishtails" or "fans" in M-A plots and challenge the subsequent analyses.

16.3 Bayesian Normalization

16.3.1 Measurement-Error Models

Normalization intends to remove effects of systematic measurement errors in a microarray experiment. The two most important effects are the dye effect and the print-tip effect. The dye effect comes from the labeling efficiencies and scanning properties of the two fluors, and it may be complicated by the use of different scanner settings. When two identical mRNA samples are labeled with different dyes and hybridized to the same slide in a self-self experiment, the red intensities often tend to be lower than the green intensities even though there is no differential expression and it is expected that the red and green intensities are equal (Smyth *et al.*[25]). Furthermore, the imbalance between the red and green intensities is usually not constant across the spots within and between arrays, and can vary according to the overall spot intensity, location on the array, slide origin, and other variables. The print-tip effect originates from the differences between the print-tips on the array printer and variations over the course of the print run.

Although normalizing microarray data is partially processed in the image analysis of the data by removing background noises, the resultant data of image analysis need to be further normalized to remove other systematic measurement errors, such as the dye and print-tip effects. Normalization of differential gene expression data usually refers to the normalization after image analysis of the data and assumes that some genes, such as housekeeping genes, are not differentially expressed. Global normalization by adjusting all differential gene expression values with a constant for each array is considered and may be incorporated into the model to identify differentially expressed genes (Chen *et al.*;[4] Newton *et al.*;[19] Kerr *et al.*[16]).

Dudoit *et al.*[7] treated the normalization as a procedure independent of identifying differentially expressed genes. Instead of global normalization, they consider within-print-tip-group local normalization (intensity-dependent normalization). Observing the obvious nonlinear relationship between the log-differential expression ratio M and the log-intensity A for each print-tip group, Dudoit *et al.* (2002) suggest to use a LOWESS (LOcally WEighted Scatterplot Smoothing) approach (Cleveland;[5] Cleveland and Devlin[6]) for each print-tip group to remove the effect of log-intensity A from M. The residuals of M are therefore employed to identify differentially expressed genes by possibly using models proposed by Chen *et al.*,[4] Newton *et al.*,[19] Efron *et al.*,[8] or Lönnstedt and Speed.[18] As the LOWESS smoother is available in many statistical packages, the ideas of Dudoit *et al.*[7] can be easily implemented. In addition, the two-step procedure for taking normalization and identification separately has been commonly accepted although the results are biased due to ignoring measurement errors in log-intensity A.

For the j-th replication of the i-th gene, assume $\log_2 R_{ij} = r_{ij} + \epsilon_{rij}$ and $\log_2 G_{ij} = g_{ij} + \epsilon_{gij}$, where r_{ij} and g_{ij} are the ideal log-intensities for individual channels, ϵ_{rij} and ϵ_{gij} represent the measurement errors. We further assume that ϵ_{rij} and ϵ_{gij} are normally distributed with $var(\epsilon_{rij}) = var(\epsilon_{gij})$. Therefore, we have $A_{ij} = (r_{ij} + g_{ij})/2 + \xi_{ij}$ and $M_{ij} = r_{ij} - g_{ij} + \epsilon_{ij}$, where $\xi_{ij} = (\epsilon_{rij} + \epsilon_{gij})/2$ is independent of $\epsilon_{ij} = \epsilon_{rij} - \epsilon_{gij}$ even though ϵ_{rij} and ϵ_{gij} may be correlated.

With the true abundance $\zeta_{ij} = (r_{ij} + g_{ij})/2$, a measurement-error model for normalizing a specific array can be taken as, following the intensity-dependent normalization by Dudoit *et al.*,[7]

$$\begin{cases} M_{ij} = h(\zeta_{ij}) + \gamma_i + \epsilon_{ij}, \\ A_{ij} = \zeta_{ij} + \xi_{ij}, \end{cases} \tag{16.1}$$

where the measurement errors ϵ_{ij} and ξ_{ij} are assumed to be independently and identically distributed as $N(0, \sigma_\epsilon^2)$ and $N(0, \sigma_\epsilon^2/4)$, respectively. Note that we can preprocess the array by averaging out the differences between print-tip groups.

A likelihood approach to estimate the gene effects in the above measurement error model is difficult since the dimension of parameter space is in the magnitude of thousands and $h(\cdot)$ is an unknown nonlinear function. We therefore develop a Bayesian framework to fit the measurement error model (16.1). Bayesian approach is a natural choice since it not only enables us to consider the above measurement error model by using some appropriate prior distributions, but also enables us to efficiently draw valid conclusions from small number of samples relative to the large number of parameters.

16.3.2 Bayesian Framework

Model (16.1) is a semiparametric measurement-error model. We use penalized splines (or simply P-splines) approach to approximate the unknown nonparametric function $h(\cdot)$. See Eilers and Marx[9] for a discussion on the benefits of this method. A convenient basis (the truncated power basis), with given knots $(t_1, t_2, \cdots, t_\kappa)$, is chosen as

$$B(x) = (1, x, x^2, \cdots, x^d, (x - t_1)_+^d, \cdots, (x - t_\kappa)_+^d)^T.$$

If $h(\cdot)$ is a smooth function, with appropriately chosen knots $(t_1, \cdots, t_\kappa)$, spline degree d, and coefficients $\beta = (\beta_0, \beta_1, \cdots, \beta_{d+\kappa})^T$, $B(x)^T \beta$ can approximate $h(x)$ well enough to ignore the estimation error. Therefore, we can focus on developing a Bayesian inference procedure for the following model,

$$\begin{cases} M_{ij} = B(\zeta_{ij})^T \beta + \gamma_i + \epsilon_{ij}, & \epsilon_{ij} \sim N(0, \sigma_\epsilon^2), \\ A_{ij} = \zeta_{ij} + \xi_{ij}, & \xi_{ij} \sim N(0, \sigma_\epsilon^2/4), \end{cases} \tag{16.2}$$

where $1 \leq j \leq m_i$, and $1 \leq i \leq n$. Full Bayesian inference on nonparametric measurement-error models was considered by Berry *et al.*[1] Here, we will establish the full Bayesian inference on our proposed semiparametric measurement-error model to simultaneously normalize the microarray data and identify the differentially expressed genes.

Prior Distributions: Let $\beta = (\beta_1^T, \beta_2^T)^T$, where β_1 includes the first $d + 1$ coefficients for the polynomial part and β_2 includes the last κ coefficients for the non-polynomial part of the spline regression. We therefore set up the prior distributions,

$$\beta_1 \sim N\big(0_{(d+1)\times(d+1)}, \lambda_1^{-1} I_{(d+1)\times(d+1)}\big),$$
$$\beta_2 \sim N\big(0_{\kappa\times\kappa}, \lambda_2^{-1} I_{\kappa\times\kappa}\big),$$

where $\lambda_1 = \alpha_1/\sigma_\epsilon^2$, and $\lambda_2 = \alpha_2/\sigma_\epsilon^2$. This prior is related to a penalized least squares (or penalized likelihood) estimation with the penalty term as

$\frac{\alpha_1}{N}\beta_1^T\beta_1 + \frac{\alpha_2}{N}\beta_2^T\beta_2$ (N is the total number of observations). For Bayesian inference with nonparametric measurement-error models, Berry et al.[1] suggest a diffuse prior on the polynomial coefficients β_1 by setting $\alpha_1 = 0$, however, it performs better to use a non-diffuse prior on β_1 for the semiparametric measurement-error models based on our experience in microarray studies.

Note that γ_i is the effect of the i-th gene and our primary interest is to test whether an individual gene effect is significantly different from zero, i.e., the following hypothesis,

$$H_0 : \gamma_i = 0, \quad vs. \quad H_1 : \gamma_i \neq 0,$$

for each i. Since most genes are not differentially expressed, we therefore assume H_0 can be true with a nonzero probability $1 - w$, i.e., a priori $P(\gamma_i \neq 0) = 1 - P(\gamma_i = 0) = w$. Although an appropriate heavy-tailed prior for nonzero γ_i may make the Bayes estimator of γ_i to be minimax, we assume that it follows a normal distribution for convenience. Hence, the distribution of γ_i is assumed to be a mixture of a point mass at zero and a normal distribution,

$$\gamma_i \overset{indep}{\sim} (1-w)\delta_0 + wN(0, \sigma_\gamma^2), \tag{16.3}$$

where δ_0 is the Dirac delta function at zero. This prior is previously used by Lönnstedt and Speed[18] to construct an empirical Bayes approach to identify differentially expressed genes. Apparently, γ_i corresponding to each housekeeping gene should be set to zero.

Since the true abundance ζ is the location parameter of the distribution of the observed abundance, the observed abundance A_{ij} is distributed as normal with mean ζ_{ij} under the usual normality assumption of the measurement errors. We therefore suggest using Jeffrey's reference prior for ζ_{ij}, i.e., $\pi(\zeta_{ij}) \propto 1$, which is an improper non-informative prior. Although Berry et al.[1] suggest a fixed proper prior for this uncontaminated latent variable, the true abundance here varies too widely to use a fixed prior.

For simplicity, conjugate priors are selected for parameters w and σ_ϵ^2, and hyperparameters σ_γ^2, λ_1, λ_2 and φ_s, $s = 1, 2, \cdots, S$. For example, the prior of w is set as $Beta(\theta_w, \phi_w)$, and the priors of σ_ϵ^{-2}, σ_γ^{-2}, λ_1, λ_2 and φ_s are set as Gamma distributions with parameters $(\theta_\epsilon, \phi_\epsilon)$, $(\theta_\gamma, \phi_\gamma)$, $(\theta_{1\lambda}, \phi_{1\lambda})$, $(\theta_{2\lambda}, \phi_{2\lambda})$ and $(\theta_{s\varphi}, \phi_{s\varphi})$, $s = 1, 2, \cdots, S$, respectively. Without loss of generality, $Gamma(0, \infty)$ is defined as the improper density function of positive τ proportional to $1/\tau$, which is Jeffrey's reference prior for the variance of a normal distribution.

Implementation of the Gibbs Sampler: Although only w and $\gamma = (\gamma_1, \cdots, \gamma_n)^T$ are of our primary interest, their posterior distributions are difficult to calculate due to many nuisance parameters involved. With the available full conditional distributions, we therefore use the Gibbs sampling algorithm. All parameters and hyperparameters, except ζ_{ij}, have full conditional distributions. Therefore, it is straightforward to get random draws of these parameters (except ζ_{ij}) from their full conditional distributions. However, the

Metropolis-Hastings algorithm nested in Gibbs samplers can be developed to get random draws of ζ_{ij} from its full conditional distribution.

Setting the number of knots, κ, to be at most 8 usually provides an excellent approximation of cubic spline functions (*i.e.*, $d = 3$) to the nonlinear function h in microarray studies. The knots can then be chosen at equally spaced quantiles of the observed log-intensities A_{ij}.

While it is feasible to choose $\theta_w = \phi_w$ to be either 0.5 or even smaller values, all other unspecified hyperparameters can be set as $\theta_\epsilon = \theta_\gamma = \theta_{1\lambda} = \theta_{2\lambda} = \theta_{j\varphi} = 0$, and $\phi_\epsilon = \phi_\gamma = \phi_{1\lambda} = \phi_{2\lambda} = \phi_{j\varphi} = \infty$ by using non-informative priors for the corresponding parameters, $j = 1, 2, \cdots, S$. The hyperparameter values can be adjusted accordingly whenever prior information is available.

Choosing $\alpha_1 = 1$ and $\alpha_2 = 0.1$, we can then roughly estimate β, γ, σ_ϵ^2 and σ_γ^2 by assuming no measurement errors on log-intensities A_{ij}. We suggest using this estimation to set up the initial values $\beta^{(0)}, \gamma^{(0)}, \eta^{(0)}, \sigma_\epsilon^{2(0)}, \sigma_\gamma^{2(0)}$ and $\lambda_1^{(0)} = \alpha_1/\sigma_\epsilon^{2(0)}$, $\lambda_2^{(0)} = \alpha_2/\sigma_\epsilon^{2(0)}$. The initial value $\varphi^{(0)}$ can be chosen based on $\eta^{(0)}$, and $p^{(0)}$ can be set as the proportion of $\gamma_i^{(0)}$ whose absolute values are larger than $3\sigma_\gamma^{(0)}$. Finally, the initial values of ζ_{ij} are set as $\zeta_{ij}^{(0)} = \max(0, A_{ij})$.

16.3.3 Identifying Differentially Expressed Genes

For each gene in model (16.2), one parameter is used to identify whether it is differentially expressed under different conditions. It is, for the i-th gene, the posterior probability $w_i = P(\gamma_i \neq 0|\mathbf{M}, \mathbf{A})$. Note that the above Gibbs sampler provides the stationary Markov chain for γ_i by sampling from

$$[\gamma_i|\beta, M, \eta, \zeta, \sigma_\epsilon^2, \sigma_\gamma^2, p] \sim \tilde{w}_i \delta_0 + (1 - \tilde{w}_i) N \left(\frac{\sum_{j=1}^{m_i}(M_{ij} - B_{ij}^T \beta)}{m_i + \sigma_\epsilon^2/\sigma_\gamma^2}, \frac{\sigma_\epsilon^2 \sigma_\gamma^2}{m_i \sigma_\gamma^2 + \sigma_\epsilon^2} \right),$$

where the weight $\tilde{w}_i$ is updated as

$$\tilde{w}_i = \frac{1 - p}{1 - p + p \left(1 + \frac{m_i \sigma_\gamma^2}{\sigma_\epsilon^2}\right)^{-1/2} \exp\left\{-\frac{1}{2} \frac{(\sum_{j=1}^{m_i}[M_{ij} - B(\zeta_{ij})^T \beta])^2}{m_i \sigma_\epsilon^2 + \sigma_\epsilon^4/\sigma_\gamma^2}\right\}}.$$

When identification of differentially expressed genes is concerned, the median value of the stationary γ_i chain serves as a natural estimate of γ_i. The chain of $\tilde{w}_i$ updated along the stationary chain of γ_i provides a natural estimate $\hat{w}_i$ of w_i, following

$$w_i = E[\tilde{w}_i|\mathbf{M}, \mathbf{A}].$$

Obviously, $1 - w_i$ is the local *false discovery rate* (FDR) at gene i in the sense of Efron *et al.*,[8] and $1 - \hat{w}_i$ is an estimate of this local FDR. Due to the measurement-error model (16.2), this local FDR is not a monotone function of γ_i. Within the Bayesian framework, we can estimate the FDR with the direct posterior probability approach suggested by Newton *et al.*,[20] that is,

$$\widehat{FDR}(\alpha) = \frac{\sum_{i=1}^n (1 - \hat{w}_i) \times I[\hat{w}_i \geq \alpha]}{\sum_{i=1}^n I[\hat{w}_i \geq \alpha]},$$

which is a monotone function of α. If none of the variables is selected at level α, the corresponding FDR is defined to be zero.

It is natural to identify differentially expressed genes on the basis of the Bayes factor, which, for gene i, is defined as

$$B_i = \frac{w_i}{1 - w_i}.$$

When choosing genes with the levels of Bayes factor higher than b, we can estimate the FDR as,

$$\widehat{FDR}(b) = \frac{\sum_{i=1}^n I[B_i \geq b]/(1 + B_i)}{\sum_{i=1}^n I[B_i \geq b]},$$

which is a decreasing function of b, and is set to zero if none of the genes is selected at level b. This calculation is valid for all $b \in (0, \infty)$, however, it is too liberal to use $b < 1$. Therefore, we suggest plotting $\widehat{FDR}(b)$ $(1 \leq b \leq 1000)$ to help control FDR.

16.4   Generalized Empirical Bayes Method for Multiple Arrays

When identifying differentially expressed genes using multiple microarrays, the estimate of γ_i for the i-th gene can be output from the above Bayesian normalization on each array. Averaging the estimated values across arrays, we can get a new set of data $\mathbf{Y}_p = (y_1, y_2, \cdots, y_p)$, with y_i representing the i-th gene. We then need to estimate a high-dimensional parameter $\boldsymbol{\mu} = (\mu_1, \mu_2, \cdots, \mu_p)$, assuming that

$$y_i - \mu_i \overset{iid}{\sim} \varphi(\cdot), \tag{16.4}$$

where $\varphi(\cdot)$ is a symmetric log-concave density function. Given that a large number of the components in $\boldsymbol{\mu}$ are zero, $\boldsymbol{\mu}$ resides in a relatively sparse subspace.

Johnstone and Silverman[12] proposed an empirical Bayes method that is adaptive to the high-dimensional sparse parameter space, and they also demonstrated some nice theoretical properties and numerical advantages of their estimator. Zhang *et al.*[35] introduced and developed a class of generalized shrinkage and thresholding estimators, which are adaptive to both sparsity and asymmetry of the parameter space. They used the Bayesian approach merely as a tool to place a measure on the sparse and asymmetric parameter space, and therefore construct better decision rules adaptive to the scenario at hand. The proposed estimators have the bounded shrinkage property under a slightly broader condition than the one given by Johnstone and Silverman[12]. An empirical Bayes construction is presented for estimating multivariate normal mean.

16.4.1 Generalized Empirical Bayes Estimators

Formally, as defined in Zhang *et al.*,[35] a function $\delta(y, \tau_-, \tau_+)$ is called a *generalized shrinkage estimator* if and only if (i) $\delta(y, \tau) = \delta(y, -\tau, \tau)$ is antisymmetric for any $\tau \geq 0$ (i.e., symmetric but with opposite sign); (ii) $\delta(y, \tau_-, \tau_+)$ is increasing on $y \in \mathbb{R}$ for each pair of non-negative (τ_-, τ_+); and (iii) $-|y| \leq \delta(y, \tau_-, \tau_+) \leq |y|$, for all $y \in \mathbb{R}$. The shrinkage estimator $\delta(y, \tau)$ is further called a *generalized thresholding estimator* with thresholds (τ_-, τ_+) if and only if

$$\delta(y, \tau_-, \tau_+) = 0 \Leftrightarrow \tau_- \leq y \leq \tau_+.$$

Zhang *et al.*[35] investigated construction of generalized shrinkage and thresholding estimators for $\boldsymbol{\mu}$ in model (16.4) using a Bayesian approach. With a unimodal and symmetric distribution function $\gamma(\cdot)$, we denote $\gamma_+(\mu) = 2\gamma(\mu)1_{[0,\infty)}(\mu)$ and $\gamma_-(\mu) = 2\gamma(\mu)1_{(-\infty,0]}(\mu)$. Consider a Bayesian estimator of $\boldsymbol{\mu}$ by assuming that the components of $\boldsymbol{\mu}$ have the following prior,

$$\mu_i \overset{iid}{\sim} (1 - w_- - w_+)\delta_0(\mu) + w_-\gamma_-(\mu) + w_+\gamma_+(\mu), \qquad (16.5)$$

where $\delta_0(\cdot)$ is Dirac's delta function. Here w_- and w_+ are the weights for the negative and positive parts with density distributions $\gamma_-(\mu)$ and $\gamma_+(\mu)$, respectively.

Given the model and prior specifications, we can obtain the posterior distribution of each parameter. Let $\tilde{w}_-$ and $\tilde{w}_+$ denote the posterior probabilities of μ_i being negative and positive respectively. Let f_- and f_+ denote the posterior conditional densities for the negative and positive parts respectively. Then the posterior distribution of the parameter μ_i, given the observed value y_i, is

$$\begin{aligned}\mu_i | y_i, w_-, w_+ \sim\ & \{1 - \tilde{w}_-(y_i; w_-, w_+) - \tilde{w}_+(y_i; w_-, w_+)\}\delta_0(\mu_i) \\ & + \tilde{w}_+(y_i; w_-, w_+)f_+(\mu_i | y_i; w_-, w_+) \\ & + \tilde{w}_-(y_i; w_-, w_+)f_-(\mu_i | y_i; w_-, w_+).\end{aligned}$$

For fixed w_- and w_+, a Bayesian estimator of μ_i (under componentwise absolute error loss) is its posterior median, i.e.,

$$\hat{\mu}(y_i; w_-, w_+) = \text{median}(\mu_i | y_i; w_-, w_+). \qquad (16.6)$$

Let $g_+(y_i) = \int_0^\infty \varphi(y_i - \mu)\gamma_+(\mu)d\mu$. Then define the constant $a = \frac{\varphi(0)}{g_+(0)+\varphi(0)} \in [0,1]$, and the simplex,

$$S(a) = \{(w_-, w_+) \in [0,1]^2 : (2a-1)w_- + w_+ \le a, w_- + (2a-1)w_+ \le a\}.$$

Then, Zhang et al.[35] proved the following theorem.

THEOREM 16.1
For the model (16.4), (i) $\hat\mu(y; w_-, w_+)$ is a generalized shrinkage estimator if and only if $(w_-, w_+) \in S(a)$; (ii) $\hat\mu(y; w_-, w_+)$ is a generalized thresholding estimator if and only if $(w_-, w_+) \in S(a)$.

This theorem states a sufficient and necessary condition for our proposed Bayesian estimator $\hat\mu(y; w_-, w_+)$ to be a generalized shrinkage/thresholding estimator, i.e., $(w_-, w_+) \in S(a)$. However, the antisymmetric Bayesian estimator developed by Johnstone and Silverman[12] essentially requires $(w_-, w_+) \in S(0)$, i.e., $w_- = w_+ \in [0,1]$. Obviously, the Bayesian estimator developed here gains more flexibility by offering a much larger space for (w_-, w_+).

We can construct the generalized shrinkage/thresholding estimators with a quasi-Cauchy prior, i.e., taking

$$\begin{cases} \gamma_+(\mu|\theta_+) = 2(\frac{1}{\theta_+} - 1)^{-1/2}\phi(\frac{\mu}{1/\theta_+ - 1})1_{[0,\infty)}(\mu), & \theta_+ \sim \text{Beta}(0.5, 1), \\ \gamma_-(\mu|\theta_-) = 2(\frac{1}{\theta_-} - 1)^{-1/2}\phi(\frac{\mu}{1/\theta_- - 1})1_{(-\infty,0]}(\mu), & \theta_- \sim \text{Beta}(0.5, 1), \end{cases} \quad (16.7)$$

or equivalently,

$$\begin{cases} \gamma_+(\mu) = \sqrt{\frac{2}{\pi}}\left(1 - \frac{\mu(1-\Phi(\mu))}{\phi(\mu)}\right)1_{[0,\infty)}(\mu), \\ \gamma_-(\mu) = \sqrt{\frac{2}{\pi}}\left(1 + \frac{\mu\Phi(\mu)}{\phi(\mu)}\right)1_{(-\infty,0]}(\mu), \end{cases}$$

which have tails similar to those of Cauchy densities, i.e., much heavier than Gaussian distribution as desired.

Assuming that $\varphi(\cdot) = \phi(\cdot)$ in model (16.4), we then have,

$$g_+(y_i) = \frac{1}{y_i^2\sqrt{2\pi}}\left(2\Phi(y_i) - \exp(-y_i^2/2) - \frac{2y_i\exp(-y_i^2/2)}{\sqrt{2\pi}}\right).$$

Since $\phi(0) = 1/\sqrt{2\pi}$ and $g_+(0) = \lim_{y\downarrow 0} g_+(y) = 1/\sqrt{8\pi}$, we have $a = 2/3$, and $S(a)$ is defined by

$$\begin{cases} w_+ + 3w_- \le 2 \\ 3w_+ + w_- \le 2. \end{cases} \quad (16.8)$$

Maximizing the marginal distribution of $\mathbf{Y}_p$ for $(w_-, w_+) \in S(2/3)$, we then construct a generalized thresholding estimator with the posterior median, which is essentially an empirical Bayes estimator.

16.4.2   Identifying Differentially Expressed Genes

With $(\hat{w}_-, \hat{w}_+) \in \mathcal{S}(2/3)$ maximizing the marginal distribution of $\mathbf{Y}_p$, we can calculate the posterior probabilities $\tilde{w}_-(y_i; \hat{w}_-, \hat{w}_+)$ and $\tilde{w}_+(y_i; \hat{w}_-, \hat{w}_+)$. The local *false discovery rate* at gene i in the sense of Efron *et al.*[8] can be estimated by $1 - \tilde{w}_-(y_i; \hat{w}_-, \hat{w}_+) - \tilde{w}_+(y_i; \hat{w}_-, \hat{w}_+)$, which is a monotone function of y_i.

Let $\hat{w}_i = \tilde{w}_-(y_i; \hat{w}_-, \hat{w}_+) + \tilde{w}_+(y_i; \hat{w}_-, \hat{w}_+)$. Then we can estimate the FDR using the method by Newton *et al.*[20] and detect differentially expressed genes with Bayes factor using similar procedure described in Section 16.3.3.

16.4.3   Reducing ChIP-Chip Data

The ChIP-chip experiment has been described in detail by Ren *et al.*[22] In summary, after cross-linked to the protein, the DNA fragments are enriched by immunoprecipitation, namely IP-enriched. Then both the IP-enriched and unenriched DNA samples are amplified using ligation-mediated-polymerase chain reaction (LM-PCR), and dyed with Cy5 and Cy3 respectively before hybridizing to DNA microarray. Finally, the ratio of fluorescence intensities between these two samples are utilized to identify the genomic binding sites for the protein of interest. Because of its high-throughput nature, ChIP-chip data share some similarities with regular microarray data, such as high dimensionality. In addition, a unique property of ChIP-chip data is that there are only two possibilities for each DNA fragment, it is either enriched or not. Such characteristic motivated the development of data reduction method in order to identify binding sites more effectively and efficiently.

For each arrayed element on the microarrays, its precise genomic location is known and a summary statistic is calculated as the difference between the logarithmic intensities of the IP-enriched and unenriched channels. Let y_i be the summary statistic for the i-th arrayed element. With appropriate normalization and transformation, we can assume the model (16.4). In a perfect ChIP-chip experiment, $\mu_i > 0$ if the i-th arrayed element corresponds to a binding site, and $\mu_i = 0$ otherwise. However, technical limitations may result in $\mu_i > 0$ at certain arrayed elements that are not related to any binding sites. Nonetheless, when the whole genome tiling arrays are employed for ChIP-chip experiments, the high-throughput biotechnology presents high-dimensional data $\mathbf{Y}_p = (y_1, y_2, \cdots, y_p)$ with the mean vector $\boldsymbol{\mu} = (\mu_1, \mu_2, \cdots, \mu_p)$ having sparse non-zero components, i.e., $\boldsymbol{\mu}$ lies in a high-dimensional but sparse parameter space. This sparse parameter space is characterized by asymmetry since all the nonzero components of $\boldsymbol{\mu}$ are positive.

A well-understood distribution of μ_i in model (16.4) certainly helps identifying the binding sites. However, as shown by Zheng *et al.*,[36] it is computationally intensive to implement the corresponding algorithm if a fine probability model is developed for the ChIP-chip data. Henceforth we propose to use the above generalized empirical Bayes estimator to effectively reduce the dimension of ChIP-chip data by efficiently estimating all non-zero μ_i before further models are utilized to identify the binding sites of interest (Zhang and Zhang[33]).

Specifically, we assume the prior of μ in (16.5) with $w_- = 0$, where $\gamma_+(\mu)$ is specified in (16.7). As the simplex $\mathcal{S}(a)$ defined in (16.8) reduces to $w_+ \leq 2/3$, we propose to maximize the marginal likelihood function for an optimal $\hat{w}_+ \in [0, 2/3]$, and therefore construct an empirical Bayes estimator $\hat{\mu}(y_i; 0, \hat{w}_+)$, which is a generalized shrinkage/thresholding estimator. Here the optimal $\hat{w}_+$ describes the degree of sparsity in μ, i.e., the degree of sparsity for the underlying parameter space. In practice, σ^2 can be estimated independently either through control probes or certain robust estimators by taking advantage of the characteristics of ChIP-chip data (i.e., most probes are not related to binding sites).

16.5　Software Packages

16.5.1　MicroBayes

Package MicroBayes can be downloaded at

http://www.stat.purdue.edu/~zhangdb/MicroBayes/

and run under MATLAB®. It implements the Bayesian inference on spotted microarray data, proposed by Zhang *et al.*[32] It also provides the multiple background correction of Zhang *et al.*[34] The local FDR will be reported for each gene, and the $\widehat{FDR}(b)$ will also be calculated for different levels of b. The main function is

MATLAB>> MicroBayes(config).

Here the input argument config is an array containing the following structures to specify the input file and its data structure,

DataFile: the name of the input file;

bBGCorrect: 1 if MBC should be applied, 0 otherwise;

GIDIdx: indicates the column of gene ID;

MIdx: indicates the column of M;

AIdx: indicates the column of A;

Cy3FgIdx: indicates the foreground reading column of Cy3 channel;

Cy3BgIdx: indicates the background reading column of Cy3 channel;

`Cy5FgIdx`: indicates the foreground reading column of Cy5 channel;

`Cy5BgIdx`: indicates the background reading column of Cy5 channel;

Note that a structure may not be specified if the input file does not include the corresponding records. The input data should be put in a plain text file, with each row representing one probe.

The input argument of `MicroBayes` can also be used to specify other optional settings, with default ones specified in the function `MBDefault`. For example, the structure `nMCMCBurnIns` sets the number of burn-in samples, and `nMCMCSamples` sets the number of samples for the Gibbs sampler after the burn-in period, both with default values being set at 5,000. The structure `ResPrefix` of `config`, together with the name of the input file, specifies the folder name containing parameter chains from the Gibbs sampler.

A text file will be created including the final report on each gene, the estimates of γ_i, Bayes factor, and local FDR. A plot will be presented for $\widehat{FDR}(b)$ against b.

16.5.2 GEBCauchy

Package `GEBCauchy` can be downloaded at

http://www.stat.purdue.edu/ ~zhangdb/GEBCauchy/

and run under MATLAB®. It implements the generalized empirical Bayes estimator, proposed by Zhang *et al.*[35] The local FDR will be reported for each gene, and the $\widehat{FDR}(b)$ will also be calculated for different levels of b. The main function is

MATLAB>> retRes = GEBCauchy(inX,inSD,inPatt).

Here the input argument `inX` should be a column vector including the data $\mathbf{Y}_p$, and `inSD` specifies the standard deviation with the default value calculated using the median absolute deviation (MAD). The input argument `inPatt` specifies the possible patterns of (w_-, w_+),

[-1,1,1]: $(w_-, w_+) \in \mathcal{S}(2/3)$ as in (16.8);

[1,1,1]: $w_- = w_+ \in [0, 0.5]$;

[1,0,1]: $w_- = w_+ = 0.5$;

[0,1,1]: $w_- = 0$ and $w_+ \in [0, 2/3]$;

[-1,1,0]: $w_+ = 0$ and $w_- \in [0, 2/3]$.

The output `retRes` is a structure containing the following structures,

mu: a vector of the estimator $\hat{\mu}$;

sd: the estimator of $|\sigma|$;

priorp: the estimator of (w_-, w_+);

postp: a matrix of $\left(\tilde{w}_-(y_i; \hat{w}_-, \hat{w}_+), \tilde{w}_+(y_i; \hat{w}_-, \hat{w}_+) \right)$;

A plot will be presented for $\widehat{FDR}(b)$ against b. A text file will be created including the final report on each gene, as well as the estimates of γ_i, Bayes factor, and local FDR.

References

[1] S. M. Berry, R. J. Carroll, and D. Ruppert. Bayesian smoothing and regression splines for measurement error problems. *Journal of the American Statistical Association*, 97:160–169, 2002.

[2] C. S. Brown, P. C. Goodwin, and P. K. Sorger. Image metrics in the statistical analysis of DNA microarray data. *Proceedings of the National Academy of Sciences of the United States of America*, 98:8944–8949, 2001.

[3] P. O. Brown and D. Botstein. Exploring the new world of the genome with DNA microarrays. *Nature Genetics*, 21:33–37, 1999.

[4] Y. Chen, E. R. Dougherty, and M. L. Bittner. Ratio based decisions and the quantitative analysis of cDNA microarray images. *Journal of Biomedical Optics*, 2:364–374, 1997.

[5] W. S. Cleveland. Robust locally weighted regression and smoothing scatterplots. *Journal of the American Statistical Association*, 74:829–836, 1979.

[6] W. S. Cleveland and S. J. Devlin. Locally weighted regression: An approach to regression analysis by local fitting. *Journal of the American Statistical Association*, 83:596–610, 1988.

[7] S. Dudoit, Y. H. Yang, M. J. Callow, and T. P. Speed. Statistical methods for identifying differentially expressed genes in replicated cDNA microarray experiments. *Statistica Sinica*, 12:111–139, 2002.

[8] B. Efron, R. Tibshirani, J. D. Storey, and V. Tusher. Empirical Bayes analysis of a microarray experiment. *Journal of the American Statistical Association*, 96:1151–1160, 2001.

[9] P. H. C. Eilers and B. D. Marx. Flexible smoothing with B-splines and penalties (with discussion). *Statistical Science*, 11:89–102, 2001.

[10] D. B. Finkelstein, J. Gollub, R. Ewing, F. Sterky, S. Somerville, and J. M. Cherry. Iterative linear regression by sector. Methods of Microarray Data Analysis. Papers from CAMDA 2000 (S. M. Lin and K. F. Johnson, eds), Kluwer Academic, 2001.

[11] N. Friedman, M. Linial, I. Nachman, and D. Pe'er. Using Bayesian networks to analyze expression data. *Journal of Computational Biology*, 7:601–620, 2000.

[12] I. M. Johnstone and B. W. Silverman. Needles and straw in haystacks: empirical Bayes estimates of possibly sparse sequence. *The Annals of Statistics*, 32:1594–1649, 2004.

[13] T. B. Kepler, L. Crosby, and K. T. Morgan. Normalization and analysis of DNA microarray data by self-consistency and local regression. Santa Fe Institute Working Paper, Santa Fe, New Mexico, 2000.

[14] M. K. Kerr and G. A. Churchill. Experimental design for gene expression microarrays. *Biostatistics*, 2:183–201, 2001.

[15] M. K. Kerr and G. A. Churchill. Statistical design and the analysis of gene expression microarrays. *Genetics Research*, 77:123–128, 2001.

[16] M. K. Kerr, M. Martin, and G. A. Churchill. Analysis of variance for gene expression microarray data. *Journal of Computational Biology*, 7:819–837, 2000.

[17] J. H. Kim, D. M. Shin, and Y. S. Lee. Effect of local background intensities in the normalization of cDNA microarray data with a skewed expression profiles. *Experimental and Molecular Medicine*, 34:224–232, 2002.

[18] I. Lonnstedt and T. Speed. Replicated microarray data. *Statistica Sinica*, 12:31–46, 2002.

[19] M. Newton, C. Kendziorski, C. Richmond, F. Blattner, and K. Tsui. On differential variability of expression ratios: improving statistical inference about gene expression changes from microarray data. *Journal of Computational Biology*, 8:37–52, 2001.

[20] M. Newton, A. Noueiry, D. Sarkar, and P. Ahlquist. Detecting differential gene expression with a semiparametric hierarchical mixture model. *Biostatistics*, 5:155–176, 2004.

[21] D. Pe'er, A. Regev, G. Elidan, and N. Friedman. Inferring subnetworks from perturbed expression profiles. *Bioinformatics*, 17, S1:S215–224, 2001.

[22] R. Ren, F. Robert, J. J. Wyrick, O. Aparicio, E. G. Jennings, I. Simon, J. Zeitlinger, J. Schreiber, N. Hannett, E. Kanin, T. L. Volkert, C. J. Wilson, S. P. Bell, and R. A. Young. Genome-wide location and function of DNA binding proteins. *Science*, 290:2306–2309, 2000.

[23] E. E. Schadt, C. Li, B. Ellis, and W. H. Wong. Feature extraction and normalization algorithms for high-density oligonucleotide gene expression array data. *Journal of Cellular Biochemistry*, Suppl 37:120–125, 2002.

[24] M. Schena, D. Shalon, R. W. Davis, and P. O. Brown. Quantitative monitoring of gene expression patterns with complementary DNA microarray. *Science*, 270:467–470, 1995.

[25] G. K. Smyth, Y. H. Yang, and T. P. Speed. Statistical issues in cDNA microarray data analysis. *Functional Genomics: Methods and Protocols* (M.J. Brownstein and A.B. Khodursky, eds.), Humana Press, Totowa, NJ, 2002.

[26] G. C. Tseng, M.-K. Oh, L. Rohlin, J. C. Liao, and W. H. Wong. Issues in cDNA microarray analysis: quality filtering, channel normalization, models of variations and assessment of gene effects. *Nucleic Acids Research*, 29:2549–2557, 2001.

[27] R. D. Wolfinger, G. Gibson, E. D. Wolfinger, L. Bennett, H. Hamadeh, P. Bushel, C. Afshari, and R. S. Paules. Assessing gene significance from cDNA microarray expression data via mixed models. *Journal of Computational Biology*, 8:625–637, 2001.

[28] Y. H. Yang, M. J. Buckley, S. Dudoit, and T. P. Speed. Comparison of methods for image analysis on cDNA microarray data. http://www.stat.Berkeley.EDU/users/terry/zarray/Html/image.html, 2000.

[29] Y. H. Yang, S. Dudoit, P. Luu, D. M. Lin, V. Peng, J. Ngain, and T. P. Speed. Normalization of cDNA microarray data: a robust composite method addressing single and multiple slide systematic variation. *Nucleic Acids Research*, 30 (4):e15, 2002.

[30] Y. H. Yang, S. Dudoit, P. Luu, and T. P. Speed. Normalization for cDNA microarray data. Microarrays: Optical Technologies and Informatics, Volume 4266 of *Proceedings of SPIE* (M.L. Bittner, Y. Chen, A.N. Dorsel and E.R. Dougherty, eds.), 2001.

[31] Y. H. Yang and T. P. Speed. Design issues for cDNA microarray experiments. *Nature Genetics*, 3:579–588, 2002.

[32] D. Zhang, M. T. Wells, C. D. Smart, and W. E. Fry. Bayesian normalization and identification for differential gene expression data. *Journal of Computational Biology*, 12:391–406, 2005.

[33] D. Zhang and M. Zhang. Estimating nonnegative location parameters: with application to ChIP-chip data. Submitted, 2008.

[34] D. Zhang, M. Zhang, and M. T. Wells. Multiplicative background correction for spotted microarrays to improve reproducibility. *Genetical Research*, 87:195–206, 2006.

[35] M. Zhang, D. Zhang, and M. T. Wells. Generalized thresholding estimators for high-dimensional location parameters. *Statistica Sinica*, in press, 2009.

[36] M. Zheng, L. O. Barrera, B. Ren, and Y. N. Wu. ChIP-chip: data, model, and analysis. *Biometrics*, 63:787–796, 2007.

Chapter 17

Bayesian Classification Method for QTL Mapping

Min Zhang

Department of Statistics, Purdue University

17.1 Introduction

Variation in quantitative traits is due to the combination of genetic and non-genetic factors. Identifying such genetic factors by associating the trait variation with genotypic differences at certain loci is referred to as quantitative trait loci (QTL) mapping. In other words, the goal of QTL mapping is to identify regions of the chromosome that are associated with the phenotype of interest. More specifically, we would like to estimate the total number of QTL, their locations on the chromosomes, and the magnitude of their effects on the phenotypic value. It has been widely accepted that genetic markers can be used to study QTL (Sax 1923, Thoday 1960, Jayakar 1970), where such markers are identifiable and presumably closely linked to genes that affect the trait under investigation. QTL mapping holds the promise of locating genes underlying quantitative traits which help further investigate and manipulate the genes to improve the trait, and thus, it has attracted considerable attention in the research community recently.

17.1.1 Non-Bayesian Methods

Traditional non-Bayesian approaches have been prevalently employed in QTL mapping for the past two decades. Such approaches map QTL on the basis of univariate or multiple regression models, and maximum likelihood methods are used to detect QTL and estimate their effects. After the earlier years of single marker analyses using t-test and regression models, Lander and Botstein (1989) proposed one of the most popular approaches in QTL mapping, namely interval mapping (IM), where the chromosomes are densely gridded using available marker information and likelihood ratio tests are performed at each of these grid points to detect QTL. As an extension of the single-QTL model assumed in IM, Zeng (1993, 1994) proposed composite interval mapping (CIM) by incorporating the markers outside the interval that may also affect the trait of interest, and similarly, Jansen (1993) and Jansen and Stam (1994) introduced multiple QTL mapping (or marker-QTL-marker, MQM). By including multiple intervals simultaneously, Kao et al. (1999) presented multiple interval mapping (MIM) to directly regress the trait on a number of markers. MIM allows the investigation of epistasis (i.e., the interaction of genes or QTL) and provides the estimates of variance components as well as heritabilities. A detailed review on the background of QTL mapping and statistical issues involved is available from Doerge et al. (1997) and Jansen (2007). Here we will give a brief overview on MIM which will be used to provide locations of the pseudomarkers prior to mapping QTL with the newly developed Bayesian method.

On the basis of Cockerham's genetic model, Kao et al. (1999) proposed MIM to identify QTL from multiple intervals simultaneously via a multiple regression model that includes m pseudomarkers located in m different marker intervals, and the digenic epistasis between these m pseudomarkers. Maximum likelihood method is employed to make inference on the locations and effects of the detected QTL. More specifically, model selection procedures (e.g., stepwise selection, chunkwise selection) using likelihood ratio test (LRT) statistic are employed, and significance is evaluated by comparing the LRT statistic with the critical value from IM (or CIM) obtained after Bonferroni adjustment. MIM can estimate the number of significant QTL as well as their locations on the chromosomes, separate linked QTL, and identify significant epistasis between a pair of pseudomarkers (Kao et al. 1999). These information, combined with the estimated heritability and variance components, provide information for marker assisted selection to further improve the trait of interest.

17.1.2 Bayesian Methods

The utility of Bayesian approaches in QTL mapping has been recognized only recently, which is largely attributed to the increasing power of modern computers and the advent of Markov chain Monte Carlo (MCMC) methods (Shoemaker et al., 1999). In addition to the general characteristics of Bayesian statistical methods described in Berger (1985) and Gelman et al. (1995), the Bayesian framework offers specific advantages in QTL mapping, such as accommodating missing genotypic and phenotypic values and handling a large number of parameters in the model. Missing genotype information is common in QTL mapping, either due to failure to genotype or selective genotyping.

Using MCMC, Bayesian methods provide a natural framework to impute the missing genotypic value based on the available flanking markers at each iteration of the Markov chain. As a result of the development of technologies that enable fast and cost-effective genotyping of thousands of individuals and hundreds or thousands of markers per individual, the total number of markers on the genetic map is getting larger and larger. Furthermore, the desire to simultaneously modeling main and epistatic effects across the entire genome further challenge the statistical analysis of QTL data. Therefore, the ability to include large number of markers in the model, as provided by Bayesian methods, is essential for the success of QTL mapping.

Earlier considerations of Bayesian methods in QTL mapping include Thomas and Cortessis (1992), Hoeschele and VanRaden (1993a, 1993b), and later, a series of Bayesian approaches proposed to map QTL. Among them, some methods are developed to estimate the number of QTL as well as their effects on the trait by using reversible-jump MCMC algorithm (Green 1995), such as Satagopan et al., (1996), Sillanpää and Arjas (1998), Stephens and Fisch (1998), and Yi and Xu (2002). To further improve the performance of MCMC mixing, Yi et al. (2003) proposed a Bayesian approach to identify multiple QTL on the basis of stochastic search variable selection methodology (George and McCulloch 1993). This method considers all markers of the entire genome simultaneously, and a mixture of two normal distributions is specified for the effect of each marker. Both normal distributions are centered at zero with one variance being small and the other variance being large, in order to capture the small effects and large effects respectively (Yi et al. 2003).

17.2 Bayesian Classification to Map QTL

The existence of epistasis has been recognized for over a decade (Bateson, 1909), however, lack of efficient statistical tools is still the major barrier to identifying the epistatic effects. The statistical challenge is "large p small n", i.e., we have a huge number of predictors but relatively small number of subjects. As shown by Broman and Speed (2002), Kao et al. (1999), Zeng et al. (1999), and Ball (2001), QTL mapping can be considered as a model selection process. A number of methods have been proposed to identify epistatic effects, such as Kao and Zeng (2002), Yi and Xu (2002), Yi et al. (2005, 2007), Bogdan (2004), Cui and Wu (2005), Żak et al. (2007), and Shi et al. (2007), among others.

To break the curse of dimensionality for models with epistases in QTL mapping, we propose a new Bayesian framework with a two-stage procedure to simultaneously investigate the main effects of each candidate marker and the epistatic effects between all possible pairs of candidate markers across the whole genome. The candidate markers can be observed markers or pseudo-markers (e.g., imputed genotypes). Here and thereafter, we will use markers to represent both markers and pseudomarkers and assume that QTL are located at the markers. With the proposed Bayesian approach, all candidates are classified into three groups, the positive-effect group that includes all QTL with positive effects on the phenotype, the negative-effect group that includes all QTL with negative effects on the phenotype, and the zero-effect group that includes all QTL with negligible effects. Therefore, the method we proposed for QTL mapping is called Bayesian classification. Taking advantage of the Bayesian framework, a mixture-of-three-components prior distribution is specified for each candidate marker on the genome. More specifically, we use positive and negative truncated normal distributions for the positive-effect and negative-effect groups respectively, and the different parameters of these two normal distributions allow the asymmetric effects between the two groups in terms of the number of QTL and the sizes of their effects. After the prior information being updated by the observed data, the posterior probabilities will be used to infer the group membership of each marker. The inference is carried out using Bayesian framework via Gibbs sampling.

17.2.1 Model and Prior Specification

We will focus the discussion on continuous phenotypes in populations with binary genotypes, such as backcross, double haploid and recombinant inbred lines. With a multiple regression model including up to digenic epistasis, the relationship between phenotypic value (Y_i) and the genetic/non-genetic factors of the i-th individual can be written as

$$Y_i = \mu + \sum_{j=1}^{d} \alpha_j Z_{ij} + \sum_{j=1}^{m_b} \beta_j X_{ij} + \sum_{j=1}^{m_g} \gamma_j W_{ij} + \epsilon_i, \quad i = 1, \ldots, n. \quad (17.1)$$

where μ is the overall mean, Z_{ij} represents the j-th non-genetic covariate of subject i and α_j is its corresponding effect, X_{ij} is the genotypic value for the j-th marker of subject i, $W_{ij} = X_{ik}X_{il}$ with X_{ik} and X_{il} representing the genotypes of the k-th and the l-th marker of the i-th individual respectively, β_j is the additive effect of the j-th marker, γ_j represents the j-th epistasis between the k-th and l-th marker, and ϵ_i is the residual error which follows $N(0, \sigma_\epsilon^2)$. Finally, d is the total number of non-genetic factors, m_b and m_g are the total numbers of main and epistatic effects respectively. Note that, if all pairwise epistases are considered, we have $m_g = m_b \times (m_b - 1)/2$.

First, a mixture-of-three-component prior distribution is specified for the main effect of each marker and for the epistatic effect of each pair of markers as follows,

$$\beta_j \overset{iid}{\sim} (1 - p_{\beta+} - p_{\beta-})\delta_{\{0\}} + p_{\beta+}F_{\beta+} + p_{\beta-}F_{\beta-}$$

$$\gamma_j \overset{iid}{\sim} (1 - p_{\gamma+} - p_{\gamma-})\delta_{\{0\}} + p_{\gamma+}F_{\gamma+} + p_{\gamma-}F_{\gamma-},$$

where $p_{\beta+}$ and $p_{\beta-}$ are the probabilities for a marker to have positive and negative effect respectively, $\delta_{\{0\}}$ is a Dirac function which equals one at zero and equals zero otherwise, then $F_{\beta+}$ and $F_{\beta-}$ refer to the population distributions of the positive and negative groups. Similarly, $p_{\gamma+}$ and $p_{\gamma-}$ represent the probabilities for a pair of markers to have positive and negative epistatic effect respectively, and $F_{\gamma+}$ (or $F_{\gamma-}$) is the population distribution of the positive (or negative) epistasis. Let $N_+(m, s)$ and $N_-(m, s)$ denote the normal distribution with moments (m, s) restricted to the positive and negative half line respectively. We choose $F_{\beta+} = N_+(0, \sigma_{\beta+}^2)$ and $F_{\beta-} = N_-(0, \sigma_{\beta-}^2)$. Then $F_{\gamma+}$ and $F_{\gamma-}$ are defined in a similar manner as $F_{\beta+}$ and $F_{\beta-}$ by just replacing β with γ.

The priors for other parameters are selected as the standard non-informative priors for location and scale parameters, i.e., $\mu \propto 1$, and $\sigma_\epsilon^2 \propto 1/\sigma_\epsilon^2$. Given a set of non-genetic covariates $\mathbf{Z}$, the corresponding α_j can be partitioned into homogeneous groups with α_j from the k-th group, *a priori*, following *iid* normal distribution $N(0, \sigma_{\alpha k}^2)$. In addition, we have $\sigma_{\alpha k}^2 \propto 1/\sigma_{\alpha k}^2$.

For the hyperparameters, we choose the following priors,

$$p_{\beta+} + p_{\beta-} \sim \text{Uniform}(0, c\sqrt{n}/m_b),$$
$$p_{\gamma+} + p_{\gamma-} \sim \text{Uniform}(0, c\sqrt{n}/m_g), \tag{17.2}$$

and

$$\sigma_{\beta+}^2 \sim \text{Inverse Gamma}(\eta_{\beta+}, \zeta_{\beta+}), \quad \sigma_{\beta-}^2 \sim \text{Inverse Gamma}(\eta_{\beta-}, \zeta_{\beta-})$$
$$\sigma_{\gamma+}^2 \sim \text{Inverse Gamma}(\eta_{\gamma+}, \zeta_{\gamma+}), \quad \sigma_{\gamma-}^2 \sim \text{Inverse Gamma}(\eta_{\gamma-}, \zeta_{\gamma-}).$$

17.2.2 Bayesian Inference and the Two-Stage Procedure

On the basis of the observed data $(Y_i, \mathbf{X}_i, \mathbf{Z}_i)$ for $i = 1, \ldots, n$, and prior specifications for each parameter in model (17.1), the joint posterior density function of the parameters conditional on the observed data can be obtained. Then we can construct a Gibbs sampler to iteratively draw samples of the missing genotypic and/or phenotypic values, as well as the model parameters from their full conditional posterior distributions.

The starting value for each parameter is obtained as follows. The overall mean μ and σ_ϵ^2 are initialized with the sample mean and variance of the phenotypic values, the genetic effects of all markers $\{\beta_j : j = 1, \ldots, m_b\}$ and $\{\gamma_j : j = 1, \ldots, m_g\}$ are initialized using the values from the univariate regression, the initial values for variance components $(\sigma_{\beta+}^2, \sigma_{\beta-}^2, \sigma_{\gamma+}^2, \sigma_{\gamma-}^2)$ and prior probabilities of each marker and pair of markers having positive or negative effect $(p_{\beta+}, p_{\beta-}$ and $p_{\gamma+}, p_{\gamma-})$ are calculated based on the initial values of β_j and γ_j, $j = 1, \ldots, n$. Given these initial values, we update each of them (i.e., μ, α_j, β_j, γ_j, σ_ϵ^2, $p_{\beta+}$, $p_{\beta-}$, $p_{\gamma+}$, $p_{\gamma-}$, $\sigma_{\beta+}^2$, $\sigma_{\beta-}^2$, $\sigma_{\gamma+}^2$, $\sigma_{\gamma-}^2$) by sampling from its corresponding posterior distributions. Details of the Gibbs sampler are reported in Zhang et al. (2005) and Zhang et al. (2008).

Bayesian inference can be carried out by calculating the statistics using the marginal distributions of the underlying parameters. New samples will be obtained at each iteration of the Markov chain. After discarding the initial iterations (i.e., burn-in samples), the subsequent parameter samples will approximately follow the posterior distributions. Let K be the total number of iterations after the burn-in period, and the Bayesian inference can be made on the basis of the K samples from the Markov chain of each parameter. Let β_{-j} denotes all components in β except the j-th component (β_j). On the basis of the following probabilities,

$$\tilde{p}_{\beta j-} = P(\beta_j < 0 | \boldsymbol{Y}_n, \boldsymbol{X}_n, \boldsymbol{Z}_n, \mu, \beta_{-j}, \sigma_\epsilon^2, \sigma_{\beta+}^2, \sigma_{\beta-}^2),$$

$$\tilde{p}_{\beta j+} = P(\beta_j > 0 | \boldsymbol{Y}_n, \boldsymbol{Y}_n, \boldsymbol{Z}_n, \mu, \beta_{-j}, \sigma_\epsilon^2, \sigma_{\beta+}^2, \sigma_{\beta-}^2),$$

we can obtain two chains, i.e., $\{\tilde{p}_{\beta j-}^{(k)} : k = 1, \ldots, K\}$ and $\{\tilde{p}_{\beta j+}^{(k)} : k = 1, \ldots, K\}$, to check whether or not marker j has a significant effect (negative or positive) on the phenotype. More specifically, we can calculate the posterior probabilities $P(\beta_j > 0 | \boldsymbol{y}_n, \boldsymbol{x}_n, \boldsymbol{z}_n)$ and $P(\beta_j < 0 | \boldsymbol{y}_n, \boldsymbol{x}_n, \boldsymbol{z}_n)$ using these two chains. The posterior probabilities for epistatic effects can be estimated in a similar manner by simply replacing β with γ.

As the total number of predictors is much larger than that of individuals in model (17.1), we employ a two-stage procedure to improve the efficiency and accuracy of the inference. In the first stage, we screen all the main and epistatic effects in model (17.1), and select the top $c\sqrt{n}$ out of m_b additive effects and the top $c\sqrt{n}$ out of m_g epistatic effects using a restrictive prior distribution for each coefficient (for example, $c = 2$), as shown in (17.2). Then, with a much smaller number of candidates selected from the first stage, we rerun the same algorithm in the second stage and stochastically search the significant main and epistatic effects out of the pool of candidates chosen from the first stage. The only difference between the two stages is that, the priors used in the second stage are non-restrictive, i.e.,

$$p_{\alpha+} + p_{\alpha-} \sim \text{Uniform}(0, 1), \qquad p_{\beta+} + p_{\beta-} \sim \text{Uniform}(0, 1). \quad (17.3)$$

Let the posterior probabilities of marker j ($j = 1, \ldots, m_b$) having non-zero additive effect be $p_{\beta j+} = P(\beta_j > 0 | \boldsymbol{y}_n, \boldsymbol{x}_n, \boldsymbol{z}_n)$ and $p_{\beta j-} = P(\beta_j < 0 | \boldsymbol{y}_n, \boldsymbol{x}_n, \boldsymbol{z}_n)$. Similarly, we have $p_{\gamma j+} = P(\gamma_j > 0 | \boldsymbol{y}_n, \boldsymbol{x}_n, \boldsymbol{z}_n)$ and $p_{\gamma j-} = P(\gamma_j < 0 | \boldsymbol{y}_n, \boldsymbol{x}_n, \boldsymbol{z}_n)$ for the posterior probabilities of the j-th epistatic effect being significantly different from zero. After estimating these quantities using the samples drawn by Gibbs samplers in the second stage, we can calculate the Bayes factor (BF) to test whether each marker has statistically significant additive effect or involved in non-zero epistatic effect. More specifically,

$$BF(\beta_j) = \frac{p_{\beta j+} + p_{\beta j-}}{1 - p_{\beta j+} - p_{\beta j-}}, \quad \text{and} \quad BF(\gamma_j) = \frac{p_{\gamma j+} + p_{\gamma j-}}{1 - p_{\gamma j+} - p_{\gamma j-}}. \quad (17.4)$$

As proposed by Jeffreys (1961), "a bare mention" evidence is provided if BF is between 1 and 3.2, "substantial" evidence is provided if BF is between 3.2

and 10, "strong" evidence for BF between 10 and 100, and "decisive" evidence is provided if BF is greater than 100. On the basis of the two Bayes factors, we can select markers that have non-zero main and epistatic effects on the phenotype.

For each marker with non-zero effect, we can further estimate the magnitude of its effect on the basis of its corresponding Markov chain. Usually the median value of the chain is utilized to estimate the size of the main (or epistatic) effects.

17.3　QTLBayes: Software for QTL Mapping with Bayesian Classification

The Bayesian classification method for QTL mapping has been implemented in QTLBayes that allows simultaneous analysis of the additive and epistatic effects of all markers across the whole genome. QTLBayes can be used to analyze data from experimental crosses, and data from natural populations where inbreeding can not be controlled. For example, QTLBayes has been employed to analyze data from several collaborative projects, including barley and *Drosophila* (Zhang et al., 2005), tomato (with S. Tanksley, Cornell University), rice (with D. Salt, Purdue University), and human (with A. Clark, Cornell University).

17.3.1　Availability

The algorithms and functions have been developed and tested in MATLAB® for Windows and Unix. QTLBayes can be freely downloaded from http://www.stat.purdue.edu/~zhangdb/QTLBayes/. The example datasets and an information file about the package ('ReadMe.txt') are also provided at this web site.

17.3.2　Model, Bayesian Inference, and Algorithm

Considering the main effects of each marker, digenic epistasis between markers, and the effects from other non-genetic variables, QTLBayes maps QTL on the basis of the multiple regression model (17.1). As discussed in Section 17.2.1, the homogeneity among the non-genetic variables is acknowledged and can be built into the analysis with QTLBayes.

Bayesian classification method is used to estimate parameters in model (17.1). Details of the Gibbs sampling algorithm are described in Zhang et al. (2005).

17.3.3   Input Data and Format

One dataset is required if only the observed markers are included in model (17.1) and no missing genotypic values are involved. However, if one also wants to impute the genotypic values of pseudomarkers between flanking markers and include them in model (17.1), a separate genetic map file is necessary for the imputation. Both data files should be provided in text format (e.g., ".txt").

- "`config.CTMFile`": This is the name of the file which includes a matrix of dimension $n \times C$. Here n is the total number of subjects in the data and thus, each row collects the genetic and non-genetic information for each subject. The C columns include the genetic line index for each individual, the phenotypic value, the values for non-genetic covariates, and all the genotypic values of each marker. Note that the only requirement here is that each row is the data for each individual, and the columns can be organized freely as long as they match the specifications in Section 17.3.4. The genotypic values for binary markers can be represented with any two numbers, and they will be transformed into -0.5 or 0.5 in the algorithm to avoid correlation between main and epistatic effects. For F2 populations, the input of genotypic values can be any three consecutive numbers representing the three possible genotypes, i.e., gg, Gg, or GG, respectively. By default, the program will use -0.5, 0, and 0.5 to code the additive effect and use 0, 1, and 0 to code the dominant effect of gg, Gg, and GG respectively. Consequently, the corresponding epistatic effects, *additive × additive, additive × dominant, dominant × additive, dominant × dominant*, will be included in the model if the user would like to identify epistatic effects as well.

- "`config.MPosFile`": The name of the file which includes a column vector to specify the genetic map of all markers. This is required if there is missing genotypic value in the dataset or one would like to add pseudomarkers between flanking markers. In this file, the order of the markers should be the same as the order of the genotypic values in "`config.MrksIdx`". Note that QTLBays could not impute the missing genotypic values for F2 populations yet, so one needs to use other packages, such as `fastPHASE` (Scheet et al. 2006), to get the complete genotype information for F2 data.

17.3.4   Specifications and Default Setting

The current version only offers standardization of the trait variables or non-genetic covariates, so if other types of transformation is necessary, users need to transform the data first before running the program. In addition to the input data described in Section 17.3.3, users have the option to specify the following variables. The data structure is specified by the following structures of array `config`,

- "`config.ChLens`": The total number of markers on each chromosome. This is a row vector with dimension equals to the total number of chromosomes, and each component of this vector represents the number of markers on each chromosome ordered from the first to the last chromosome.

- "`config.LIDIdx`": The column index of genetic line ID for each observation. Observations from the same line should share an identical ID.

- "`config.TraitIdx`": The column index of the trait value for each individual.

- "`config.MrksIdx`": A row vector specifying the markers' column indexes in the file "`config.CTMFile`".

- "`config.CovIdx`": The column indexes of non-genetic variables in the file `config.CTMFile`. There is no non-genetic variables by default.

- "`config.ZGrps`": This specifies the homogeneous groups of non-genetic covariates. Users can group the similar covariates together by this specification. Each non-genetic covariate forms a homogeneous group by default.

- "`config.EpiFile`": The name of the file which specifies epistatic effects in the model. If only main effects are of interest, the user does not need to define it or let it be an empty character, i.e., `config.EpiFile = ''`;. If all pairwise interactions are considered, one can use `config.EpiFile = 'ALL'`. If only a subset of epistases are considered, they can be specified in the corresponding file which is a matrix with dimension equals to $m_g \times 2$. Each row is a pair of indexes indicating the two markers.

- "`config.bNormT`": This equals to 0 if no standardization of the phenotypic values is required, and equals to 1 if the phenotypic values need to be standardized. The default value is 0.

- "`config.bNormZ`": This equals to 1 if the non-genetic covariates need to be standardized and equals to 0 otherwise. The default value is 0.

- "`config.bF2`": This equals to 1 if the data are from F2 population, and 0 otherwise.

- "`config.nCstCov`": This specifies the cut-off value for highly unbalanced markers to stay in the model. For example, the two genotypes of a marker are coded as -0.5 and 0.5, if the genotypes of most individuals in the population are 0.5, then ``config.nCstCov'' = 4 means that at least 4 individuals need to have genotype -0.5 in order to keep this marker in the model. Otherwise, this marker will be removed from the model after the initial screening of all available markers in the dataset prior to analysis. The default value is 0.

- "config.nMCMCBurnIns": This specifies the total number of burn-in samples of the Markov chain. Since the chains usually converge after $5,000$ iterations based on our experience, the default value is $5,000$.

- "config.nMCMCSamples": This specifies the total number of samples of the Markov chain used to make inference. The default value is $5,000$. On the basis of our experience, we can keep all samples after the burn-in period and there is no thinning procedure necessary. Therefore, all samples after burn-in will be saved, however, users can still thin it later if necessary.

- "config.MapFunc": This specifies the mapping function used to impute the genotypic values for pseudomarkers. "config.MapFunc = 0" corresponds to Haldane's mapping function (1919), "config.MapFunc = 1" indicates Kosambi's mapping function (1944), and "config.MapFunc = 2" corresponds to Morgan mapping function. The default value is 1.

17.3.5　Output and Results

For each parameter in model (17.1), we have the corresponding Markov chains from the Gibbs sampler which consist of a reasonable large number of samples. The number specified in "config.nMCMCSamples" should be large enough such that there is a relative small Monte Carlo error while estimating the summary statistics, such as the posterior mean or median, the variance, and the posterior distribution etc. The algorithm will create two folders that save the Markov chains for each parameter generated from the first and second stages respectively. Results from the second stage will be used to estimate the parameters in model (17.1).

Before making inference on the basis of these Markov chains, users need to check the convergence of the chain by using diagnostic tools described in Cowles and Carlin (1996). There are some available packages specially designed for convergence assessment and posterior inference of MCMC output, such as boa in R (Smith 2007).

Users can obtain the results for QTL with significant main and epistatic effects on phenotype by using the two functions, "ReportMainEffects" and "ReportEpiEffects". The input for both functions is the value specified for thinning. For example, $thin = 5$ means that every 5-th sample will be used for estimation. On the basis of our experience, the estimates from different values of thinning are similar to those without thinning (i.e., $thin = 1$). The default value for both functions is 1. "ReportMainEffects" provides the information of the markers with significant main effects, which includes: the marker index, the point estimate of the magnitude of the effect as well as its standard deviation, and the corresponding Bayes factor. "**ReportEpiEffects**" reports similar information for the identified significant epistatic effects.

With the standardized phenotypic values and no non-genetic covariates in model (17.1), one can estimate the heritability by using $1 - \hat{\sigma}_\epsilon^2$ where $\hat{\sigma}_\epsilon^2$ can be obtained from the chain of σ_ϵ^2.

17.3.6 Summary

QTLBayes is still under continuous development. We are trying to develop a user friendly interface and adding other features to make it more widely applicable. Please contact Dabao Zhang (zhangdb@stat.purdue.edu) or Min Zhang (minzhang@stat.purdue.edu) for the most recent version.

17.4 An Example of Data Analysis Using QTLBayes

17.4.1 The Data

The example dataset is originally presented in Liu et al. (1996) and Zeng et al. (2000), and the phenotype as well as genotype information are made publicly available at the QTL Cartographer web site (http://statgen.nscu.edu /qtlcart/cartographer.html). In summary, four backcross populations were generated from two inbred lines, *Drosophila simulans* and *Drosophila mauritiana*. The phenotype is the morphometric of the posterior lobe and there are 45 markers across all 3 chromosomes. Liu et al. (1996) has more details about this dataset. Here we take the *mauritiana* backcross II data as an example.

17.4.2 QTL Analysis

First, we select 9 putative markers based on the results from multiple interval mapping (Zeng et al., 2000) and the genotypes of these pseudomarkers will be imputed on the basis of their flanking markers. Then, we put all 54 main effects (45 markers and 9 pseudomarkers), and 1431 epistasis (all pairwise interactions between 54 markers/pseudomarkers) into the model. The phenotype and genotype information of all 54 markers/pseudomarkers are organized into a matrix, where the genotypes of all 9 pseudomarkers are missing and indicated by 'NaN' in the dataset. The name of the file containing this matrix is specified in "**config.CTMFile**". The standardized phenotypic values are used in the analysis. The genetic map estimated from the gametes is available (Liu et al. 1996, Zeng et al. 2000) and the corresponding file name is specified by "**config.MPosFile**". We analyze the data by using the following command in MATLAB®,

TABLE 17.1:   Partial List of the QTL
with Main Effect BF>1 for BM2 Dataset

Index:	beta:	sd:	bf:
1	-0.2180	0.1448	4.5036e+015
15	-0.0561	0.1071	1.2878e+000
16	-0.1473	0.1161	1.7689e+003
25	-0.1145	0.0879	7.9064e+001
33	-0.1749	0.1357	4.9999e+004
⋮	⋮	⋮	⋮

```
≫ QBRunBM2;
```
where the function QBRunBM2 is written as follows,

```
config.CTMFile = 'BM2.txt';
config.MPosFile = 'BM2MPos.txt';
config.ChLens = [6 19 29];
config.LIDIdx = 1;
config.TraitIdx = 56;
config.MrksIdx = 2:55;
config.EpiFile = 'ALL';
config.bNormT = 1;
QTLBayes(config);
```

After the algorithm is done at each stage, a folder will be created under the current directory where all the parameter chains are saved. For example, 'pqtlp.txt' and 'pqtln.txt' will save the posterior probabilities of each marker having positive and negative main effects respectively, 'pepip.txt' and 'pepin.txt' correspond to the posterior probabilities of each pair of markers having positive and negative effects respectively. Users can make inference about any parameter of interest on the basis of these chains.

To check which marker/pseudomarker has significant positive or negative effect on the phenotype, one can use
```
≫ ReportMainEffects;
```
There are 20 markers with BF>1 in the output, and a partial list of these QTL are given in Table 17.1. The first column of Table 17.1 is the marker index that corresponds to the order specified by "config.MrksIdx". The second column (beta) is the estimated magnitude of the effect, and the third column (sd) is the standard deviation of the estimate. The last column is the BF calculated using (17.4).

Similarly, for epistatic effects, one can use
```
≫ ReportEpiEffects;
```
to list all pairs of markers whose BF for epistatic effects are greater than 1. Part of the list is shown in Table 17.2. Table 17.2 is similar to Table 17.1 except that there are two indexes for the marker, Index 1 and Index 2, representing

TABLE 17.2: Partial List of the Pairs of QTL with Epistatic Effect BF>1 for BM2 Dataset

Index 1:	Index 2:	gamma:	sd:	bf:
3	47	0.2231	0.1533	1.1630e+002
9	50	0.3304	0.0750	5.0000e+004
12	42	0.2688	0.0964	1.0461e+003
25	30	0.0896	0.0926	2.6083e+000
49	23	0.1678	0.0921	1.2229e+001
⋮	⋮	⋮	⋮	⋮

the two markers involved in epistasis. Note that the values of BF vary a lot in both tables, ranging from 1 to 10^{15}. On the basis of these BF, users can be more or less conservative when selecting the total number of significant main and epistatic effects. Finally, the heritability can be estimated by using

```
>> sigmae = load('sigmae.txt');
>> 1- median(sigmae)
```

and the estimated heritability is 0.9023 for this dataset.

17.5 Conclusion

We focus our discussion on the trait that follows a normal distribution. For other types of traits, such as categorical or limited traits, similar Bayesian classification strategy can be proposed for the generalized linear model framework. In addition, the gene by environment interactions can be easily incorporated into the model if the user is interested in modeling phenotype plasticity. Although the method is developed for QTL mapping, it can be easily extended to other applications of large p small n with minimal modifications. The disadvantage of Baysian methods is the intensive computation as the computation burden is much heavier than non-bayesian methods, however, the gain can be significant and thus make it worthwhile in practice.

17.6 Acknowledgment

I would like to thank an anonymous reviewer for the very helpful comments. This work is partially supported by NSF DBI-0701119 and Purdue Alumni Research Grant.

References

Ball, R. D. 2001. Bayesian methods for quantitative trait loci mapping based on model selections: approximate analysis using the Bayesian information criterion. *Genetics* 159:1351-1364.

Bateson, W. 1909. *Mendel's Principles of Heredity*. Cambridge: Cambridge University Press.

Berger, J. O. 1985. *Statistical Decision Theory and Bayesian Analysis* (2nd edn). Springer-Verlag.

Bogdan, M. Ghosh, J. K., Doerge, R. W. 2004. Modifying the Schwartz Bayesian information criterion to locate multiple interacting quantitative trait loci. *Genetics* 167:989-999.

Broman, K. W., Speed, T. P. 2002. A model selection approach for the identification of quantitative trait loci in experimental crosses. *Journal of the Royal Statistical Society Series B* 64:641-656.

Cockerham, C. C. 1954. An extension of the concept of partitioning hereditary variance for analysis of covariances among relatives when epistasis is present. *Genetics* 39:859-882.

Cowles, M. K., and Carlin, B. P. 1996. Markov chain Monte Carlo convergence diagnostics: a comparative review. *J. Am. Stat. Assoc.* 91:883-904.

Cui, Y. H., Wu, R. 2005. Mapping genome-genome epistasis: a high dimensional model. *Bioinformatics* 21:2447-2455.

Doerge, R. W., Zeng, Z-B., Weir, B. S. (1997). Statistical issues in the search for genes affecting quantitative traits in experimental populations. *Statistical Science* 12:195-219.

Gaffney, P. J. 2001. An efficient reversible jump Markov chain Monte Carlo approach to detect multiple loci and their effects in inbred crosses. *PhD thesis* Department of Statistics, University of Wisconsin, Madison, WI.

Gelman, A., Carlin, J. B., Stern, H. S et al. 1995. *Bayesian Data Analysis*. Chapman & Hall/CRC.

George, E.I. and McCulloch, R.E. 1993. Variable selection via Gibbs sampling. *J. Am. Stat. Assoc.* 88, 881-889.

Green, P.J. 1995. Reversible jump Markov Chain Monte Carlo computation and Bayesian model determination. *Biometrika* 82, 711-732.

Haldane, J. B. S. 1919. The combination of linkage values and the calculation of distance between the loci of linked factors. *J. Genet.* 8:299-309.

Hoeschele, I., and VanRaden, P. M. 1993a. Bayesian analysis of linkage between genetic markers and quantitative trait loci. I. Prior knowledge. *Theor Appl Genet* 85:953-960.

Hoeschele, I., and VanRaden, P. M. 1993b. Bayesian analysis of linkage between genetic markers and quantitative trait loci. II. Combining prior knowledge with experimental evidence. *Theor Appl Genet* 85:946-952.

Jansen, R. C. 1993. Interval mapping of multiple quantitative trait loci. *Genetics* 135:205-211.

Jansen, R. C., and Stam, P. 1994. High resolution of quantitative traits into multiple loci via interval mapping. *Genetics* 136:1447-1455.

Jansen, R. C. 2007. Quantitative trait loci in inbred lines. In *Handbook of Statistical Genetics*, Third Edition. Edited by Balding, D. J., Bishop, B. M., and Cannings, C. John Wiley & Sons, Ltd.

Jayakar, S. D. 1970. On the detection and estimation of linkage between a locus influencing a quantitative trait character and a marker locus. *Biometrics* 26:451-464.

Jeffreys, H. 1961. *Theory of Probability.* Oxford: Clarendon Press.

Kao, C.-H., Zeng, Z-B., and Teasdale, R. D. 1999. Multiple interval mapping for quantitative trait loci. *Genetics* 152:1203-1216.

Kao, C.-H., Zeng, Z.-B. 2002. Modeling epistasis of quantitative trait loci using Cockerham's model. *Genetics* 160:1243-1261.

Kosambi, D. D. 1944. The estimation of map distance from recombination values. *Ann. Eugen* 12:172-175.

Lander, E. S., and Botstein, D. 1989. Mapping mendelian factors underlying quantative traits using RFLP linkage maps. *Genetics* 121:185-199.

Liu, J., Mercer, J. M., Stam, L. F. et al., 1996. Genetic analysis of a morphological shape difference in the male genitalia of *Drosophila simulans* and *D. mauritiana. Genetics* 142:1129-1145.

Satagopan, J. M., Yandell, B. S., Newton, M. A. et al. 1996. A Bayesian approach to detect polygene loci using Markov Chain Monte Carlo. *Genetics* 144:805-816.

Sax, K. 1923. The association of size difference with seed-coat pattern and pigmentation in *Phaseolus vulgarls. Genetics* 8:552-560.

Scheet, P. and Stephens, M. 2006. A fast and flexible statistical model for large-scale population genotype data: applications to inferring missing genotypes and haplotypic phase. *Am J Hum Genet* 78:629-644.

Shi, W., Lee, K. E., and Wahba, G. 2007. Detecting disease-causing genes by LASSO-Patternsearch algorithm. *BMC Proceedings*1(Suppl 1):S60.

Shoemaker, J. S., Painter, I. S., and Weir, B. S. 1999. Bayesian statistics in genetics. *Trends in Genetics* 15:354-358.

Sillanpää, M.J. and Arjas, E. 1998. Bayesian mapping of multiple quantitative trait loci from incomplete inbred line cross data. *Genetics* 148, 1373-1388.

Smith, B. J. 2007. boa: An R package for MCMC output convergence assessment and posterior inference. *Journal of Statistical Software* 211-37.

Stephens, D.A. and Fisch, R.D. 1998. Bayesian analysis of quantitative trait locus data using reservible jump Markov chain Monte Carlo. *Biometrics* 54, 1334-1347.

Thoday, J. M. 1960. Location of polygenes. *Nature* 191:368-370.

Thomas, D. C., Cortessis, V. 1992. A Gibbs sampling approach to linkage analysis. *Hum Hered* 42:63-76.

Yi, N., George, V. and Allison, D.B. 2003. Stochastic search variable selection for identifying multiple quantitative trait loci. *Genetics* 164, 1129-1138.

Yi, N., Xu, S. 2002. Mapping quantitative trait loci with epistatic effects. *Genetical Research* 79:185-198.

Yi, N., Yandell, B. S., Churchill, G. A. et al. 2005. Bayesian model selection for genome-wide epistatic quantitative trait loci analysis. *Genetics* 170:s1333-1344.

Yi, N., Banerjee, S. Pomp, D., et al. 2007. Bayesian mapping of genomewide interacting quantitative trait loci for ordinal traits. *Genetics* 176:1855-1864.

Żak, M., Baierl, A., Bogdan, M. et al. 2007. Locating multiple interacting quantitative trait loci using rank-based model selection. *Genetics* 176:1845-1854.

Zeng, Z-B. 1993. Theoretical basis of separation of multiple linked gene effects on mapping quantitative trait loci. *Proc. Natl. Acad. Sci. USA* 90:10972-10976.

Zeng, Z-B. 1994. Precision mapping of quantitative trait loci. *Genetics* 136:1457-1468.

Zeng, Z-B., Kao, C. H., and Basten, C. J. 1999. Estimating the genetic architecture of quantitative traits. *Genetical Research* 74:279-289.

Zeng, Z-B., Liu, J., Stam, L. F., et al., 2000. Genetic architecture of a morphological shape difference between two *drosophila* species. *Genetics* 154:299-310.

Zhang, M., Montooth, K. L., and Wells, M. T. et al., 2005. Mapping multiple quantitative trait loci by Bayesian classification. *Genetics* 169:2305-2318.

Zhang, M., Zhang, D., and Well, M. T. 2008. Variable selection with large p small n regression models: mapping QTL with epistasis. *BMC Bioinformatics* 9:251.

Index